T0301785

Biohydrometallurgical Processes

Extensive industrialization has led to an increased release of toxic metals into the soil and air. Industrial waste can include mine overburden, bauxite residue, and E waste, and these can serve as a source of valuable recoverable metals. There are relatively simple methods to recycle these wastes, but they require additional chemicals, are expensive, and generate secondary waste that causes environmental pollution. Biohydrometallurgical processing is a cost-effective and ecofriendly alternative where biological processes help conserve dwindling ore resources and extract metals in a nonpolluting way. Microbes can be used in metal extraction from primary ores, waste minerals, and industrial and mining wastes. *Biohydrometallurgical Processes: Metal Recovery and Remediation* serves as a useful guide for microbiologists, biotechnologists, and various industrialists dealing with mining, metallurgy, chemical engineering, and environmental sciences.

Features:

- Examines advances in biohydrometallurgy, biomineralization, and bioleaching techniques
- Discusses the importance of bacteria in biohydrometallurgical processes and microbial interventions for waste cleanup and upgradation of minerals
- Presents the latest techniques for biosynthesis related to different metals, along with recent developments in alternative procedures using extremophiles and leaching bacteria

Microbial Biotechnology for Food, Health, and the Environment Series
Series Editor: Ashok Kumar Nadda

Plant-Microbial Interactions and Smart Agricultural Biotechnology
Edited by Swati Tyagi, Robin Kumar, Baljeet Saharan, Ashok Kumar Nadda

Microbial Products
Applications and Translational Trends
Edited by Mamtesh Singh, Gajendra Pratap Singh, Shivani Tyagi

Biohydrometallurgical Processes
Metal Recovery and Remediation
Edited by Satarupa Dey

For more information about this series, please visit: www.routledge.com
https://www.routledge.com/Microbial-Biotechnology-for-Food-Health-and-the-Environment/book-series/CRCMBFHE

Biohydrometallurgical Processes

Metal Recovery and Remediation

Edited by

Satarupa Dey

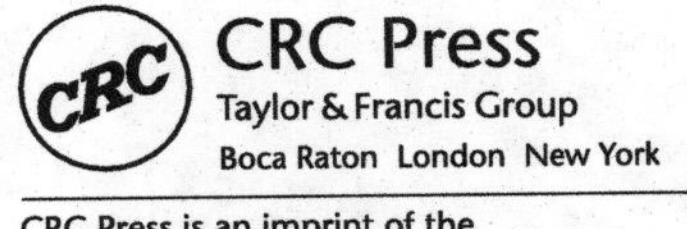

CRC Press is an imprint of the
Taylor & Francis Group, an **informa** business

Designed cover image: Shutterstock

First edition published 2024
by CRC Press
2385 NW Executive Center Drive, Suite 320, Boca Raton FL 33431

and by CRC Press
4 Park Square, Milton Park, Abingdon, Oxon, OX14 4RN

CRC Press is an imprint of Taylor & Francis Group, LLC

ISBN: 978-1-032-58781-3 (hbk)
ISBN: 978-1-032-58782-0 (pbk)
ISBN: 978-1-003-45145-7 (ebk)

DOI: 10.1201/9781003451457

Typeset in Times
by codeMantra

Contents

Preface

Extensive industrialization and urbanization have led to extensive pollution and release of toxic metals into the soil and air. These industries produce several wastes such as mine overburden, bauxite residue, and E-waste, which have been dumped in huge amounts and can serve as a source of metal values. They can also be recycled to extract metals from them. There exist several physicochemical methods that are simple and fast, but they require additional chemicals, are therefore expensive, and generate secondary waste that causes environmental pollution. Biohydrometallurgical processing of metals is considered a cost-effective and ecofriendly alternative where biological processes and microbes help to conserve dwindling ore resources and help in the extraction of metals in a non-polluting way. Microbes can be used in metal extraction from primary ores, waste minerals, and industrial and mining wastes. This book not only deals with microbes which play a crucial role in biohydrometallurgical processes, but also deals with the concepts of microbial diversity in various operations and molecular biology involved in such systems (extremophiles).

This book comprises a total of 13 chapters which address in detail the aspects of biohydrometallurgy, biomineralization, and bioleaching techniques and cover in detail topics like biofouling and corrosion produced by biofilms. Few other aspects such as bioremediation of industrial waste using biohydrometallurgical processes are also discussed in brief. The importance of bacteria in biohydrometallurgical processes is addressed in this book, which can serve as a valuable source of information for scientists and students in a number of disciplines, including geology, chemistry, metallurgy, microbiology, and chemical engineering.

Satarupa Dey
Shyampur Siddheswari Mahavidyalaya, West Bengal, India

Acknowledgments

I am grateful and acknowledge my in-depth gratitude to all the contributors and reviewers who trusted me and supported immensely in shaping this book. This book would not have been a success without the vital contributions from the authors. I wholeheartedly express our thanks to all the authors for their efforts in writing wonderful chapters for this book. I also express my thanks to all the reviewers who have given their vital inputs to improve the chapters. I am thankful to CRC Press and all the team members for giving me this opportunity to edit this book. I would like to express my gratitude to my mentors, teachers, and students who have always guided and motivated me. I convey my thanks to them for giving me the confidence and strength to accomplish this task. Also, I would like to thank our colleagues, family, and friends who provided encouragement and support during the work of this book.

Satarupa Dey
Shyampur Siddheswari Mahavidyalaya, West Bengal, India

Editor

Dr. Satarupa Dey is an Assistant Professor in the Department of Botany at Shyampur Siddheswari Mahavidyalaya, affiliated to the University of Calcutta, West Bengal, India. She earned her BSc in Botany from the University of Calcutta, West Bengal, India. She holds also an MSc in Botany with Microbiology specialization from the same university. She completed her PhD in 2012, and her field of expertise is Bioremediation of Toxic Metals. She was selected as a Department of Biotechnology, Government of India research fellow and later worked as a visiting scientist in the Agricultural and Ecological Research Unit of the Indian Statistical Institute, Kolkata. She is the recipient of the Woman Scientist (WOS A) fellowship of the Department of Science and Technology, Government of India. She has published research and review articles in several journals of national and international repute. She has co-edited three books and is an active reviewer for many high-impact journals published by Elsevier, Springer Nature, and Nature Publishers.

Contributors

Robina Aziz
Department of Botany
Government College Women University
Sialkot, Pakistan

Neha Bhadauria
Amity Institute of Environmental Toxicology, Safety and Management
Amity University
Noida, India

Somya Bhandari
Department of Microbiology
Modern College of Arts, Science & Commerce
Pune, India

Pranjal Bharali
Applied Environmental Microbial Biotechnology Laboratory, Department of Environmental Science
Nagaland University
Zunheboto, India

Arunima Bhattacharya
Postgraduate Department of Microbiology
St. Xavier's College (Autonomous)
Kolkata, India
and
Institut Européen de Chimie et Biologie, Université de Bordeaux
Institut National de la Santé et de la Recherche Médicale (U1212) and Centre National de la Recherche Scientifique (UMR 5320)
Pessac, France

Ratul Bhattacharya
Raja Rammohun Roy Mahavidyalaya
Hoogly, India

Roumi Bhattacharya
Department of Civil Engineering
Indian Institute of Engineering, Science and Technology
Howrah, India

Arunima Biswas
Department of Microbiology
Raidighi College
Parganas, India

Rajeswari Chakraborty
Department of Microbiology
St Xavier's College (Autonomous)
Kolkata, India

Riddhi Chakraborty
Department of Microbiology
St Xavier's College (Autonomous)
Kolkata, India

Niti Choudhury
Department of Microbiology
St Xavier's College (Autonomous)
Kolkata, India

Deeksha Dave
Environmental Studies, School of Inter Disciplinary and Trans Disciplinary Studies
IGNOU
New Delhi, India

Satarupa Dey
Department of Botany
Shyampur Siddheswari Mahavidyalaya (affiliated to University of Calcutta)
Howrah, India

Subhankar Dey
Dept. of Zoology, Block L
New Alipore College
Kolkata, India

Sayak Ganguli
Post Graduate Department of Biotechnology
St. Xavier's College (Autonomous)
Kolkata, India

Shakuntala Ghorai
Department of Microbiology
Raidighi College
Parganas, India

Aindri Ghosh
Department of Microbiology
St Xavier's College (Autonomous)
Kolkata, India

Jaydip Ghosh
Department of Microbiology
St Xavier's College (Autonomous)
Kolkata, India

Mahashweta Mitra Ghosh
Post Graduate Department of Microbiology
St. Xavier's College (Autonomous)
Kolkata, India

Bhagyudoy Gogoi
Applied Environmental Microbial Biotechnology Laboratory, Department of Environmental Science
Nagaland University
Zunheboto, India

Vartika Jain
Department of Botany
Govt. Meera Girls College
Udaipur, India

Harshita Jonwal
Department of Biotechnology
Dr. B. Lal Institute of Biotechnology Malviya Industrial Area
Jaipur, India

Nivedhitha Kabeerdass
Department of Microbiology
Muthayammal College of Arts and Science
Namakkal, India

Rupsha Karmakar
Post Graduate Department of Microbiology
St. Xavier's College (Autonomous)
Kolkata, India

Muhammad Majeed
Department of Botany
University of Gujrat
Gujrat, Pakistan

Madhumita Majumder
Department of Botany
Raidighi College
Raidighi, India

Maghimaa Mathanmohun
Department of Microbiology
Muthayammal College of Arts and Science
Namakkal, India

Arup Kumar Mitra
Department of Microbiology
St Xavier's College (Autonomous)
Kolkata, India

Biplob Kumar Modak
Sidho Kanho Birsha University
Purulia, India

Ayan Mondal
Department of Zoology
Government General Degree College
Paschim Medinipur, India

Subhamita Sen Niyogi
Central Water Testing Laboratory, New Town Kolkata Water Supply Division 1
Administrative Building, Water Treatment Plant site
Kolkata, India

Gunadhor Singh Okram
UGC-DAE Consortium for Scientific Research
University Campus, Khandwa Road
Indore, India

Preeti Pradhan
Department of Engineering Sciences and Humanities (ESH)
Indore Institute of Science & Technology
Indore, India

Anjalika Roy
Department of Botany (UGC DRS (SAP-II), DST-FIST)
Visva-Bharti
Santiniketan, India

Bedaprana Roy
Department of Microbiology
St Xavier's College (Autonomous)
Kolkata, India

Lemzila Rudithongru
Applied Environmental Microbial Biotechnology Laboratory, Department of Environmental Science
Nagaland University
Zunheboto, India

Debanjana Sengupta
Postgraduate Department of Microbiology
St. Xavier's College (Autonomous)
Kolkata, India

Sonia Sethi
Department of Biotechnology
Dr. B. Lal Institute of Biotechnology Malviya Industrial Area
Jaipur, India

Meesha Singh
Post Graduate Department of Microbiology
St. Xavier's College (Autonomous)
Kolkata, India

Dwaipayan Sinha
Department of Botany
Government General Degree College
Paschim Medinipur, India

Viphrezolie Sorhie
Applied Environmental Microbial Biotechnology Laboratory, Department of Environmental Science
Nagaland University
Zunheboto, India

Arjun Suresh
Save Vibrant Earth Foundation
Greater Noida West, India

Rakesh Vinayagam
Department of Microbiology
Muthayammal College of Arts and Science
Namakkal, India

1 Formation Mechanisms and Prevention Strategies of Microorganism-Influenced Corrosion Biofilms

Maghimaa Mathanmohun, Nivedhitha Kabeerdass, and Rakesh Vinayagam
Muthayammal College of Arts and Science

Gunadhor Singh Okram
UGC-DAE Consortium for Scientific Research

1.1 INTRODUCTION

Austin, in the year 1788, coined first the very word '*corrosion*' after noticing that water, originally neutral, tends to become alkaline in the presence of iron. Later, Thenard suggested it as an electrochemical phenomenon and published it in 1819 (Roberge, 2000). The surveillance of the microorganisms becomes common on metals since iron does not rust in the anaerobic condition (Miller and Tiller, 1970). The corrosion rates indicate the difference between the chemical action and the generation of electric currents by the two famed laws of electrochemical action of Michael Faraday. In simple words, the deterioration of the metals or compounds into a stable form is a natural process according to their environment; it may be by microorganisms, chemical reactions, or electrochemical reactions. There are various types of corrosion as reported (Roberge, 2000). They are galvanic, pitting, and uniform; some of the microorganisms influence corrosion (MIC). Pitting corrosion is a destructive and localized form of corrosion. Degradation forms a pit or hole on the small metal surface (Davis Joseph, 1998), especially in aluminum (Balaji Ganesh and Radhakrishnan, 2006) or steel, which finally leads to cracks, flaws, and scratches on the surface. Pitting corrosion occurs mainly due to the damage, stress, and nonuniform coating of the cathode's surface protecting layer or film, which leads to a small exposure of the metal and turns it anodic. The silver bridge that collapsed in West Virginia (USA) is an example of pitting corrosion, as reported by Lutzenkirchen-Hecht and Strehblow (2006). This corrosion is commonly seen

DOI: 10.1201/9781003451457-1

in marine environments and is also misnamed as electrolysis. Galvanic corrosion occurs due to the voltage differences between two metals in galvanic series and the size of the exposed cathode metal relative to their anodic metal. According to Dexter Stephen (1999), the diatoms (Nashima and Palaniasmy, 2016), marine microorganisms, and slime film might increase the potential differences. Uniform corrosion, also known as general corrosion, causes dulling of the metal surface, such as silver, and fogging of nickel, finally leading to a rough surface. The deterioration of the metals can be due to the metabolic activities of the microorganisms, in which they play a dual role of corrosion and anticorrosive activity. In order to protect the metal surfaces, many researchers coated the surface using the sol–gel method (Jhobalia et al., 2005), in which many protective metabolic products of harmless microorganisms can act as antibacterial and antifungal agents. In the past, due to a lack of awareness about MIC, many internal corrosions related to pipeline systems and underground storage systems were misunderstood to be due to hydrogen sulfide, oxygen, and carbon dioxide. In the case of the oil industry (TPC-3, 1982), many recent findings suggested that the antagonistic activity of the nonpathogenic *Bacillus* strain (B21) might be a consortium and act as a biocontrol agent in the field of corrosion. Research proved that there is a reduction in the production of sulfides when sulfate-reducing bacteria (SRB) are co-cultured with the bacterial strain (B21) (Gana et al., 2011). According to Bano and Qazi (2011), the minimal inhibitory concentration of the noncorrosive *Bacillus* is labeled as SN-8 and SS-14 against steel corrosion influenced by SNB4 and SS-5, respectively. Therefore, the intensity of the microbially influenced corrosion is reduced, and consequently, selected potential microbial strains might be biocontrolled strategies to protect the metals from corrosion (Neoh and Kang, 2011). Thus, the main aim of this section is to focus on the dual role of the microbiota in inducing and controlling corrosion.

1.2 WHAT IS MIC?

The acronym MIC for microorganism-influenced corrosion (Tatnal, 1993) is generally accepted and is used as corrosion induced by the activities and presence of microorganisms (Javaherdashti, 2017). Notably, since there are many similarities among microbes, microorganisms, and microbiologies, they are used here interchangeably for MIC. However, ‘biocorrosion’ is also progressively used as a synonym for MIC in Latin America and Europe to represent the corrosion of implants within a living body because of the involvement of both biotic and abiotic processes (Hansen, 2008; Milošev, 2017). The MICs, or all these terms collectively, denote the influence of the microorganisms in the aerobic and anaerobic conditions as a result of the secondary metabolite they form as a slime layer on the metals, which finally leads to destruction (Hansen, 2008; Milošev, 2017). In chemical terms, corrosion is an interfacial process that directly depends on environmental parameters like pH, concentration of dissolved oxygen, physical and chemical nature of the salts, and redox potential of oxygen (Little et al., 2020). Each one of these can be influenced, in a much more localized fashion, by microbial biofilms that colonize the interfaces (Little et al., 2020). Activities of Archaea, bacteria, and fungi in colonies that can form biofilms on the surfaces of materials or in local environments with direct contact with materials can lead to MIC, and most metals, including few nonmetals, can be affected by this kind of corrosion. The MIC plays a vital role in many environments like freshwater, marine (APHA, 1996), food stuffs, soils, demineralized water, sewage, aircraft, petrol, human plasma, and chemical processes.

The formation of the slime layer, or biofilm, by the multiple microorganisms is a significant example of how it accelerates the corrosion. In the case of the food industry, the high concentration of chlorides present in different ranges of temperature and humidity leads to galvanic corrosion in poultry and meat. The tin and steel cans in the food industry are sometimes corroded by the atmospheric pollutant, so the cans are thrown out as garbage, which leads to an economic decline in the meat, dairy and beverage industries. In marine environments, the synergistic action of both anaerobic SRB (Enos and Taylor, 1996) and iron-oxidizing aerobic bacteria accelerates the corrosion of materials. In soil, the corrosion of metals is linked with the biogeochemical cycles of iron, nitrogen, and extracellular electron transfer. In sewage treatment plants, the microbiota plays a significant role in the synthesis of the sulfur cycle, which finally leads to the synthesis of sulfuric acid. In aircraft, many researchers suggested that the fungus *Hormoconis resinae*, also called kerosene fungus, creates pitting corrosion and finally ends up in the contaminated fuel tank. In oil refineries and petrol bunks, the oil storage tanks (Magot et al., 1997; Javaherdashti et al., 2016) and pipes are colonized by aerobic and anaerobic organisms that vary in their morphology according to their environmental parameters. In repairing and replacing tissues in dental, orthopedic, and medical devices using titanium-based, molybdenum-based, silver-based, and gold-based materials for many decades, MICs may take place. In the case of humans, the body fluids, which are the electrolytes, cause degradation of the devices when in contact with metals.

1.3 SIGNIFICANT MICROORGANISMS IN CORROSION

The roles of many iron oxidizers, sulfur-reducing bacteria, manganese oxidizers, and iron reducers are well known in natural and artificial environments. Many scrapings from the soil, petroleum industry, marine, and sewage plants contain colonies of many aerobic and anaerobic organisms. They are SRB such as *Desulfovibrio* (McNeil et al., 1991; McNeil and Mohr, 1993; McNeil and Odom, 1994; Javaherdashti, 1999), iron-oxidizing bacteria (IOB) (Javaherdashti, 2010), and manganese-oxidizing bacteria (Tables 1.1, 1.2, and Figures 1.1, 1.2).

The image explores the diverse bacteria and fungi involved in MIC studies.

1.3.1 Sulfate-Reducing Bacteria (SRB)

SRB are probably the most notorious type of chromium-reducing bacteria known to industry. However, this is in fact one of the "myths" of MIC repeatedly put forward by MIC experts. SRB that grow anaerobically (no oxygen) are known for their capability of reducing sulfate (SO_4^{-2}) to sulfide (S^{-2}) where, in the absence of metallic ions such as iron, they produce H_2S gas (Javaherdashti, 1999), and, with ferrous ions, iron sulfide is generated.

TABLE 1.1
Microorganism-Influenced Corrosion (MIC) Using Diverse Bacteria and Materials

S. No.	Type of Organism	Name of the Organisms	Mode of Action	Materials	References
1	Bacteria Genetically engineered microorganisms (Bacteria)	*Escherichia coli* and *Pseudomonas* sp. *Bacillus brevis* *B. subtilis* WB600 and Bacillus subtilis BE1500	Protective biofilm, (oxygen removal) SRB inhibition by the production of antimicrobials	Carbon steel (SAE 1018)	Jayaraman et al. (1997a, b, c) Jayaraman et al. (1999a) Jayaraman et al. (1999b)
2	Bacteria	*Pseudomonas* sp.9 and *Serratia marcescens* EF190	Metabolic activity (oxygen removal)	Steel	Pedersen and Hermansson (1991)
3	Bacteria	*Bacillus brevis* 18 and *Pseudomonas fragi* K	Protective biofilm (oxygen removal)	Unalloyed copper and aluminum alloy 2024	Jayaraman et al. (1999c)
4	Bacteria	*Pseudomonas cichorii* Pseudomonas flava	Protective biofilm, phosphate precipitation layer	Mild steel	Chongdar et al. (2005) Gunasekaran et al. (2004)
5	Bacteria	*Shewanella algae* and *Shewanella* sp.	Protective biofilm	Aluminum, mild steel and brass	Nagiub and Mansfeld (2002)
6	Bacteria	*Shewanella oneidensis* strain MR-1	Protective precipitation layer	Mild steel 1018	Dubiel et al. (2002)
7	Bacteria	Bacillus pasteurii	Calcium carbonate precipitation	Cement-based building material	Chunxiang et al. (2009)
8	Bacteria	*Bacillus licheniformis* and *B. subtilis*	Production of anionic corrosion inhibitors	Aluminum 2024-T3 and C26000 brass	Mansfeld et al. (2002); Örnek et al. (2002a, b)
9	Bacteria	Bacillus sphaericus	MICP (microbial-induced carbonate precipitation)	Concrete and mortar	De Muynck et al. (2008)
10	Bacteria	*Bacillus pseudofirmus* DSM 8715 and *B. cohnii* DSM 6307	MICP (microbial-induced carbonate precipitation)	Concrete	Jonkers et al. (2010)

SRB, sulfate-reducing bacteria.

TABLE 1.2
Microorganism-Influenced Corrosion (MIC) Using Diverse Fungi, Cyanobacteria and Materials

S. No.	Type of Organism	Name of the Organisms	Mode of Action	Materials	References
1	Cyanobacteria	Spirulina platensis (Arthrospira)	Protective biofilm	Carbon steel	Mert et al. (2011)
2	Fungi	*Aspergillus niger*, *Aspergillus alliaceae*, *Penicillium* sp., *Beauveria bassiana*, and *Fusarium* sp.	Protective oxalate layer	Bronze and copper	Joseph et al. (2011)

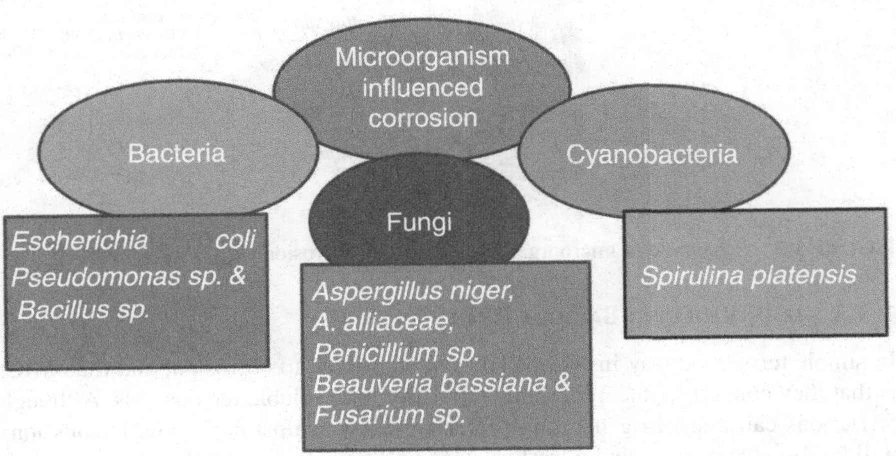

FIGURE 1.1 Microorganism-influenced corrosion (MIC) using diverse microorganisms.

1.3.2 Iron-Oxidizing Bacteria (IOB)

IOB utilizes the ferrous ion Fe^{+2} as a source of energy and is characterized by the deposition of ferric Fe^{+3} hydroxide. A common example of IOB is *Gallionella* sp. These bacteria will oxidize ferrous to ferric, and therefore this will be the initial "food" for IRB. In addition, we should not forget that the oxidation of iron and the conversion of iron atoms into their ionic form (ferrous) is in fact an anodic reaction (that is, corrosion). Therefore, these bacteria will in fact enhance corrosion in this way too.

IOB are not just a member of chromium-reducing bacteria in its negative sense; these bacteria can have very useful applications in mining, particularly in a method that is called "bioleaching" (Javaherdashti et al., 2016).

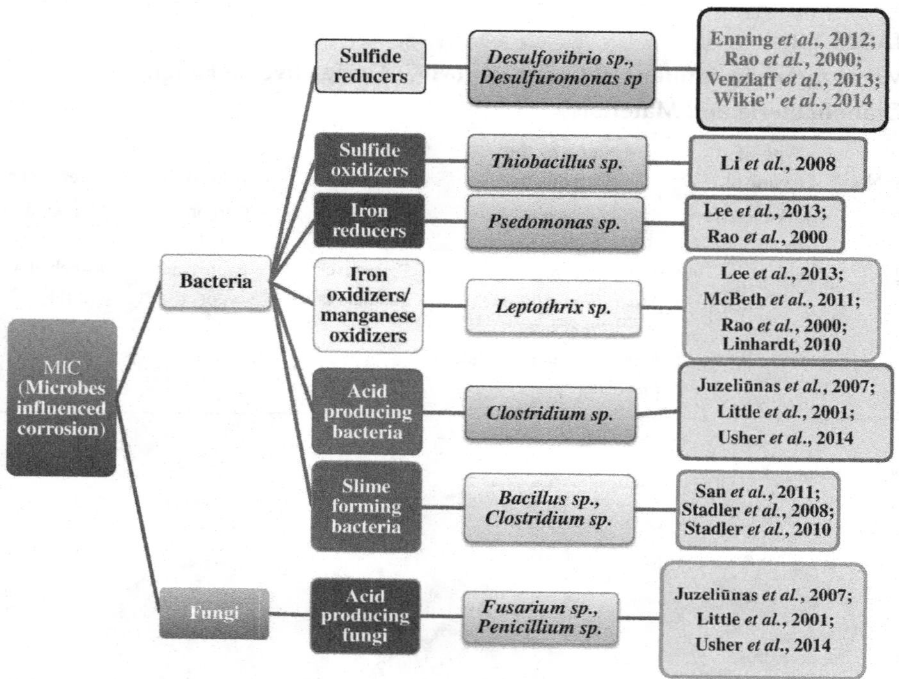

FIGURE 1.2 Schematic of microorganism-influenced corrosion (MIC) studies.

1.3.3 Iron-Reducing Bacteria (IRB)

In simple terms, the way in which IRB can contribute to corrosion, and thus MIC, is that they convert (reduce) insoluble ferric iron into soluble ferrous ions. Although ferric ions can assist in generating protective films against corrosion, ferrous ions will be dissolved easily, and a fresh surface will be presented to the corrosive environment surrounding the material (Javaherdashti, 2010).

1.4 MINERALOGICAL TECHNIQUES

The presence and activities of microorganisms cause the MIC (Little and Lee, 2007), which is best understood as microbiologically mediated reactions with metal oxides, disruptions of passivating layers, or the formation/dissolution of minerals. The following sections will review the mineralogy associated with MIC. The relationships between microbiologically mediated reactions and metal oxides define many of the mechanisms for MIC. For example, MIC can involve disruption of a passivating oxide (e.g., under deposit corrosion), conversion of a protective oxide to a less protective layer (e.g., sulfide derivitization), or removal of an oxide layer (e.g., metal oxide reduction). Microorganisms can also produce minerals, particularly sulfides, which cannot be produced abiotically in most near-surface aquatic environments. Consequently, in some cases, sulfides are mineralogical indicators of MIC.

As alternatives to phenotype and genotype and to identify the corrosion products and the nature of the MIC, some mineralogical techniques are used. This technique is used to identify the causative agent and the condition under which the corrosion occurred. Proliferation of the microbiota within the biofilm, which develops a microenvironment with the biofilm substratum interface, shows variation in environmental parameters like pH, temperature, and also with the concentration of organic and inorganic species (Little and Lee, 2007). The corrosion product obtained from biomineralization (Jroundi et al., 2012) is different from that of abiotic environments. For example, the corrosion of steel under abiotic conditions produces iron oxides and oxyhydroxides such as magnetite, maghemite, and hematite. In other phases, the dominant bacteriogenic iron oxides produced from both freshwater and marine waters are like two-line ferrihydrite, which is found in many sheaths and stalks and produced by IOB, e.g., *Leptotherix* and *Gallionella* (Little and Lee, 2007; McBeth et al., 2011). Ferrihydrite is a poorly ordered mineral that can be changed into hematite. According to Chan et al. (2011), the Fe-encrusted protein provides a robust biosignature for lithiotropic iron-oxidizing-based metabolism.

Under aerobic conditions, the presence of manganese and IOB can be used as possible indicators for MIC (Neville and Hodgkiess, 2000; Ray et al., 2011; Gerke et al., 2012). Dissimilatory microbial sulfate reduction results in sulfides and is deficient in heavy sulfur isotopes. Sulfur-rich copper corrosion products were concentrated in the residual sulfate in the culture medium (Webster et al., 2000). In the case of sewer systems, the microbial sulfate reduction produces sulfides, which are oxidized by sulfur-oxidizing bacteria to sulfuric acid, which reacts with the concrete to produce gypsum.

1.5 LABORATORY IDENTIFICATION OF MIC

The presence of specific groups of corrosion-inducing pathogens is identified in bulk medium (planktonic cells) or sessile cells through phenotypic and genotypic studies. It can be classified as follows:

- Isolation and culture of the microorganisms in liquid or solid media.
- Extract and enumerate a meticulous cell constituent.
- Measure/demonstrate several cellular actions.
- Characterization study of the corrosion product.

1.5.1 Culture Techniques

Inoculated isolates in a chamber contain 5% carbon dioxide, 5% hydrogen, and 90% nitrogen. Later, the SRB (Cowan, 2005) are enriched by inoculation in a standard post-gate medium in which sodium lactate is used as a carbon source. Then, the aliquots containing modified post-gate media are incubated at 38°C in an anaerobic environment for 120 hours to develop the optimal growth of the isolates. Later, the blackish color change in the media might indicate the presence of iron sulfides and progress in growth (Parker, 1990).

The composition of the standard post-gate medium per liter comprises 1.0g NH_4Cl, 0.5g KH_2PO_4, 1.0g Na_2SO_4, 1.0g $CaCl_2$, 1.83g $MgCl_2.6H_2O$, 1.0g yeast extract, 0.1g ascorbic acid, 0.03g sodium thioglycolate, 6.3g sodium citrate, 1.75mL sodium lactate, 2.0 ml resazurin, 0.5g $FeSO_{4.}7H_2O$, and 3.5 ML NaCl. These components were later dissolved in the distilled water by maintaining a pH of 7.5–8.0 with 5M HCl. Then, the homogenized solution is sterilized at 120°C for 30 minutes (Bernardez and de Andrade Lima, 2015).

1.6 CHARACTERIZATION OF THE CORROSION PRODUCT

2.8mL of aliquots are inoculated into the post-gate medium present in a 250mL Erlenmeyer flask. Then, these aliquots are incubated at 38°C in an environmental chamber (anaerobic condition) to monitor the optimal growth for 120hours. 45mL of the aliquots can be mixed with 2mL of HCL in each case. Then, the culture is centrifuged at 11,000rpm at 10°C for 20 minutes. The upper layer (supernatant) can be removed and rinsed in distilled water. Again, the supernatant is centrifuged at 14,310rpm, and the obtained biomass is again resuspended in 10mL of distilled water that might act as a precursor for the bacterial suspension.

Subsequent dilutions of the bacterial suspension can be made by dissolving 1mL of the bacterial suspension in 9mL of distilled water, and the following duplicates can be made in the ratios of 1:5, 1:10, 1:15, 1:20, and 1:25. Later, the optical density and dry weight of the samples are determined.

In an oven, a membrane filter made up of cellulose acetate with a pore size of 0.22μm is placed in an empty aluminum weighing pan and dried. Then, it is later stored in a desiccator made up of calcium sulfate for 24hours. About 5mL of the culture is poured into the filter membrane, which was held by a reservoir, after applying a vacuum. This helps pull the liquid into the membrane. Distilled water is used to rinse the reservoir and then remove the paste adhered to the glassware. After all the water droplets are exhausted or pulled through, a wet weight is calculated. The cell paste is dried in the oven at 105°C. A calculation is done using the weight of the pan or filter cell paste, which was monitored periodically until no decrease in the weight is realized. The whole process of a search takes 24hours to dry completely.

The appropriate concentrations of the dried samples and their absorbance can be measured using a spectrophotometer at 600nm. The curve corresponding to its dry-weight cell can be calculated. According to the turbidity measurement, the optical density of 1unit corresponds to 1.0g/L of a dry cell. Inoculated in the modified post-gate medium, the SRB cells are harvested in the late exponential phase. About 1.5mL of the aliquots are added to a clean, dry test tube and centrifuged at 10,000rpm for 10 minutes each. Using phosphate solubilization buffer, the cells are washed twice and then re-suspended in a 1mL stock solution containing acridine orange. The solution was covered by an aluminum foil for 30 minutes and left it in the dark, and then observed under the fluorescence microscope immediately by spreading it on a slide.

For the determination of the kinetics, 200mL of the autoclaved medium and 50mL of the inoculum placed inside the anaerobic chamber are co-inoculated in a

250 mL sterile screw-capped bottle. Then, such bottles are placed in a mechanical shaker for continuous shaking at 120 rpm at 38°C. After certain time intervals, one bottle is chosen and used for biological analysis (Larson and Robert, 1994). The amounts of sodium sulfate and ferrous sulfate are adjusted correspondingly to get a concentration of 1,790 mg/L (Bernardez et al., 2012). After evaluating the biomass concentration and acidification at suitable time intervals, a calibrated curve is obtained.

1.7 INHIBITION OF BIOFILM FORMATION BY THE MIC

Biofilm formation is a serious threat since it possesses serious health hazards and is very difficult to combat. Plant-based nanoparticle synthesis opens a new epoch to fight against the biofilm caused by the microbiota of the MIC. Plants are recognized for their therapeutic properties in nanoparticle synthesis (Maghimaa and Alharbi, 2020; Kabeerdass et al., 2021; Sun et al., 2021; Vakayil et al., 2021d; Kabeerdass et al., 2022) against *E. coli*, *Salmonella* sp., and *S. aureus*. All these pathogens attach to the walls of the food and beverage plants, forming a ring-like slime layer on the pipeline system (Parkar et al., 2004; Romero et al., 2005; Zhu et al., 2005; Simões et al., 2010). Biofilm inhibition against these microorganisms is achieved in the dairy, food, and beverage industries through chemical, enzymatic, and biosurfactants disruptions. An interesting example is that nanocomposite materials reduce the adhesive property of bacterial pathogens by inhibiting the ica genes (extracellular matrix of biosynthesis) (Galie et al., 2018). Nanoparticles synthesized via plant extracts like *Bucida buceras*, *Viminalis callistemon*, and green tea phenols are effective and act as an antibiofilm agent against *Shewanella balitica*. In marine environments, the antibiofilm agents are small molecules, enzymes, phage particles, and degrading agents. These soil biofilm formations increase the metabolic activities of microorganisms. In sewage plants, the antiadhesive properties of *Psidium guajava* extract are shown against the fungal pathogens *Candida albicans* (Díaz-de-Cerio et al., 2017). Recent studies (Abirami and Maghimaa, 2019; Vakayil et al., 2019; Abirami et al., 2021; Vakayil et al., 2021a, b, c, d; Baburam et al., 2022; Vakayil et al., 2022a, b) suggest that many phytochemical-mediated nanoparticle syntheses have the capacity to inhibit the adhesion property and loss the cell motility. Victoria et al. (2021) explored the inhibitory action of *Azadirachta indica* extract on the slime layer formed on steel pipe (API5LX80) due to SRB such as *Desulfovibrio alaskensis*, *Desulfovibrio africanus*, and *Desulfovibrio microbium* (McNeil et al., 1991; McNeil and Mohr, 1993; McNeil and Odom, 1994; Javaherdashti, 1999). In this case, the neem extract showed a 50% reduction and propensity in cell growth, sulfide production, sessile cell density, and slime layer production. This biofilm inhibitory effect of the neem extract on MIC is confirmed by field-effect scanning electron microscopy, energy dispersive X-ray spectroscopy, linear polarization resistance method, and open circuit potential (Ravikumar et al., 2021). Organic derivatives provide fine production control in nanoparticle synthesis, resulting in customizable nanocomposites with analyzed properties that improve corrosion resistance and modify corrosion reactions on metals.

1.8 ACTIVE PROTECTION OF THE CORROSION

Active protection is the process by which a potential and reliable system is used to protect against corrosion-induced organisms. Implementing nanocoatings possesses numerous merits, like surface thickness and long-term surface hardness, that allow flexibility, lower maintenance, and improve efficiency. The nanocoating possesses coatings either at the nanoscale level or at lower than 100 nm. Nanocomposite coating provides good coating and better physical coverage due to its fine structures (Rathish et al., 2013).

The particles used in nanocoating are of a fine size, which fills the space and blocks the corrosive elements from diffusing into the substrate surface more efficiently. The boundaries of the nanocomposites have a better adhesive property, so the time period and durability of the coating are increased. It also provides superior mechanical and electronic properties. It is harder and stronger. So, it has the capacity to withstand the corrosion environment and the availability of a replacement chromium-toxic coating. Reducing biofouling and biocorrosion and also responding to environmental parameters like pH, humidity, temperature, heat, stress, radiation, and distortion are the advantages of nanocoatings. In the case of traditional coatings, they possess a different corrosion behavior than nanocoatings. For example, the zinc oxide nanoparticles show better results and have the capacity to overcome corrosion than weldability, which results in poor failure in the peculiar finish. The other extraordinary properties of nanocoatings can be seen in our daily lives, such as in cell phones, eyeglasses, clothes, and computers. In the field of construction, they are used in paints, air filters, windows, flooring, walls, ceramics, and tiles. Nanolayer coatings are used on scratch-resistant, antigraffit, and self-cleaning surfaces. They also possess good antifogging, antifouling, and antiadhesive properties. It prevents many corrosions in biomedical devices, drug delivery, and automobile industries.

Nanocomposite coatings reduce corrosion in two ways. First, it forms a passive film on the surface of the nanocomposite coating. Second is the addition of electrical conductivity in the polymer matrix, which enables anticorrosion activity. The depletion of oxygen on the polymer surface reduces the potential areas and the corrosion effect on the surfaces. The synthetic inhibitors (inorganic polymers) used to nanocomposite test for anticorrosion activity are polythiophene, polypyrrole, and polyaniline coated with oxides like ZnO, $CaCO_3$, and TiO_2. Among these inorganic polymers, TiO_2-polyaniline (TiO_2-PANI) shows a high anticorrosion effect against MIC. Similarly, in the case of plant exudates, inhibitors (organic polymers) are eco-friendly and do not contain excessively deleterious elements. The natural raw material along with the nanoparticles prevents or reduces the destruction of the element. For example, eucalyptus oil-mediated silver, lawsonia extract-mediated silver, and aqueous extract of the tobacco plant and its parts mediated zinc oxide nanoparticles show 50% effectiveness against the adherence of the slime layer caused by *Pseudomonas aeruginosa* and SRB like *Desulfovibrio* and *E. coli* in urine catheters, steelpipes, tanks, and other metal containers in the food, marine, and automobile industries. Among these natural inhibitors or nanocomposite natural polymers, the active ingredient of eucalyptus oil is monomtrene1,8-cineole-mediated Ag nanoparticles, which show better effectiveness and might also be an anticorrosive agent against these MICs in the upcoming days.

1.9 PASSIVE PROTECTION OF CORROSION

Passive corrosion protection means all measures to shield the corrosive media that take place very slowly, even those made of metal components, under the influence of the corrosive media. The main principle of this coating is to prevent corrosion attacks on the metal (Potekhina et al., 1999; Videla and Herrera, 2005; Zuo, 2007; Videla and Herrera, 2009) through physical blocking. In the same way as active protection, passive protection also includes nanocomposites instead of inorganic polymers. According to the Food and Drug Administration, the approved biopolymer is xanthan gum, which is an anionic extracellular polysaccharide that is nontoxic and biocompatible and commonly found in the food sector. Quorum sensing is the communication system that occurred as a result of the density of the MIC. Green synthesis-mediated zinc oxide nanoparticles are used as nanocomposites and have antiquorum sensing potential against biosensor strains of *P. aeruginosa*. Quroum sensing controls the violacein production and has identified the chitinase activity in *C. violaceum* (Venkatramanan et al., 2020). For example, Quroum sensing regulated virulence traits of *S. marscescens* such as protease and prodigiosin production at subinhibitory concentrations of xantham gum-mediated zinc oxide nanoparticles (Husain et al., 2020). The biofilm inhibitory action of ZnOXG against both pathogens like *C. violaceum* and *S. marcescens* exposed its significant reduction through microtiter plate assay. As per Galie et al. (2018), a 64 µg/mL concentration of biofilm formed on the pipeline and steel by the pathogen in the food industry shows a 54% inhibition at 128 µg/mL. Further confirmation and the quantitative analysis of the biofilm inhibition are confirmed through confocal laser scanning microscopy, in which an uncontrolled image of dense aggregations of cells is found (Al-Shabib et al., 2020). It shows reduced surface coverage, a scattered appearance of cells, and disturbed integrity of the biofilm. According to the root extract, *Vetiveria zizanioides*-mediated silver nanoparticles show an inhibitory action at 67 µg/mL concentration against the slime layer formation adhered to the pipes and other automobile and steel industries.

1.10 CONCLUSION

Corrosion resistance in a material is defined as its stability, durability, and importance in identifying the nature of the material. Corrosion research provides information regarding the kinetics and mechanism of corrosion. Nanocoating provides information regarding its structure, size, surface area, grain size, and intermetallic particle distribution. These constituents provide small and dense grain boundaries and also develop new corrosion theories and their interactions with the surface. Nanocomposite coatings using organic and inorganic polymers provide a finer, smoother coating than organic polymers. Biocorrosion is an immensely complicated task. The involvement of microbiological agents in which the environmental factors show drastic change throughout the process. The time factor and succession in microbiology have a great impact on biocorrosion. Using natural plant fibers as composites in nanocoating presents a significant challenge as an antibiocorrosive agent because they have a complex lattice structure. If rigorous studies are done on the quality and pharmacokinetics of the plant fiber-based nanomaterial synthesis, it might be a novel natural antibiofilm agent and also provide safety and efficacy in controlling the microorganism-induced corrosion.

ACKNOWLEDGMENTS

The authors are grateful to the UGC-DAE-CSR, through a Collaborative Research Scheme (CRS) project in Indore, Madhya Pradesh, for the financial support (CRS/2021-22/01/461). This work was partially carried out using the facilities of the UGC-DAE CSR, Indore, and the DST–FIST Centralized Laboratory, Muthayammal College of Arts and Science, Rasipuram, Namakkal, Tamil Nadu, India, for executing this work.

REFERENCES

Abirami, K., and M. Maghimaa. "Phytochemical screening and bioactivity of zingiber officinale to combat the multidrug-resistant bacterial pathogens using foldscope." *Uttar Pradesh Journal of Zoology*, 40 (2019): 67–74.

Abirami, K., Karthikeyan Murugesan, and Maghimaa Mathanmohun. "Phytochemical screening and antibacterial potential of *Piper nigrum* seed extract against the bacterial pathogens." *Research Journal of Agricultural Sciences an International Journal* 12, no. 6 (2021): 2029–2033.

Al-Shabib, Nasser A., Fohad Mabood Husain, Faizan Abul Qais, Naushad Ahmad, Altaf Khan, Abdullah A. Alyousef, Mohammed Arshad et al. "Phyto-mediated synthesis of porous titanium dioxide nanoparticles from Withania somnifera root extract: broad-spectrum attenuation of biofilm and cytotoxic properties against HepG2 cell lines." *Frontiers in Microbiology* 11 (2020): 1680.

APHA. *Standard Methods for the Examination of Water and Wastewater*. 19th ed., American Public Health Association Inc., Washington, DC (1996).

Baburam, Sharmila, Srinivasan Ramasamy, Gnanendra Shanmugam, and Maghimaa Mathanmohun. "Quorum sensing inhibitory potential and molecular docking studies of *Phyllanthus emblica* phytochemicals against *Pseudomonas aeruginosa*." *Applied Biochemistry and Biotechnology* 194, no. 1 (2022): 434–444.

Balaji Ganesh, A., and T. K. Radhakrishnan. "Estimation of microbiologically influenced corrosion of aluminium alloy in natural aqueous environment." *Nature and Science* 4, no. 3 (2006).

Bano, Arjumand Shah, and Javed Iqbal Qazi. "Soil buried mild steel corrosion by *Bacillus cereus*-SNB4 and its inhibition by *Bacillus thuringiensis*-SN8." *Pakistan Journal of Zoology* 43, no. 3 (2011): 555–562.

Bernardez, L. A., and L. R. P. de Andrade Lima. "Improved method for enumerating sulfate-reducing bacteria using optical density." *MethodsX* 2 (2015): 249–255.

Bernardez, L. A., L. R. P. de Andrade Lima, C. L. S. Ramos, and Paulo Fernando de Almeida. "A kinetic analysis of microbial sulfate reduction in an upflow packed-bed anaerobic bioreactor." *Mine Water and the Environment* 31, no. 1 (2012): 62–68.

Chan, Clara S., Sirine C. Fakra, David Emerson, Emily J. Fleming, and Katrina J. Edwards. "Lithotrophic iron-oxidizing bacteria produce organic stalks to control mineral growth: implications for biosignature formation." *The ISME Journal* 5, no. 4 (2011): 717–727.

Chongdar, Shobhana, G. Gunasekaran, and Pradeep Kumar. "Corrosion inhibition of mild steel by aerobic biofilm." *Electrochimica Acta* 50, no. 24 (2005): 4655–4665.

Chunxiang, Qian, Wang Jianyun, Wang Ruixing, and Cheng Liang. "Corrosion protection of cement-based building materials by surface deposition of $CaCO_3$ by *Bacillus pasteurii*." *Materials Science and Engineering: C* 29, no. 4 (2009): 1273–1280.

Cowan, Jennifer K. "Rapid enumeration of sulfate-reducing bacteria." In *Corrosion 2005, Paper No. 05485*. NACE, Houston, TX, p. 16 (2005).

Davis Joseph R. *Metals Handbook*. Desk Edition. ASM International, Member/Customer Service Center, Materials Park, OH, p. 1521 (1998).

De Muynck, Willem, Dieter Debrouwer, Nele De Belie, and Willy Verstraete. "Bacterial carbonate precipitation improves the durability of cementitious materials." *Cement and Concrete Research* 38, no. 7 (2008): 1005–1014.

Dexter, Stephen C. "Galvanic corrosion." (1999).

Díaz-de-Cerio, Elixabet, Vito Verardo, Ana María Gómez-Caravaca, Alberto Fernández-Gutiérrez, and Antonio Segura-Carretero. "Health effects of *Psidium guajava* L. leaves: an overview of the last decade." *International Journal of Molecular Sciences* 18, no. 4 (2017): 897.

Dubiel, M., C. H. Hsu, C. C. Chien, F. Mansfeld, and D. K. Newman. "Microbial iron respiration can protect steel from corrosion." *Applied and Environmental Microbiology* 68, no. 3 (2002): 1440–1445.

Enos, D. G., and S. R. Taylor. "Influence of sulfate-reducing bacteria on alloy 625 and austenitic stainless steel weldments." *Corrosion* 52, no. 11 (1996): 831–842.

Galie, Serena, Coral García-Gutiérrez, Elisa M. Miguélez, Claudio J. Villar, and Felipe Lombó. "Biofilms in the food industry: health aspects and control methods." *Frontiers in Microbiology* 9 (2018): 898.

Gana, Mohamed Lamine, Salima Kebbouche-Gana, Abdelkader Touzi, Mohamed Amine Zorgani, André Pauss, Hakim Lounici, and Nabil Mameri. "Antagonistic activity of *Bacillus* sp. obtained from an Algerian oilfield and chemical biocide THPS against sulfate-reducing bacteria consortium inducing corrosion in the oil industry." *Journal of Industrial Microbiology and Biotechnology* 38, no. 3 (2011): 391–404.

Gerke, T. L., Kirk G. Scheckel, Richard I. Ray, and Brenda J. Little. "Can dynamic bubble templating play a role in corrosion product morphology?" *Corrosion, the Journal of Science and Engineering* 68, no. 2 (2012): 025004–1.

Gunasekaran, G., Shobhana Chongdar, S. N. Gaonkar, and Pradeep Kumar. "Influence of bacteria on film formation inhibiting corrosion." *Corrosion Science* 46, no. 8 (2004): 1953–1967.

Hansen, Douglas C. "Metal corrosion in the human body: the ultimate bio-corrosion scenario." *The Electrochemical Society Interface* 17, no. 2 (2008): 31.

Husain, Fohad Mabood, Imran Hasan, Faizan Abul Qais, Rais Ahmad Khan, Pravej Alam, and Ali Alsalme. "Fabrication of zinc oxide-xanthan gum nanocomposite via green route: attenuation of quorum sensing regulated virulence functions and mitigation of biofilm in gram-negative bacterial pathogens." *Coatings* 10, no. 12 (2020): 1190.

Javaherdashti, Reza. "A review of some characteristics of MIC caused by sulfate-reducing bacteria: past, present and future." *Anti-Corrosion Methods and Materials* 46, no. 3 (1999): 173–180.

Javaherdashti, Reza. *MIC and Cracking of Mild and Stainless Steels*. VDM Verlag Dr Muller Aktiengesellschaft & Co KG, Saarbrucken (2010).

Javaherdashti, Reza. "Microbiologically influenced corrosion (MIC)." In *Microbiologically Influenced Corrosion*. Springer, Cham, pp. 29–79 (2017). https://doi.org/10.1007/978-3-319-44306-5_4

Javaherdashti, Reza, Chikezie Nwaoha, and Henry Tan, eds. *Corrosion and Materials in the Oil and Gas Industries*. CRC Press, Boca Raton, FL (2016).

Jayaraman, A., E. T. Cheng, J. C. Earthman, and T. K. Wood. "Axenic aerobic biofilms inhibit corrosion of SAE 1018 steel through oxygen depletion." *Applied Microbiology and Biotechnology* 48, no. 1 (1997a): 11–17.

Jayaraman, A., E. T. Cheng, J. C. Earthman, and T. K. Wood. "Importance of biofilm formation for corrosion inhibition of SAE 1018 steel by axenic aerobic biofilms." *Journal of Industrial Microbiology and Biotechnology* 18, no. 6 (1997b): 396–401.

Jayaraman, A., J. C. Earthman, and T. K. Wood. "Corrosion inhibition by aerobic biofilms on SAE 1018 steel." *Applied Microbiology and Biotechnology* 47, no. 1 (1997c): 62–68.

Jayaraman, A., P. J. Hallock, R. M. Carson, C.-C. Lee, F. B. Mansfeld, and T. K. Wood. "Inhibiting sulfate-reducing bacteria in biofilms on steel with antimicrobial peptides generated in situ." *Applied Microbiology and Biotechnology* 52, no. 2 (1999a): 267–275.

Jayaraman, A., F. B. Mansfeld, and T. K. Wood. "Inhibiting sulfate-reducing bacteria in biofilms by expressing the antimicrobial peptides indolicidin and bactenecin." *Journal of Industrial Microbiology and Biotechnology* 22, no. 3 (1999b): 167–175.

Jayaraman, A., D. Ornek, D. A. Duarte, C.-C. Lee, F. B. Mansfeld, and T. K. Wood. "Axenic aerobic biofilms inhibit corrosion of copper and aluminum." *Applied Microbiology and Biotechnology* 52, no. 6 (1999c): 787–790.

Jhobalia, Chintan M., An Hu, Tingyue Gu, and Srdjan Nesic. "Biochemical engineering approaches to MIC." In *Corrosion 2005, Paper No. 05500*. NACE International, Houston, TX, p. 12 (2005).

Jonkers, Henk M., Arjan Thijssen, Gerard Muyzer, Oguzhan Copuroglu, and Erik Schlangen. "Application of bacteria as self-healing agent for the development of sustainable concrete." *Ecological Engineering* 36, no. 2 (2010): 230–235.

Joseph, Edith, Anaële Simon, Silvia Prati, Marie Wörle, Daniel Job, and Rocco Mazzeo. "Development of an analytical procedure for evaluation of the protective behaviour of innovative fungal patinas on archaeological and artistic metal artefacts." *Analytical and Bioanalytical Chemistry* 399, no. 9 (2011): 2899–2907.

Jroundi, Fadwa, Patricia Gómez-Suaga, Concepción Jimenez-Lopez, Maria Teresa González-Muñoz, and Maria Antonia Fernandez-Vivas. "Stone-isolated carbonatogenic bacteria as inoculants in bioconsolidation treatments for historical limestone." *Science of the Total Environment* 425 (2012): 89–98.

Kabeerdass, Nivedhitha, Selvam Kandasamy, Gadah Albasher, Ohoud Alamri, Nouf Alsultan, Selvankumar Thangaswamy, and Maghimaa Mathanmohun. "*Limonia acidissima* leaf mediated gold nanoparticles synthesis and their antimicrobial and wound healing properties." *Materials Letters* 314 (2022): 131893.

Kabeerdass, Nivedhitha, Karthikeyan Murugesan, Srinivasan Ramasamy, and Maghimaa Mathanmohun. "Phyto mediated silver nanoparticle synthesis coated fabrics towards antibacterial potential." (2021).

Larson, Allen C., and Robert B. Von Dreele. "Gsas." *Report lAUR*, pp. 86–748 (1994).

Little, B. J., D. J. Blackwood, J. Hinks, F. M. Lauro, E. Marsili, A. Okamoto, S. A. Rice, S. A. Wade, and H.-C. Flemming. "Microbially influenced corrosion-any progress." *Corrosion Science* 170 (2020): 108641.

Little, Brenda J., and Jason S. Lee. *Microbiologically Influenced Corrosion*. John Wiley & Sons, Washington, DC (2007).

Lutzenkirchen-Hecht, D., and H. Strehblow. "Synchrotron methods for corrosion research." In Marcus, P., and Mansfeld, F., eds. *Analytical Methods in Corrosion Science and Engineering*. CRC Press, New York, pp. 169–235 (2006).

Maghimaa, M., and Sulaiman Ali Alharbi. "Green synthesis of silver nanoparticles from *Curcuma longa* L. and coating on the cotton fabrics for antimicrobial applications and wound healing activity." *Journal of Photochemistry and Photobiology B: Biology* 204 (2020): 111806.

Magot, M., G. Ravot, X. Campaignolle, Bernard Ollivier, B. K. C. Patel, M.-L. Fardeau, P. Thomas, J.-L. Crolet, and J.-L. Garcia. "*Dethiosulfovibrio peptidovorans* gen. nov., sp. nov., a new anaerobic, slightly halophilic, thiosulfate-reducing bacterium from corroding offshore oil wells." *International Journal of Systematic and Evolutionary Microbiology* 47, no. 3 (1997): 818–824.

Mansfeld, F., H. Hsu, D. Örnek, T. K. Wood, and B. C. Syrett. "Corrosion control using regenerative biofilms on aluminum 2024 and brass in different media." *Journal of the Electrochemical Society* 149, no. 4 (2002): B130.

McBeth, Joyce M., Brenda J. Little, Richard I. Ray, Katherine M. Farrar, and David Emerson. "Neutrophilic iron-oxidizing "Zetaproteobacteria" and mild steel corrosion in near-shore marine environments." *Applied and Environmental Microbiology* 77, no. 4 (2011): 1405–1412.

McNeil, M. B., J. M. Jones, and B. J. Little. "Mineralogical fingerprints for corrosion processes induced by sulfate reducing bacteria." In *Proc. NACE Corrosion `91, Paper No 580*. NACE, Houston, TX (1991).

McNeil, M. B., and D. W. Mohr. "Formation of copper-iron sulfide minerals during corrosion of artifacts and possible implications for pseudogilding." *Geoarchaeology* 8, no. 1 (1993): 23–33.

McNeil, Michael B., and A. L. Odom. "Thermodynamic prediction of microbiologically influenced corrosion (MIC) by sulfate-reducing bacteria (SRB)." *ASTM Special Technical Publication* 1232 (1994): 173–173.

Mert, Başak Doğru, M. Erman Mert, Gülfeza Kardaş, and Birgül Yazıcı. "The role of *Spirulina platensis* on corrosion behavior of carbon steel." *Materials Chemistry and Physics* 130, no. 1–2 (2011): 697–701.

Miller, John David Allan, and A. K. Tiller. *Microbial Aspects of Metallurgy*. American Elsevier Pub. Co. Inc., NewYork (1970).

Milošev, Ingrid. "Biocorrosion special issue." *Corrosion* 73, no. 12 (2017): 1399–1400.

Nagiub, A., and F. Mansfeld. "Evaluation of microbiologically influenced corrosion inhibition (MICI) with EIS and ENA." *Electrochimica Acta* 47, no. 13–14 (2002): 2319–2333.

Nashima, K., and A. Palanisamy. "Prevalence and distribution of diatoms in the paddy fields of Rasipuram area, Namakkal Dt, Tamilnadu, India." *International Journal of Current Microbiology and Applied Sciences* 5, no. 8 (2016): 402–413.

Neoh, Koon Gee, and En-Tang Kang. "Combating bacterial colonization on metals via polymer coatings: relevance to marine and medical applications." *ACS Applied Materials & Interfaces* 3, no. 8 (2011): 2808–2819.

Neville, A., and T. Hodgkiess. "Study of passive film on stainless steels and high grade nickel base alloy using X-ray photoelectron spectroscopy." *British Corrosion Journal* 35, no. 3 (2000): 183–188.

Örnek, D., A. Jayaraman, B. C. Syrett, C-H. Hsu, F. B. Mansfeld, and T. K. Wood. "Pitting corrosion inhibition of aluminum 2024 by *Bacillus* biofilms secreting polyaspartate or γ-polyglutamate." *Applied Microbiology and Biotechnology* 58, no. 5 (2002a): 651–657.

Örnek, D., T. K. Wood, C. H. Hsu, and F. Mansfeld. "Corrosion control using regenerative biofilms (CCURB) on brass in different media." *Corrosion Science* 44, no. 10 (2002b): 2291–2302.

Parkar, S. G., S. H. Flint, and J. D. Brooks. "Evaluation of the effect of cleaning regimes on biofilms of thermophilic bacilli on stainless steel." *Journal of Applied Microbiology* 96, no. 1 (2004): 110–116.

Parker, C. H. J. *The Influence of Sulphate-Reducing Bacteria on Hydrogen Absorption by Steel during Microbial Corrosion*. Cranfield University, Swindon, United Kingdom, 1990.

Pedersen, Amelie, and Malte Hermansson. "Inhibition of metal corrosion by bacteria." *Biofouling* 3, no. 1 (1991): 1–11.

Potekhina, J. S., N. G. Sherisheva, L. P. Povetkina, A. P. Pospelov, T. A. Rakitina, F. Warnecke, and G. Gottschalk. "Role of microorganisms in corrosion inhibition of metals in aquatic habitats." *Applied Microbiology and Biotechnology* 52, no. 5 (1999): 639–646.

Rathish, R. Joseph, R. Dorothy RM Joany, M. Pandiarajan, and Susai Rajendran. "Corrosion resistance of nanoparticle-incorporated nano coatings." *European Chemical Bulletin* 2, no. 12 (2013): 965–970.

Ravikumar, Arjun, Paul Rostron, Nader Vahdati, and Oleg Shiryayev. "Parametric study of the corrosion of API-5L-X65 QT steel using potentiostat based measurements in a flow loop." *Applied Sciences* 11, no. 1 (2021): 444.

Ray, R. I., J. S. Lee, B. J. Little, and T. L. Gerke. "The anatomy of tubercles on steel." In *Corrosion 2011*. OnePetro (2011).

Roberge, Pierre R. "Library of congress cataloging-in-publication data." (2000).

Romero, Juan Manuel, Elia Velazquez, Jose L. Garcia-Villalobos, Manuel Amaya, and Sylvie Le Borgne. "Genetic monitoring of bacterial populations in a sea water injection system, identification of biocide resistant bacteria and study of their corrosive effect." In *Corrosion 2005*. OnePetro (2005).

Simões, Manuel, Lúcia C. Simões, and Maria J. Vieira. "A review of current and emergent biofilm control strategies." *LWT-Food Science and Technology* 43, no. 4 (2010): 573–583.

Sun, Xiaopeng, Vishnu Priya Veeraraghavan, Krishna Mohan Surapaneni, Sardar Hussain, Maghimaa Mathanmohun, Sulaiman Ali Alharbi, Aref Ali Mohammed Aladresi, and Arunachalam Chinnathambi. "Eugenol-piperine loaded polyhydroxy butyrate/polyethylene glycol nanocomposite-induced apoptosis and cell death in nasopharyngeal cancer (C666-1) cells through the inhibition of the PI3K/AKT/mTOR signaling pathway." *Journal of Biochemical and Molecular Toxicology* 35, no. 4 (2021): e22700.

Tatnal, R. E. Introduction. In Kobrin, G., ed. *Microbiologically Influenced Corrosion*. National Association of Corrosion Engineers, Houston, TX (1993).

TPC-3. *The Role of Bacteria in the Corrosion of Oil Field Equipment*. National Association of Corrosion Engineers, Third printing (1982).

Vakayil, Ramsi, T. Abdul Nazeer, and Maghimaa Mathanmohun. "Evaluation of the antimicrobial activity of extracts from *Acorus calamus* rhizome against multidrug-resistant nosocomial pathogens." *Research Journal of Agricultural Sciences* 12 (2021a): 1613–1617.

Vakayil, Ramsi, Murugesan Anbazhagan, Gnanendra Shanmugam, Srinivasan Ramasamy, and Maghimaa Mathanmohun. "Molecular docking and in vitro analysis of phytoextracts from B. serrata for antibacterial activities." *Bioinformation* 17, no. 7 (2022a): 667. Antibacterial activity and molecular characteristics of Indian olibanum (*Boswellia Serrata*) phytochemicals: an in silico approach. *Applied Ecology and Environmental Research* 20, no. 2, 919–929.

Vakayil, Ramsi, Nivedhitha Kabeerdass, Abirami Kuppusamy, and Maghimaa Mathanmohun. "Phytochemical screening and antibacterial properties of *Punica granatum* extracts against gastrointestinal infection an in-vitro study." *Uttar Pradesh Journal of Zoology* 40 (2019): 25–32.

Vakayil, Ramsi, Nivedhitha Kabeerdass, Rajkumar Srinivasan, Gnanendra Shanmugam, Srinivasan Ramasamy, and Maghimaa Mathanmohun. "Invitro and insilico studies on antibacterial potentials of phytochemical extracts." *Materials Today: Proceedings* 47 (2021b): 453–460.

Vakayil, Ramsi, Sivakumar Krishnamoorthy, Murugesan Anbazhagan, Nachimuthu Senthil Kumar, and Maghimaa Mathanmohun. "Antibacterial potential of *Acorus calamus* extracts against the multi-drug resistant nosocomial pathogens." *Uttar Pradesh Journal of Zoology* 42 (2021c): 144–150.

Vakayil, Ramsi, Sankareswaran Muruganantham, Nivedhitha Kabeerdass, Manikandan Rajendran, Srinivasan Ramasamy, Tahani Awad Alahmadi, Hesham S. Almoallim, Velu Manikandan, and Maghimaa Mathanmohun. "*Acorus calamus*-zinc oxide nanoparticle coated cotton fabrics shows antimicrobial and cytotoxic activities against skin cancer cells." *Process Biochemistry* 111 (2021d): 1–8.

Vakayil, Ramsi, Srinivasan Ramasamy, Tahani Awad Alahmadi, Hesham S. Almoallim, Nandakumar Natarajan, and Maghimaa Mathanmohun. *Boswellia serrata*-mediated zinc oxide nanoparticles-coated cotton fabrics for the wound healing and antibacterial applications against nosocomial pathogens. *Applied Nanoscience* 12, no. 10 (2022b): 2873–2887. https://doi.org/10.1007/s13204-022-02573-9.

Venkatramanan, Mahendrarajan, Pitchaipillai Sankar Ganesh, Renganathan Senthil, Jeyachandran Akshay, Arumugam Veera Ravi, Kulanthaivel Langeswaran, Jamuna Vadivelu, Samuthira Nagarajan, Kaliaperumal Rajendran, and Esaki Muthu Shankar. "Inhibition of quorum sensing and biofilm formation in *Chromobacterium violaceum* by fruit extracts of *Passiflora edulis*." *ACS Omega* 5, no. 40 (2020): 25605–25616.

Victoria, S. Noyel, Akansha Sharma, and R. Manivannan. "Metal corrosion induced by microbial activity-mechanism and control options." *Journal of the Indian Chemical Society* 98, no. 6 (2021): 100083.

Videla, Héctor A., and Liz K. Herrera. "Microbiologically influenced corrosion: looking to the future." *International Microbiology* 8, no. 3 (2005): 169.

Videla, Hector A., and Liz Karen Herrera. "Understanding microbial inhibition of corrosion. A comprehensive overview." *International Biodeterioration & Biodegradation* 63, no. 7 (2009): 896–900.

Webster, B. J., S. E. Werner, D. B. Wells, and P. J. Bremer. "Microbiologically influenced corrosion of copper in potable water systems-pH effects." *Corrosion* 56, no. 9 (2000): 942–950.

Zhu, Xiangyang, John Kilbane, Alvin Ayala, and Hetal Modi. "Application of quantitative, real-time PCR in monitoring microbiologically influenced corrosion (MIC) in gas pipelines." In *Corrosion 2005*. OnePetro (2005).

Zuo, Rongjun. "Biofilms: strategies for metal corrosion inhibition employing microorganisms." *Applied Microbiology and Biotechnology* 76, no. 6 (2007): 1245–1253.

2 Impacts of Biofouling on Marine Environments and Its Prevention

Neha Bhadauria
Amity University Noida

Preeti Pradhan
Indore Institute of Science and Technology

Arjun Suresh
Save Vibrant Earth Foundation

2.1 INTRODUCTION

Biofouling is a broad expression often used to refer to the adhesion and proliferation of aquatic creatures on manufactured solid structures. It is an unwelcome growth that interferes with the surface. It is a worldwide crisis for oyster, finfish, and seaweed farming and involves the establishment and growth of undesirable aquatic life on both natural and manufactured surfaces (Bannister et al., 2019). One of the primary obstacles to effective and sustainable production in marine aquaculture is biofouling. Ecological catastrophes are driven by biofouling in the shipping sector, which costs billions each year. When creatures adhere to a ship's hull, drag and the outer layer's smoothness deteriorate. This causes the hydrodynamic mass to rise, which lowers the maximum speed and reduces dexterity (Lindholdt et al., 2015). However, because aquaculturists deploy a wide range of management strategies and costing procedures, the expense of biofouling usually fluctuates throughout farms, creatures, and organizations. Problems with maritime buildings like pier pilings and oil platforms, as well as shipping, are the main effects of fouling in the marine environment on humans. On aquaculture farms, the nets that make up the cages are also subject to fouling. The significant growth of fouling organisms reduces the flow of water through a cage, which leads to oxygen depletion. This bioprocess has an impact on structures like pipes, water intake systems, desalination equipment, probes, and other objects in the aquatic domain. Biofouling is one of the major issues that have made it difficult to use the seas for commercial purposes (Archana and Sundaramoorthy, 2019; Figure 2.1).

 DOI: 10.1201/9781003451457-2

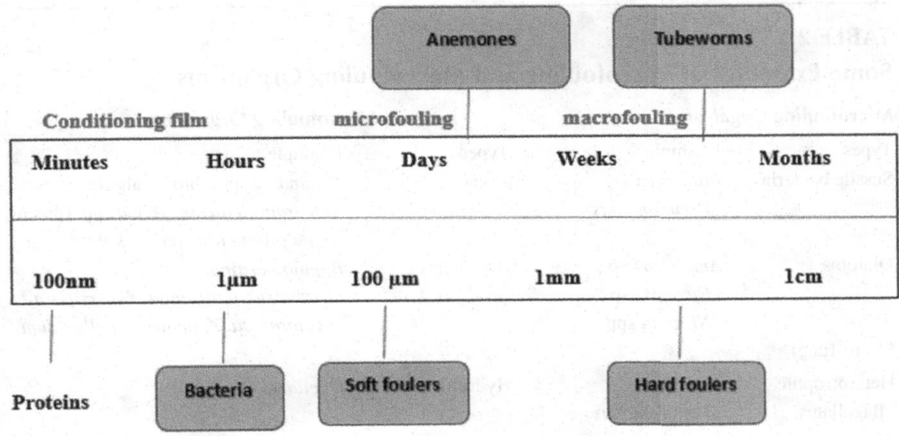

FIGURE 2.1 The biofilm formation process and biofouling.

2.2 BIOFOULING: MICRO- AND MACRO-FOULING

When a clean surface is immersed in water for a few minutes, it adsorbs a molecular conditioning film composed of solubilized organic material. Bacteria, unicellular algae, and cyanobacteria can colonize within hours (blue-green algae). These initial small colonizers form a biofilm, which is an agglomeration of attached cells also known as "microfouling" or "slime." A macrofouling community (either 'soft fouling' or 'hard fouling') may form and outgrow the microfouling. Algae and invertebrates such as soft corals, sponges, anemones, tunicates, and hydroids are examples of soft fouling, whereas barnacles, mussels, and tubeworms are examples of hard fouling. The organisms that grow in a fouling community are determined by the substrate, geographical location, season, and other variables such as competition and predation. The process of fouling is extremely dynamic (Table 2.1).

2.3 BIOFOULING EFFECTS ON AQUACULTURE

Clusters of foulants, which can also comprise parasites and diseases, may compete directly with cultured species for nutrition. Form (shellfish) with fouling problems is difficult to sell due to aesthetic concerns or since the fouling impedes generic product manufacturing or wrapping operations. In biofouling habitats, a variety of bottom-sea creatures produce secondary compounds that act as organic antifoulants or ant predators; however, biofouling can also have direct hazardous effects. Other marine organisms, including cultivated organisms, may be harmful to these substances. Deoxygenation and degradation products play a significant role in the significant losses of cultures when biofouling populations perish or simply overwhelm the cultures/cages, blocking the interchange.

Shellfish aquaculture is linked to a diverse range of biofouling microbes worldwide. There are three major categories into which these species effects on farm infrastructure and shellfish can be divided: (1) added weight; (2) physical injury; and (3) a decline in shellfish fitness, and those effects could result in a significant decrease in farm productivity, in addition to the expense of biofouling treatment.

TABLE 2.1
Some Examples of Microfouling and Macrofouling Organisms

Microfouling Organisms		Macrofouling Organisms	
Types	Example	Types	Example
Sessile bacteria	*Micrococcus*, *Pseudomonas*	Algae	*Laminaria* spp. (brown alga) *Enteromorpha* spp., *Ulva* spp. (green algae) *Ahnfeltia* spp. (red alga)
Diatoms	*Amphora* spp., *Navicula* sp., *Nitschia* spp.	Bryozoans	*Bugula neritina*, *Cryptosulapallasiana*, *Watersipora subtorquata*, *Zoobotryon pellucidum*
Micro-fungi		Corals	
Heterotrophic flagellates	*Monosiga*, *Pteridomonas*	Hydroids	*Obelia* sp.
Sessile ciliates		Sea cucumbers	
Barnacles	*Amphibalanus amphitrite*, *Amphibalanus reticulatus*, *Balanus amphitrite*	Sponges	Acanthella cavernosa

2.3.1 Added Weight

Because of the drastically increased load of the stock and culture equipment due to fouling, stock segregation in long line culture results in higher buoyancy and anchorage costs (Fitridge et al., 2012). The consequences of shellfish's excess weight are less frequently evaluated than those of finfish and seaweed culture, leaving yield consequences largely unexplored.

2.3.2 Physical Injury

Boring creatures like clionid sponges and polychaete worms of the class Polydora can enter shellfish and create boils, fissures, and holes. This leads to an unorganized evolution of the shells and joints, as well as thinner coatings that are vulnerable to illnesses, pathogens, and predation (Fitridge et al., 2012). Cultured seaweeds are significantly heavier due to biofouling, which makes them more likely to break and become dislodged. The lamina of certain fouling varieties makes grown kelp brittle, which enhances breakage vulnerability (Kim et al., 2017).

2.3.3 Decline in Shellfish Fitness

One of the biggest consequences of foulants in shellfish farming is the decline in shellfish health. Usually, these reductions directly conflict with other organisms for nutrients or indirectly by suffocating or conflicting with the correct working of valves (Woods et al., 2012). Using exploratory mussel lines, Sievers et al. (2013) conducted a research study to add biofouling and discovered distinctive proof of

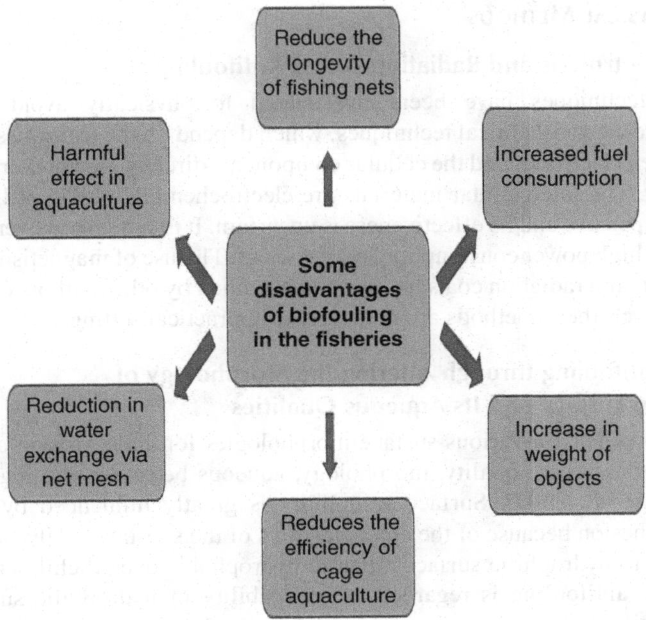

FIGURE 2.2 Impacts of biofouling in aquaculture.

industrially significant decreases in the densities of fat and development mussels. Although there is very limited evidence available to counter physiological declines (Sievers et al., 2017).

2.3.4 Farm Productivity

Estimating the overall negative effects of biofouling on farm output has garnered significant interest due to the above-mentioned effects and the high cost of eradication. The New Zealand economy would lose US$16.4 million a year, or 10% of its business, as a result of *M. galloprovincialis* mussel farms. This cost was based on impairments in yield and seed-stock availability (Forrest and Atalah, 2017) (Figure 2.2).

2.4 ANTIFOULING METHODS

High water temperatures are the primary factor influencing the breeding seasons and rates of growth of biofouling species. Unfortunately, there is nothing that can be done to significantly change these crucial elements. From a different perspective, research into how microscale physical parameters affect biofouling during the development of microtechnology has prompted the creation of some fresh anticorrosive techniques that include altering physical characteristics, and some concentrated extracts of these metabolites may work well as antifouling agents. Three categories—chemical, physical, and biological—can be used to classify antifouling techniques (Bannister et al., 2019).

2.4.1 Physical Methods

2.4.1.1 Electrolysis and Radiation-Based Antifouling

Numerous techniques have been investigated to physically avoid biofouling. Microcosmic electrochemical techniques, which depend on the transmission of electrons between electrodes and the cellular components directly are another way to prevent fouling. The intercellular materials are electrochemically oxidized as a result, but this is expensive and its effectiveness is uncertain. It is difficult to overcome these approaches' high power consumption, nevertheless. The use of magnetism, Exposure to ultraviolet, and radiation coats has also been studied by other authors (Yebra et al., 2004); however, these methods are not relevant in practical settings.

2.4.1.2 Antifouling through Altering the Morphology of the Surface and Its Aqueous Qualities

The latest evidence on various surface morphologies for anticorrosion applications has focused on surface quality, morphology, aqueous behavior, and higher viscosity (Bowen et al., 2007). Surface wettability is greatly influenced by biofouling organism adhesion because of the characteristics of the substance. Ulva spores cling more firmly to hydrophilic surfaces than to hydrophobic ones (Schilp et al., 2007). Additionally, antifouling is regarded as a capability of hydrophilic surfaces. For instance, antifouling behaviors are demonstrated when metal nanoparticles, such as TiO_2, are added. This is because the surface becomes more hydrophilic due to the photocatalytic activities brought about by solar ultraviolet light, making it easier to remove the biofilm that forms (Dineshrama et al., 2009). The development of an interface that provides both hydrophilic and hydrophobic zones to settle (attach) organisms and bacteria was encouraged by the disparities in settlement and adhesion behavior. Such surfaces' various patterns have also been assessed (Grozea et al., 2009). The experimental phase of surface topography modification research to produce antifouling is currently ongoing. Instead of using predictive models, the discovery of efficient antifouling topographies often involves trial and error.

2.4.1.3 Changes in Zeta's Potential for Antifouling

It is envisaged that electrostatic attraction will be crucial for microbiological adhesion, which is complicated and impacted by numerous physical reactions (). Assessing the opportunities of the substrates or microfouling species has thus been employed to evaluate their correlation. Microbes are charged negatively in pure saltwater (pH 7–8) and across a large pH range (Herrwerth et al., 2003). As a result, the adhesion process is facilitated by electrostatic contact between cells and metal cations. Other industries have found success in reducing biofouling by altering the pH of the solution, which shifts the surface charges of the bacterial cells (Mølbak et al., 2007). However, in the case of marine biofouling, it is not viable. As a result, efforts have been undertaken to alter the surfaces' charge potential in an effort to partially prevent cells from adhering to them.

2.4.2 Chemical Methods

The first improvement in the form of lead and copper sheets to shield their wooden vessels from biofouling is credited to the ancient Phoenicians. By the late 18th and early nineteenth centuries, vessels were regularly painted with metal coatings (Dafforn et al., 2011). Since the mid-20th century, organotin chemicals and their variants have been extensively used as antifouling coverings due to their effectiveness against a range of fouling agents. Tributyltin oxide ($C_{24}H_{54}OSn_2$) and tributyltin fluoride ($C_{12}H_{27}SnF$) are the most widely used antifoulants. These substances are potent fungicides and entirely prohibit most fouling organisms from growing at their lowest concentrations (Amara et al., 2018).

2.4.3 Traditional Technique: Organotin-Based Antifouling

Depending on the chemistry of the paint matrix and the processes used to release poisonous chemicals, these coatings can be divided into three groups. Traditional antifouling paints often have a resin matrix made of water that is present on the coverings. The biocides have a lifespan of between 6 and 12 months and are oxides of lead, arsenic, mercury, or copper. The matrix for long-lasting antifouling coatings needs to be impermeable to water. After 18–24 months, coatings must be renewed because high levels of toxicants are still retained in the paints, but there is no longer a sufficient quantity at the surface to be useful. The biocide, primarily tributyltin (TBT), is attached to the polymer network in self-polishing anticorrosive painters and freed by breakdown at the surface. The discharge rate remains constant despite being influenced by fluid velocity. Many governments have restricted the use of organotins as a result of their negative effects on marine species. For example, France restricted the application of TBT-based antifouling coatings in 1982 following large decreases in the number of oyster farms as a result of reduced oyster production, anomalies in the larval stage, and shell deformation afflicting 80%–100% of adult oysters (Terlizzi et al., 2001). The intersex effects of TBT exposure have been identified as yet another negative effect (Bryan et al. 1986).

2.4.4 Biological Methods

In recent years, successful research has been done on the use of enzymes as antifouling agents. Numerous enzyme types have been reported to exhibit antifouling properties, including hydrolases, lyases, isomerases, and ligases (Jakob et al., 2008). Biofouling issues are brought on by the development and growth of biofilms as well as the attachment of microorganism spores and larvae, according to enzymatic antifouling technology. Consequently, the following four categories can be used to group the roles of enzymes in antifouling applications:

2.4.4.1 Enzymes That Break Down Settling Adhesives

The attachment mechanism throughout macrofouling is heavily dependent on proteins and proteoglycans. Proteases are a sort of enzyme that can be used to disintegrate mucilage composed of peptides and thereby minimize biofouling since they have the ability to solubilize polypeptide chains at different locations. For example, serine protease can effectively stop Ulva spores, barnacle cyprids, and bryozoans from attaching (Pettitt et al., 2004).

The procedure is more difficult in the case of microfouling (Leroy et al., 2008), as polysaccharide-based adhesives are equally crucial to secondary adhesion as proteins are. The degradation of polysaccharides is often carried out by glycosylase. However, it is harder to breakdown polysaccharides due to the complexity of the process and the fact that glycosylase can only target a specific subset of links. As a result, selecting a suitable glycosylase for broad-spectrum antifouling would be difficult (Leroy et al., 2008).

Biofilms are extremely complicated due to the diversity of extracellular polysachharides (EPS), and the breakdown of their polymer films would require a very diverse array of hydrolase and lyase pairings (Jakob et al., 2008). Additionally, due to the high adaptability of biofilms to environmental conditions, the degeneration of a key component will result in the production of substitute components, which will take the place of the originals and create a new network for the proliferation of the organisms. Consequently, the biofilm won't degrade. It has been found that alginase can dissolve a thin biofilm, but it has no influence on a biofilm that has previously reached its full capacity. (Joao et al., 2008). As a result of the biofilm's intricacy and flexibility, the antifouling technique of rupturing the biofilm matrix may not be appropriate or successful.

2.4.4.2 Enzymes That Produce Deterrents and Biocides

Glucose and hexose oxidase are employed to produce oxidative damage in living cells caused by hydrogen peroxide (Imlay, 2008), while haloperoxidase catalyzes the synthesis of hypohalogenic acids, which are typically utilized in water treatment systems as disinfectants. Additionally, hydrogen peroxide will decompose into oxygen and water at a rather fast pace in seawater (Carman et al., 2006).

2.4.4.3 Enzymes That Obstruct Cell-to-Cell Communication

According to studies, some Gram-negative genera require *N*-acyl homoserine lactones (AHL) in order to sense the presence of a quorum (Waters, 2005). Eliminating AHL could therefore stop the growth of bacterial fouling. AHL acylase breaks down AHL, and when this enzyme's concentration rises, biofilm formation is hindered as well as the colonization of Ulva spores and Polychaeta larvae (Huang et al., 2007). As a result, AHL acylases can partially prevent microorganisms from settling.

2.5 PREVENTION OF BIOFOULING IN SHELLFISH AQUACULTURE

There are several benefits to preventing dangerous organisms from impacting aquaculture as opposed to treating them. A higher emphasis has been given on preventing marine biofouling and increasing productivity by lowering the frequency and severity of treatments as well as the direct effects of fouling organisms. Spatiotemporal restraint of biofouling can be possible in areas with predictable seasonal fouling patterns (Holthuis et al., 2015), and recently efforts have been made to study biofouling formation at aquaculture farms around the globe. Complex avoidance tactics are also being made possible by modeling biofouling settlement and development. For instance, in New Zealand, simulation of mussel spat-fall and the development of an easily usable interactive application have enabled workers to enthusiastically determine where to place lines to prevent mussel fouling (Fletcher et al., 2017). These findings may be useful in making additional informed decisions regarding the frequency of spat selection,

TABLE 2.2
Common Antifouling Methods Adopted to Prevent Biofouling

Antifouling Methods	Types	Limitations
Physical	• Surface modification • Electrolysis and radiation	Surface modifications are a good approach but more studies are needed to understand the adhesion mechanism of the organisms.
Chemical	• Organotin, e.g., TBT (Banned for marine Application) • Copper with booster biocides • Self-Polishing Copolymers (SPCs) • Foul release coating • Chlorination (Desalinization and power plants)	Many countries have regulations for the use of toxic biocides in antifouling coatings
Biological	• Enzyme-based coating • Natural product	Alternative to toxic biocides in antifouling coatings

Antifouling (AF) remedies, and husbandry practices that periodically reduce fouling, such as re-socking and sorting. Data from models and observation can also be utilized to prohibit stock culture in appropriate places during times when there is heavy fouling by particularly harmful invaders (Sievers et al., 2014).

As aquaculture production transitions into advanced culture developments, futuristic studies will necessitate solving the unresolved problems in current aquaculture processes and additionally, to broadening the scope to include a larger diversity of microbial environments (e.g., offshore aquaculture and closed culture systems). It is possible to design advanced substances and/or AF coats for arrangement as well as technological advancements such as machinery for more effective assessment, modeling, and automatic washing by studying how biofouling settles and develops in these situations. Aquaculture can contribute to world food security in the modern age, but biofouling is a bigger obstacle that needs to be removed (Table 2.2).

2.6 CONCLUSION

Biofouling in marine aquaculture is a massive problem in which both the target culture organisms and/or infrastructural facilities are subjected to a vast range of microbes, resulting in significant production impacts. It is a major management issue that leads to increased operational costs and negative effects on the organisms being cultured. Remarkably, for a concern with such a large impact in a rapidly growing global industry, information on its effects and costs is scarce. Despite some benefits, fouling is typically an expensive problem for today's aquaculturist. Numerous toxic and nontoxic antifouling techniques have been invented to mitigate biofouling problems. Unfortunately, the majority of biofouling control methods are either environmentally damaging or, in certain cases, deemed effective. Maintenance is frequently required, and expensive antifouling procedures are essential to farming techniques. Clearly, there is an imperative necessity for effective, ecologically sound antifouling agents and procedures, especially considering the explosive expansion of the fish farming industry. Furthermore, reducing marine biofouling necessitates the development of green, sustainable, feasible, appropriate, and nontoxic antifouling technology.

REFERENCES

Amara, Intissar, Wafa Miled, Rihab Ben Slama, and Neji Ladhari. "Antifouling processes and toxicity effects of antifouling paints on marine environment. A review." *Environmental Toxicology and Pharmacology* 57 (2018): 115–130.

Archana, S., and B. Sundaramoorthy. "Review on biofouling prevention using nanotechnology." *Journal of Entomology and Zoology Studies* 7 (2019): 640–648.

Bannister, Jana, Michael Sievers, Flora Bush, and Nina Bloecher. "Biofouling in marine aquaculture: a review of recent research and developments." *Biofouling* 35, no. 6 (2019): 631–648.

Bowen, J., M. E. Pettitt, Kevin Kendall, G. J. Leggett, J. A. Preece, M. E. Callow, and J. A. Callow. "The influence of surface lubricity on the adhesion of *Navicula perminuta* and *Ulva linza* to alkanethiol self-assembled monolayers." *Journal of the Royal Society Interface* 4, no. 14 (2007): 473–477.

Bryan, G. W., P. E. Gibbs, L. G. Hummerstone, and G. R. Burt. "The decline of the gastropod Nucella lapillus around south-west England: evidence for the effect of tributyltin from antifouling paints." *Journal of the Marine Biological Association of the United Kingdom* 66, no. 3 (1986): 611–640.

Carman, Michelle L., Thomas G. Estes, Adam W. Feinberg, James F. Schumacher, Wade Wilkerson, Leslie H. Wilson, Maureen E. Callow, James A. Callow, and Anthony B. Brennan. "Engineered antifouling microtopographies–correlating wettability with cell attachment." *Biofouling* 22, no. 1 (2006): 11–21.

Dafforn, Katherine A., John A. Lewis, and Emma L. Johnston. "Antifouling strategies: history and regulation, ecological impacts and mitigation." *Marine Pollution Bulletin* 62, no. 3 (2011): 453–465.

Dineshram, R., R. Subasri, K. R. C. Somaraju, K. Jayaraj, L. Vedaprakash, Krupa Ratnam, S. V. Joshi, and R. Venkatesan. "Biofouling studies on nanoparticle-based metal oxide coatings on glass coupons exposed to marine environment." *Colloids and Surfaces B: Biointerfaces* 74, no. 1 (2009): 75–83.

Fitridge, Isla, Tim Dempster, Jana Guenther, and Rocky De Nys. "The impact and control of biofouling in marine aquaculture: a review." *Biofouling* 28, no. 7 (2012): 649–669.

Fletcher, Lauren M., Anastasija Zaiko, Javier Atalah, Ingrid Richter, Celine M. Dufour, Xavier Pochon, Susana A. Wood, and Grant A. Hopkins. "Bilge water as a vector for the spread of marine pests: a morphological, metabarcoding and experimental assessment." *Biological Invasions* 19 (2017): 2851–2867.

Forrest, Barrie M., and Javier Atalah. "Significant impact from blue mussel *Mytilus galloprovincialis* biofouling on aquaculture production of green-lipped mussels in New Zealand." *Aquaculture Environment Interactions* 9 (2017): 115–126.

Grozea, Claudia M., Nikhil Gunari, John A. Finlay, Daniel Grozea, Maureen E. Callow, James A. Callow, Zheng-Hong Lu, and Gilbert C. Walker. "Water-stable diblock polystyrene-*block*-poly (2-vinyl pyridine) and diblock polystyrene-*block*-poly (methyl methacrylate) cylindrical patterned surfaces inhibit settlement of zoospores of the green alga *Ulva*." *Biomacromolecules* 10, no. 4 (2009): 1004–1012.

Herrwerth, Sascha, Wolfgang Eck, Sven Reinhardt, and Michael Grunze. "Factors that determine the protein resistance of oligoether self-assembled monolayers– internal hydrophilicity, terminal hydrophilicity, and lateral packing density." *Journal of the American Chemical Society* 125, no. 31 (2003): 9359–9366.

Holthuis, Thomas Dunér, Per Bergström, Mats Lindegarth, and Susanne Lindegarth. "Monitoring recruitment patterns of mussels and fouling tunicates in mariculture." *Journal of Shellfish Research* 34, no. 3 (2015): 1007–1018.

Huang, Yi-Li, Sergey Dobretsov, Jang-Seu Ki, Lai-Hung Yang, and Pei-Yuan Qian. "Presence of acyl-homoserine lactone in subtidal biofilm and the implication in larval behavioral response in the polychaete *Hydroides elegans*." *Microbial Ecology* 54, no. 2 (2007): 384–392.

Imlay, James A. "Cellular defenses against superoxide and hydrogen peroxide." *Annual Review of Biochemistry* 77 (2008): 755–776.

Kim, Jang K., Charles Yarish, EunKyoung Hwang, Miseon Park, Youngdae Kim, Jang K. Kim, Charles Yarish, EunKyoung Hwang, Miseon Park, and Youngdae Kim. "Seaweed aquaculture: cultivation technologies, challenges and its ecosystem services." *Algae* 32, no. 1 (2017): 1–13.

Leroy, C., C. Delbarre, F. Ghillebaert, Chantal Compere, and D. Combes. "Effects of commercial enzymes on the adhesion of a marine biofilm-forming bacterium." *Biofouling* 24, no. 1 (2008): 11–22.

Lindholdt, A., K. Dam-Johansen, D. M. Yebra, S. Kiil, C. E. Weinell, and S. M. Olsen. "Fuel efficiency and fouling control coatings in maritime transport-DTU Orbit (13/12/2018)." (2015).

Mølbak, Lars, Molin Søren, and Niels Kroer. "Root growth and exudate production define the frequency of horizontal plasmid transfer in the rhizosphere." *FEMS Microbiology Ecology* 59, no. 1 (2007): 167–176.

Pettitt, M. E., S. L. Henry, M. E. Callow, J. A. Callow, and A. S. Clare. "Activity of commercial enzymes on settlement and adhesion of cypris larvae of the barnacle *Balanus amphitrite*, spores of the green alga *Ulva linza*, and the diatom *Navicula perminuta*." *Biofouling* 20, no. 6 (2004): 299–311.

Schilp, Soeren, Alexander Kueller, Axel Rosenhahn, Michael Grunze, Michala E. Pettitt, Maureen E. Callow, and James A. Callow. "Settlement and adhesion of algal cells to hexa (ethylene glycol)-containing self-assembled monolayers with systematically changed wetting properties." *Biointerphases* 2, no. 4 (2007): 143–150.

Sievers, Eric L., and Peter D. Senter. "Antibody-drug conjugates in cancer therapy" *Annual Review of Medicine* 64 (2013): 15–29.

Sievers, Michael, Tim Dempster, Isla Fitridge, and Michael J. Keough. "Monitoring biofouling communities could reduce impacts to mussel aquaculture by allowing synchronisation of husbandry techniques with peaks in settlement." *Biofouling* 30, no. 2 (2014): 203–212.

Sievers, Michael, Isla Fitridge, Samantha Bui, and Tim Dempster. "To treat or not to treat: a quantitative review of the effect of biofouling and control methods in shellfish aquaculture to evaluate the necessity of removal." *Biofouling* 33, no. 9 (2017): 755–767.

Terlizzi, Antonio, Simonetta Fraschetti, P. Gianguzza, M. Faimali, and Ferdinando Boero. "Environmental impact of antifouling technologies: state of the art and perspectives." *Aquatic Conservation: Marine and Freshwater Ecosystems* 11, no. 4 (2001): 311–317.

Waters, Christopher M., and Bonnie L. Bassler. "Quorum sensing: cell-to-cell communication in bacteria." *Annual Review of Cell and Developmental Biology* 21, no. 1 (2005): 319–346.

Woods, Chris, Oliver Floerl, and Barbara J. Hayden. "Biofouling on Greenshell™ mussel (*Perna canaliculus*) farms: a preliminary assessment and potential implications for sustainable aquaculture practices." *Aquaculture International* 20, no. 3 (2012): 537–557.

Yebra, Diego Meseguer, Søren Kiil, and Kim Dam-Johansen. "Antifouling technology—past, present and future steps towards efficient and environmentally friendly antifouling coatings." *Progress in Organic Coatings* 50, no. 2 (2004): 75–104.

3 Metagenomic Analysis of Acid Mine Drainage, Presence of Acidometallophiles, and Their Possible Role in Biomining

Bedaprana Roy, Riddhi Chakraborty, Niti Choudhury, Aindri Ghosh, Rajeswari Chakraborty, Jaydip Ghosh, and Arup Kumar Mitra
St Xavier's College (Autonomous)

3.1 INTRODUCTION

'Metagenomics' is the study of the genome of the microbes of a particular environment, which involves the analysis of the dominant microbial strains along with the rare ones and also their metabolic potentials by comparing DNA databases. It is a new and developing field that deals with the genomic analysis of the microbiota of a certain habitat. It involves the direct isolation of the genetic material from a place, followed by its cloning and amplification in order to reconstruct the entire microbiome for further study (Sleator et al., 2008). The major metabolic capabilities of the dominant microbes in the acid mine drainage (AMD) include sulfur and iron oxidation, carbon, and nitrogen metabolisms, some of which share a syntropic relationship and play a vital role in energy conservation. Most of the rare taxa of the AMD environment are found to perform sulfur oxidation and nitrogen fixation. These metabolic capabilities not only contribute to the AMD generation but also help its microbiota adapt to harsh environmental conditions like low pH, heavy metals, and oxidative stress by reactive oxygen species.

The acid mines are generally characterized by high levels of iron, sulfur, and other heavy metals like Zn, Cu, Cr, Co and Pb, and a pH below 3. The total organic carbon content is also quite high. Pyrosequencing analysis has revealed that 99.7% of the AMD consists of bacteria and 0.3% archaea. Proteobacteria, including (1) *Betaproteobacteria*, (2) *Gammaproteobacteria* and (3) *Alphaproteobacteria, Firmucutes,* and *Nitrospora,* account for the majority of the bacterial community.

 DOI: 10.1201/9781003451457-3

The rare microbial strains with a relative abundance of less than 1% include *Acidithiobacillus* closely related to *At. ferrooxidans* and *At. thiooxidans*, strains similar to *Leptospirillum rubarum*, *Alicyclobacillus acidocaldarius*-like strains, and unclassified strains with 96% similarity to *Ferrovum myxofaciens,* most of which are acidophilic. The metabolic potential analysis revealed many central pathways like carbon metabolism, nitrogen metabolism, purine and pyrimidine metabolism and even amino acid biosynthesis. By using transcriptional analysis, the study of certain enzymes used by the AMD microorganisms for their survival in extreme environmental conditions like heavy metals and low pH has been made possible. Other survival strategies adopted by the AMD microbes include (1) high impermeability of the cell membrane to protons, (2) protons are pumped out using antiporters Na^+/H^+ or K^+/H^+, (3) maintaining positive membrane potential through uptake of potassium, and (4) degrading organic acids. This indicates the huge stress factors due to which they have to adopt certain survival strategies (Hua et al., 2015). The AMD microbial community can contribute to biomining, which is of both scientific and commercial importance. Bacteria and Archaea have certain mechanisms to detoxify the metal ions through several transformations. They have the ability to transform the toxic metals into less volatile forms by intracellular and extracellular complexation in the periplasm. Certain chemolithotrophs can be used as fuels by aerobic mineral oxidation, which would also contribute to the extraction of metals. The microbial solubilization of metals can be commercially used in industries for bioleaching or biomining of mineral ores. The acidophilic microbes of the AMD can pose a serious threat to the soil's quality by mobilizing metals as well. But on the other hand, Bioremediation can be made possible by using the specific properties of the microbe-metal interaction (Jerez, 2009).

3.2 CONDITIONS IN MINES: HEAVY METAL CONCENTRATION AND LOW pH CONDITIONS

Our environment naturally contains heavy metals, but anthropogenic activities like mining also release heavy metals in an unrestrained way that leads to the production of huge amounts of heavy metal-laden waste, which is responsible for polluting our surroundings. Some heavy metals are essential for normal physiological life processes, but higher concentrations might pose health hazards for humans. Mining and processing of ores of different metals like gold, copper, etc. significantly contaminates the environment. When sulfide-rich wastes generated due to such mining, extraction of minerals and processing, come into contact with the environment, they generate toxic compounds with low pH (Ilhwan Park et al., 2019) that drain into ecosystems and pose a serious threat to life. The study by Prafulla Kumar Sahoo et al. for over 15 years elucidated the risk associated with iron ore mining and its environmental impact. They showed that the concentrations of elements like cadmium, arsenic, manganese, copper, lead, nickel, cobalt, and zinc—most of which are essential micronutrients for the proper functioning of a cell—increase to very high levels after the mining processes, which ultimately become very toxic to any living organism inhabiting the region. The types of locations studied include soil with overburden, sediments and tailings.

3.2.1 Copper Mining

Copper mine waste sediments were sampled across mine tailings and normal soil from the top layer. Samples were analyzed for heavy metals such as barium, cadmium, lead, arsenic, cobalt, copper, nickel, vanadium and zinc. The physico-chemical properties of the soil were also analyzed. The average concentrations of heavy metals were found to be elevated in mine waste sediments compared to those in normal soil. In the mine tailings and overburdens, the concentration of cobalt was found to be the highest (after copper), followed by barium, nickel, arsenic, zinc, lead, chromium, and vanadium. The concentration of cadmium was minimal in both.

3.2.2 Gold Mining

Wastes generated from gold extraction consist of high concentrations of heavy metals like arsenic, cadmium, nickel, lead, copper, zinc, cobalt, and mercury. Table 3.1 elucidates the comparison of the amounts of metals in soil under normal conditions and those near gold mines.

Gold mining has also increased copper, arsenic, and mercury concentrations significantly in the environment.

3.2.3 Coal Mining

Coal mining is a core industry in many countries that helps in building up the economy but degrades the environment. Pollutants such as oil, grease, and heavy metals are found in the waste effluents of coal mines.

A few examples as reported by RK Tiwary (2001), concentration of iron in the Western coal mines, Northern coal mines, and Eastern coal mines of India falls in ranges of 2.0–51.03 mg/L, 32.70–84.30 mg/L, and 2.86–50.79 mg/L, respectively. Similarly, for manganese and copper, the range of concentrations varies from 0.030–3.062 mg/L, 0.02–0.40 mg/L, 0.03–2.60 mg/L, and 0.001–0.037 mg/L, 0.001–0.020 mg/L, and 0.001–0.015 mg/L, respectively. The heavy metal concentrations reported in these Indian coal mines are much higher compared to the normal levels of these heavy metals in soil. Many other heavy metals also occur in high concentrations in coal mines.

TABLE 3.1
The Heavy Metal Concentrations in Gold Mines Near Places Like Tanzania, Ghana, South Africa

Heavy Metals	Normal Concentration in Soil (mg/kg) (Fashola et al., 2016)	Concentration in Gold Mine Tailings Near Places Like Tanzania, Ghana, South Africa (Experimentally Found) (mg/kg) (Fashola et al., 2016)
Cadmium	1	Between 6.4 and 11.7
Zinc	70	177.56
Chromium	Between 2 and 60	486
Nickel	3 to 100	583
Lead	32 with a range of 10–67	Between 80 and 510

3.3 ACIDIC CONDITIONS IN MINES AND ACID MINE DRAINAGE

AMD from mining debris, mining ditches, and also from mine tailings and underground operations is mainly a function of the mineralogy of native mineral resources and the availability of oxygen and water. Known for its very low pH value and increased concentration of different heavy metals and other toxic substances, AMD might pollute surface-water, groundwater and soil. AMD is formed when water flows over sulfur-containing substances, producing acidic solutions where the sulfide-carrying solution is exposed to water and oxygen. AMD mainly comes from disused coal mines and currently active mines.

AMD releases have considerably lower values of pH, higher values of specific conductivity, and an increased concentration of heavy metals such as aluminium, manganese and iron. Concentrations of other toxic heavy metals are usually much lower.

3.3.1 Formation of Acid Mine Drainage

When exposed to oxidizing conditions, sulfide minerals get oxidized to form acidic, sulfated aqueducts in the presence of oxygen and water. The contamination by metals involved with AMD is completely dependent on the type and concentration of oxidized sulfide minerals, and also on the different types of gangue minerals found in rocks. Oxidizing minerals of sulfide and the corresponding acidity that is obtained can be explained through a number of reactions.

The first significant reaction involves the oxidation of the sulfide mineral to form sulfate, hydrogen, and dissolved iron.

$$FeS_2 + 7/2O_2 + H_2O \rightarrow Fe^{2+} + 2SO_4^{2-} + 2H^+ \quad (3.1)$$

The dissolved sulfate, ferrous ion and hydrogen ion denote an increase in the quantity of dissolved solids and the acidic nature of the water. If not neutralized, they will bring about a lowering of the pH. In an oxidizing environment (this depends on oxygen concentration, pH and microbial activity), the majority of Fe^{2+} will be oxidized to Fe^{3+}. This reaction has been shown below:

$$Fe^{2+} + 1/4O_2 + H^+ \rightarrow Fe^{3+} + 1/2H_2O \quad (3.2)$$

For values of pH within 2.3–3.5, Fe^{3+} precipitates as Ferric hydroxide and jarosite, leaving a small amount of Ferric ion in the solution while reducing the pH at the same time:

$$Fe^{3+} + 3H_2O \rightarrow Fe(OH)_3 \text{ solid} + 3H^+ \quad (3.3)$$

Any Fe^{3+} in Equation 3.2 that won't precipitate from the solution following Equation 3.3 might be used to oxidize the extra pyrite. This has been shown in the following reaction:

$$FeS_2 + 14Fe^{3+} + 8H_2O \rightarrow 15Fe^{2+} + 2SO_4^{2-} + 16H^+ \quad (3.4)$$

On the basis of simple reactions, acid production forms iron, which is further precipitated as Ferric hydroxide. This might be expressed by combining Equations 3.1 and 3.3:

$$FeS_2 + 15/4O_2 + 7/2H_2O \rightarrow Fe(OH)_3 + 2SO_4^{2-} + 4H^+ \tag{3.5}$$

Again, the overall equation for obtaining stable Fe^{3+}, which is used to oxidize extra pyrite, is that by combining Equations 3.1 and 3.3, we get:

$$FeS_2 + 15/8O_2 + 13/2Fe^{3+} + 17/4H_2O \rightarrow 15/2Fe^{2+} + 2SO_4^{2-} + 17/2H^+ \tag{3.6}$$

In all of the aforementioned equations, except Equations 3.2 and 3.3, the mineral that is oxidized is pyrite, and oxygen gas is the oxidizing agent. Some sulfide minerals, like pyrrhotite (FeS) and chalcocite (Cu_2S) have different metal sulfide and metal ratios apart from iron. Extra-oxidizing agents and sulfide minerals have various other reactions, stoichiometries and rates, but all of these things have not been studied extensively.

Thus, mining effluents are generated when pyrite, an iron sulfide, is produced and its reaction occurs with air and water, leading to the formation of dissolved Fe and H_2SO_4. Part or all of this iron might settle down and result in the formation of red, yellow, or orange sediments under streams containing the mine effluent. Heavy metals like copper, mercury, and lead get further dissolved by acid runoff in the groundwater or surface water.

The main factors that determine the acid production rate are: temperature, pH, oxygen content in the gas phase (in conditions where the saturation is <100%), degree of water saturation, concentration of oxygen in the water phase, exposed metal sufide surface area, the chemical activation energy needed to begin the production of acid, bacterial activity (in many instances, bacteria have an important role in acceleration of acid production; inhibition of bacterial activity can impair acid production), water or moist air, and sulfide minerals.

3.3.2 Acid Mine Drainage: Treatment

AMD is generally left untreated because most of the technologies involved in treating AMD are either very expensive or insufficient.

The most common process used to treat AMD is the inclusion of an alkalinity source in order to increase the pH more than the threshold level needed by iron (Fe) oxidizing bacteria, thus significantly decreasing acid production.

Some results that arise out of these actions are as follows: elimination of the acidic condition and an increase in the pH. The elimination of all heavy metals takes place. Fe^{2+} is oxidized to Fe^{3+} very quickly under acidic conditions. Once the solubility of $CaSO_4$ exceeds the addition of an adequate amount of Ca, only sulfate can be eliminated.

Shirin et al. conducted a characterization of the AMD treated with fly ash and reported the results in 2021. As per their results, the fly ash was successful in neutralizing the AMD pH and reducing the concentration of elements and sulfate compounds.

A comprehensive review has been written by Joshua O. Ighalo et al. and published in 2022, which touched upon not only the neutralization and adsorption techniques using fly ash, zeolites and bio-activated carbon, but also sophisticated methods of nanofiltration and reverse osmosis, along with biological methods of using substrate-specific microorganisms to utilize the waste. It turns out that such microorganism-aided AMD treatments are much more effective and broad-spectrum.

Independent literature reviews by Benjamin C. Hedin et al. (2020) and Guan Chen et al. (2021) have elucidated the fact that by products of AMD treatment may also serve as sources to extract rare earth elements.

3.3.3 Control of Acid Mine Drainage

3.3.3.1 Dumping Wastes Underwater

Since the availability of moisture and oxygen are essential components for oxidizing pyrite ore and releasing acids, the best approach appears to be to create an anaerobic environment that can be achieved by submerging the tailings to limit acid production, which has been successfully tried in the Solbek mine, Canada (Amyot et al., 1997). The pH was found to be neutral after a series of experiments, and abnormal concentrations of elements like iron, zinc and copper were notably decreased. This practice is assisted by a reduction in the oxidizing microbial population, a reduction in their oxidizing activity, and primarily by the presence of sulfate-reducing bacteria. These bacteria help in the inverse oxidation method by reducing sulfate to sulfide ions, further causing the precipitation of the metals found in the more stable types of metal sulfide.

3.3.3.2 To Cover and Seal with Clay

Acid formation might be prevented by covering and sealing the tailing material. Keeping aside the high price involved in laying the earthen cover, it is always difficult to find the precious clay to cover the mine spoils of India if it is not preserved at the initiation of mining.

Limestone inhibits the oxidation of iron disulfides and also continuously prevents acid generation, but only when the rate of application is high enough that the pH of the system is around 6.7 (Nairan et al., 1991). But this method results in the formation of iron precipitate that adheres to the limestone and inhibits the activities. Other acidic ameliorants like fly ash and blast furnace slag, which are rich in Al_2O_3, SiO_2, CaO and MgO, can also work well to reduce AMD production. In order to control waste emissions from dumping sites, equipment is sometimes supplemented with free lime (CaO) or portlarolite and placed under the future water table. This system has been reported to be very effective in relocating the waste dump.

3.3.3.3 Total Suspended Solids

The high amount of solid debris suspended in mining water and workshop effluents can be eliminated by allowing discharged water enter a series of settling tanks. To test the flow of runoff water from the dumps, a garland drainage system can be built around the spoil heaps. A proper coagulant has to be added to make sure that solids get settled in the garland drainage itself.

As suggested by a study done by Xin Hu et al. (2022), the iron in AMD, along with trace amounts of copper and manganese, can be removed by chemical oxidation technology in a pilot-scale, step-by-step process. This method, being very cost-efficient, was predicted to result in a significant amount of annual savings as compared to lime neutralization for treating AMD.

3.3.3.4 Removal of Dissolved Solids and Hardness

The concentration of dissolved solids and hardness can be reduced by taking the following steps: evaporation of water from large shallow retention ponds, a process that involves the exchange of ions, membrane filtration, the removal of cations by means of a cation exchange resin and the removal of anions through a multistep process.

Besides effective treatment of AMD, the challenges and avenues for beneficiation and valorization of the same in a sustainable manner, which would include water reclamation, life cycle assessment, and also incorporating the 3R's of waste management (reuse, reduce, and recycle), were evaluated by V. Masindi et al. and reported in October 2022.

3.3.4 Problems Associated with Acid Mine Drainage

Some adverse effects of AMD on the environment include contamination of drinking water, hindrance to the growth and reproduction of aquatic flora and fauna, and corrosion due to acid on different parts of infrastructure.

In the 1800s, mines were developed so that they could utilize gravity drainage in order to avoid the accumulation of excess water in the mines. This led to the drainage of water polluted by acids and heavy metals like iron, sulfur and aluminium from the mines into streams.

3.3.5 Results of Acid Mine Drainage

AMD can be considered a severe water pollution problem that can cause contamination of groundwater drinking supplies and death of valued recreational fish like trout. This also leads to a decrease in outdoor recreation and tourism. All this makes AMD an ecological as well as an economic concern of ours.

3.4 METAGENOMICS OF ACID MINES

Metagenomics involves the isolation, sequencing, and classification of the set of DNA of microbes in an environment that allows clarification of the phylogeny and metabolism of mixed populations (Liljeqvist et al., 2015). The discharge of contaminated water from the sulfide and microbe-rich environments of abandoned mines results in AMD (Natarajan, 2008). Contamination of ecosystems by AMD occurs where oxidants are available to sulfide minerals, enhanced further by mining. The rate of release of sulfur and metals is eventually determined by the catalysis of iron and sulfur oxidation by autotrophic and heterotrophic archae and bacteria (Baker & Banfield, 2003).

There are reports that the multiphase distribution of the microbial community is significantly affected by the geochemical gradient. Moreover, the toxic metal fraction is influenced by the soil particle fraction, which, along with chemical properties, cooperatively affects the microbial community structure in polluted soils (Pan et al., 2021).

AMD can be the result of waster rock and tailings, spent heap leach dumps, stock and spoil piles, and underground and open-cast mines.

3.4.1 Presence of Organisms Detected from the Acid Mine Drainage System

As reported by She et al. (2021), there were distinct geochemical gradients along with increased microbial diversity with the depth of the water column in an AMD. Heterogenous selection was most prevalent along the surface, while the significance of stochastic processes increased with depth.

The pattern of distribution of AMD microbiota in the lake is affected by the diversity of species coexisting in the environment and the acid stress that they are exposed. These reports verify the heterogeneous nature of the prokaryotic microbial community and also aid in determining the pathways that contribute to the survival of the respective species under those extremely acidic conditions.

3.4.1.1 Bacteria

From acid mines, divisions of proteobacteria, nitrospira, firmicutes, and acidobacteria have been reported. The γ-proteobacteria, especially *Acidiothiobacillus* spp. and *Thiobacillus* spp., are the most widely studied group. Among the two β-proteobacterial groups detected are *Thiomonas* spp. (Strain names: Ynys1, Ynys3). Heterotrophic α-proteobacteria of *Acidiphilum cryptum* (strain JF-5), isolated from a lake that is rich in iron sediment and acidity and is near a coal mine in eastern Germany, and another group that was detected from clone libraries of Iron Mountain AMD environments ($pH < 1.0$) were reported from acid mines. However, inhabitation at near neutral pH inside acidophilic protists is more preferable for them. Organisms belonging to *Leptospirillum ferrooxidans* are frequently detected in bioleaching and AMD systems.

Acidimicrobium ferrooxidans and *Ferromicrobium acidophilus* are two polyphyletic lineages of *Sulfobacillus* that have low G+C contents and are gram-positive in nature. These are some distinct groups from the firmicutes division that were found in AMD systems.

3.4.1.2 Archae

The archaeal lineages found in AMD systems are limited to the *Thermoplasmatales* and the *Sulfolobales*. The presence of some members of the *Metallosphaeraprunae* and *Sulfolobales* has also been spotted in acid mine drainage environments.

3.4.1.3 Eucarya

Acidophilic bacteria that oxidize minerals are grazed upon by protists, which were identified to belong to the *Cinetochilum* and *Vahlakampfia* genera.

3.4.2 Bioremediation of Metals by Microbes

Some of the prevalent bacteria accountable for acid generation through sulfide mineral oxidation are *Acidithiobacillus ferrooxidans* (capable of oxidizing iron and sulfur), *Leptospirillum ferrooxidans* (oxidizes iron), *Acidithiobacillus thiooxidans* (oxidizes sulfur), *Thiobacillus thioparus* (a neutrophile that is capable of oxidizing sulfides), and sulfate-reducing bacteria (may play a role in bioremediation of metal sulfides).

The generation of acid along with the dissolution of copper, zinc, and iron is hastened by the presence of *Acidithiobacillus ferrooxidans*. It is noticeable that the greater the number and activity of iron-oxidizing cells present, the greater the sulfide mineral dissolution (Baker et al., 2003).

Bioremediation of arsenic by microbes can be done (1) by arsenite oxidation to arsenate as established by *Thiomonas* spp. and *Bacillus* spp., both of which exhibited efficiently oxidized arsenite within 8 days; or (2) by adsorption of arsenic on jarosites getting precipitated in the course of growth of the two bacterial strains *Acidithiobacillus ferrooxidans* and *Leptospirillum ferrooxidans*. This phenomenon was examined, and the outcomes were presented as the variable conditions of pH and their corresponding percentage adsorption of arsenic. pH 2 corresponded to 50% arsenic adsorption, which was the lowest. pH 5 corresponded to 62% arsenic adsorption, pH 8 corresponded to 65%, and lastly, the highest adsorption of arsenic was observed at pH 9.5, which was 72%. A probable application of this particular approach might be the removal of toxic arsenite from acid mine tailings (Natarajan, 2008).

In anaerobic conditions, sulfate-reducing prokaryotes, or SRP, use organic compounds or hydrogen as electron donors and reduce sulfate to hydrogen sulfides.

$$SO_4^{2-} + 2CH_2O \rightarrow H_2S + 2HCO_3 \tag{3.7}$$

Hydrogen sulfide contributes to the precipitation of metals as sulfides:

$$H_2S + M^{2+} \rightarrow MS(s) + 2H^+ \tag{3.8}$$

where M^{2+} is a metal ion such as As (III or V).

Relatively negligible amounts of fixed carbon and nitrogen are fed into the subsurface AMD sites. The *Thiobacilli*, including *Acidithiobacillus ferrooxidans*, serve as a governing flora that contributes to CO_2 fixation in environments that have temperatures below 30°C and pH above 2 (Baker et al., 2003).

3.5 BIOMINING

It is the method of extraction of metals that have economic importance from rock ores and mine waste using microorganisms (microbes). Biomining techniques could also be used to clean up metal-polluted locations. Solid minerals are frequently embedded in valuable metals.

These metals can be oxidized by some microorganisms, allowing them to dissolve in water.

This is the core technique underpinning most biomining, and it's employed for metals that are easier to recover when dissolved than when extracted from solid rocks.

For metals that are not dissolved by microorganisms, a separate biomining process uses microbes to break down the surrounding minerals, making it easier to collect the metal of interest directly from the remaining rock.

3.5.1 Mechanisms of Biomining

A variety of fungus species can be utilized for biomining. Different investigations have shown that fungi such as *Penicillium simplicissimum* and *Aspergillus niger* can mobilize Sn and Cu by percentages as high as 65 and aluminium, lead, nickel, and zinc by almost 95%, respectively. Phytomining, on the other hand, is based on the functionality of certain plant species to accumulate large amounts of metals from their individual host rocks. Hyperaccumulators are specific plants that produce bio-ore, which is used as a pure metal source when cultivated on highly mineralized soils or post-mine areas. Due to the length of the operations, and consequently, their unprofitability, these methods are less widespread than bacterial mining. Metals are required by all living creatures to carry out basic enzyme activities. Metals are obtained by humans in trace amounts from their food. Metals, on the other hand, are obtained by microbes dissolving them from the minerals in their surroundings. Organic acids and metal-binding molecules are produced to do this. Scientists take advantage of these characteristics by combining microorganisms with ores in a solution and collecting the metal precipitated by these microbes on the top layer. Temperature, concentration of sugars, the stirring rate in the tank, pH, levels of carbon dioxide, and oxygen are all carefully supervised and adjusted so as to provide optimum conditions for leaching (Di Marco et al., 2020).

Figure 3.1 depicts the process of metal extraction from mine tailings.

In bioleaching, or the use of microorganisms in the industry for the process of leaching, a big heap of low-grade ore is made first, also called the leach dump. After that, dilute sulfuric acid is used to soak this large heap, and it is allowed to percolate down the heap. The pH of the acid is maintained at 2. The acid dissolves the minerals in the heap, and the liquid at the bottom of the heap, which is now rich in minerals, should be collected and relocated to the precipitation equipment. The re-precipitation of the metal is carried out, and the metal is also purified. The liquid is again forced to the top of the heap, and the same process is repeated.

Fe^{2+} may be converted to Fe^{3+} by *Acidithiobacillus ferrooxidans.*

Microbial oxidation of ferrous ions produces ferric (Fe^{3+}) ions, which are used to chemically oxidize copper ore. The oxidation of copper ore can take three different forms.

$$Cu_2S + 1/2O_2 + 2H^+ \rightarrow CuS + Cu^{2+} + H_2O \tag{3.9}$$

$$CuS + 2O_2 \rightarrow Cu^{2+} + SO_4^{2-} \tag{3.10}$$

$$CuS + 8Fe^{3+} + 4\ H_2O \rightarrow Cu^{2+} + 8Fe^{2+} + SO_4^{2-} + 8H^+ \tag{3.11}$$

The copper metal is subsequently extracted from steel cans using Fe^0:

$$Fe^0 + Cu^{2+} \rightarrow Cu^0 + Fe^{2+} \tag{3.12}$$

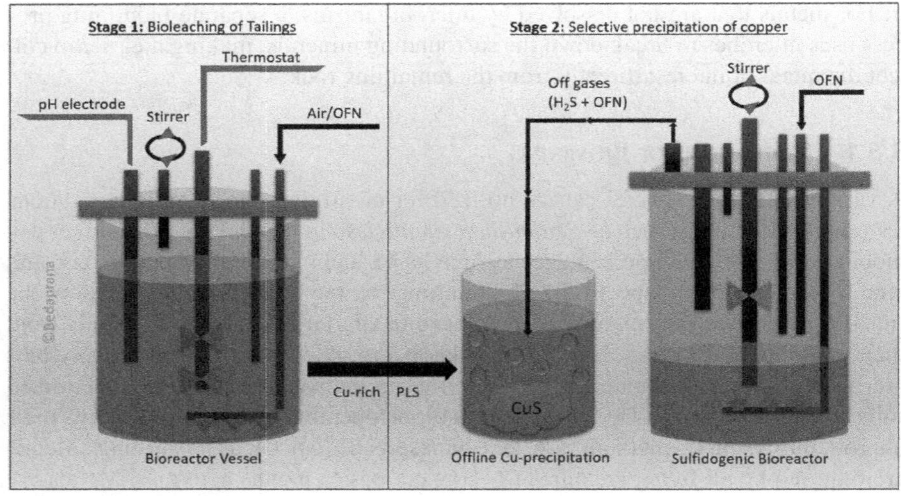

FIGURE 3.1 Bioleaching was used to recover cobalt from copper mine tailings that contain large amounts of cobalt left for decades at a disused copper mine in Kasese, Uganda, and to eliminate the deposits' environmental concern. These types of reactors were used. Image: New methods for the extraction and recovery of metals from mine tailings are being developed. (Illustration adapted from Falagán et al., 2017.)

The microbial activity causes the temperature within the leach dump to rise. As a result, the temperature shoots above 40°C. Thermophiles that are capable of oxidizing iron and are chemolithotrophs in nature, such as *Leptospirillum*, *Acidithiobacillus* sp. and the thermoacidophilic archaeon *Sulfolobus*, may become critical in the process of bioleaching. *Acidithiobacillus ferrooxidans*, like copper, can oxidize U^{4+} to U^{6+} using O_2 as an electron acceptor. However, Uranium oxidation (chemical) by Fe^{3+} is anticipated to play a larger role in the uranium leaching process, with *A. ferrooxidans* mostly carrying out the reoxidation of Fe^{2+} to Fe^{3+}, as mentioned earlier.

$$UO_2 + Fe(SO_4)_3 \rightarrow UO_2SO_4 + 2\ FeSO_4 \quad (3.13)$$

Metal sulfides oxidize when exposed to water and oxygen, resulting in a rich, acidic effluent. The environment might become contaminated and acidic with significant amounts of heavy metals if the acidity is high enough to surpass the natural buffer system capability of the neighboring rocks. Though acidophiles play a significant role in the sulfur and iron biogeochemical cycles, low pH conditions are almost entirely manmade, resulting from the cessation mining operations in areas where sulfide minerals like pyrite are found.

Acid-mine drainage can happen on the spoil heap, in the mines, or as a result of another venture that uncovers metal sulfides in high concentrations, such as large construction projects. A simple overview of the aforementioned process would be:

$$2FeS_2 + 2H_2O + 7O_2 \rightleftharpoons 2Fe^{2+} + 4SO_4^{2-} + 4H^+ \quad (3.14)$$

3.5.2 Bacterial Effects on Acid Mine Drainage

The process of metal sulfide oxidation is sluggish in the absence of acidophilic microbes, especially *Acidithiobacillus ferrooxidans*, which is a synonym for *Thiobacillus ferrooxidans*. The oxidation of pyritic acid can be sped up by 106 times using these microorganisms. The capability of using ferrous iron for the production of ferric iron catalysts was an important factor in *A. ferrooxidans'* ability to oxidize pyrite in that study:

$$Fe^{2+} + 1/4O_2 + H^+ \rightarrow Fe^{3+} + 1/2H_2O \quad (3.15)$$

Under the low pH circumstances described above, ferric iron (Fe^{3+}) is a more effective oxidant than oxygen, and results in quicker pyrite oxidation rates.

A. ferrooxidans is notably a chemolithotroph because acidic conditions are oligotrophic (low dissolved organic carbon content) and lack light for phototrophy. *A. ferrooxidans* may survive in vadose conditions if the rock holds on to moisture and the mine has enough air. In reality, the circumneutral pH of the environment, which prohibits the growth of many acidophiles, is likely to be the limiting factor for pioneer bacteria. However, when bacteria and mineral surfaces have an acidic interface, favorable geochemical conditions arise quickly, and pH drops to a level closer to the acidophilic optimum.

When *A. ferrooxidans* reaches a certain quorum level, AMD is triggered. When metal sulfides are first colonised, there are no AMDs, and AMD stays absent as the bacterial strains develop into bacterial microcolonies. When a population reaches a certain size, it causes a visible shift in water chemistry, which causes AMD to rise. This suggests that while pH isn't a good predictor of AMD risk in a mine, growing *A. ferrooxidans* (or other AMD-causing bacteria) is.

Leptospirillum ferrooxidans, *Acidithiobacillus thiooxidans*, and *Sulfobacillus thermosulfidooxidans* are among the microorganisms linked to AMD (Baker & Banfield, 2003).

3.5.2.1 Acidithiobacillus ferrooxidans

Acidithiobacillus ferrooxidans is a key member of microbe consortia that are utilized in the industrial retrieval of copper in processes of bioleaching or biomining. It thrives in environments with a very low pH and fixes both carbon and nitrogen from the air. In acidic settings, it solubilizes copper and other metals from rocks and plays a vital role in the nutrient and metal biogeochemical cycles. Extreme acidophiles, *Acidithiobacillus ferrooxidans* strains ATCC 23270 and ATCC 53993, can survive extremely high amounts of heavy metals.

A strain of *A. ferrooxidans* called ATCC 53993 has several genes that provide better metal resistance compared to another strain called ATCC 23270. These genes are grouped together in a 160 kb genomic island (GI) that is not present in the ATCC 23270 strain. These genes include those involved in detoxifying heavy metals, transporting metals into the cell, and regulating gene expression, as well as a protein called CopA3Af that may play a role in copper resistance. These genes were found to work in another type of bacteria, suggesting they are functional in *A. ferrooxidans* ATCC 53993 as well.

Future research will be very interesting to see if all the remaining potential and hypothetical genes discovered in the GI of *A. ferrooxidans* ATCC 53993 have a role in providing additional resistance to copper and other metals prevalent in bioleaching settings. This much reduced reactivity to the metal is consistent with the previously observed substantially greater copper resistance of *A. ferrooxidans* ATCC 53993.

Minor differences in their genomes caused by genetic exchanges might be used to understand why two strains of the same acidophilic microbe are resistant to copper.

3.5.2.2 Acidithiobacillus caldus

Acidophilic bacteria, such as *Acidithiobacillus caldus* ATCC 51756, have mechanisms to regulate metal levels that are similar to those used by neutrophils. A recent study found that there were many zinc transporters in *A. caldus* ATCC 51756 and other microorganisms that may be involved in importing or exporting zinc.

A strain of *A. ferrooxidans* called ATCC 53993 has better copper resistance compared to another strain, ATCC 23270, when exposed to similar copper concentrations. This is not likely due to the formation of metal complexes under the same solution conditions.

Bioinformatics analysis showed that the genes associated with zinc resistance in *A. caldus* ATCC 51756 were also present in a different strain of *A. caldus* called SM-1, specifically in its megaplasmid. Some of these genes may also be involved in copper transport. The study found that some RND-type efflux systems and P-type ATPases, which have been linked to copper resistance in *A. ferrooxidans* ATCC 23270, were also present in both strains of *A. caldus* and could be involved in copper resistance. Figure 3.2 depicts those copper resistance genes in the two organisms.

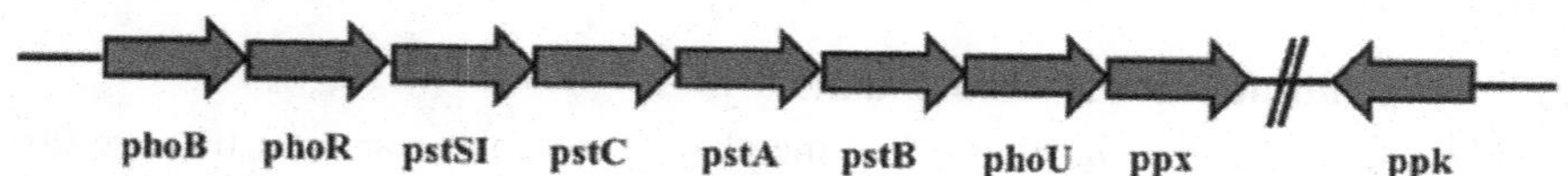

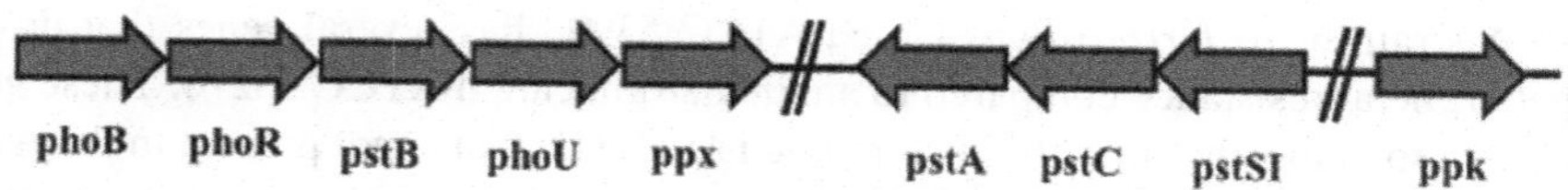

FIGURE 3.2 Putative PHO regulon in *A. caldus* and *A. ferrooxidans*. The genomes of *A. ferrooxidans* strains ATCC 23270 and 53993 both contain the same genes in the same order. The putative PHO regulon genes are found in the same genomic context in both *A. caldus* strains ATCC 51756 and SM-1, unlike *A. ferrooxidans* strains. (Navarro et al., 2013)

3.5.3 Usefulness of Biomining

The traditional mining processes include the use of harsh chemicals, a lot of energy, and a lot of pollutants. Biomining, on the other hand, uses extremely small amounts of energy and generates relatively low microbiological wastes like organic acids and gases.

Biomining may be able to effectively use poor-grade metal sources (e.g., mine tailings), which would be unprofitable and nonviable using customary methodologies otherwise due to its low cost and simplicity. Biomining is gaining popularity in countries like Finland, Chile and Uganda. Chile used biomining to recover cobalt from copper mine tailings after depleting much of its copper-rich ores, while Uganda has successfully been collecting cobalt from copper mine tailings for more than 10 years.

Iron- and sulfur-containing minerals can be easily oxidized by a variety of mineral-oxidizing bacteria. *Acidithiobacillus caldus*, *Acidithiobacillus thiooxidans*, *Acidithiobacillus ferrooxidans*, and the iron-oxidizing *Leptospirillum ferriphilum* and *Leptospirillum ferrooxidans* are among them. Biomining can be done with a variety of fungus species. Investigations have divulged that fungal strains like *Aspergillus niger* and *Penicillium simplicissimum* have the capacity to mobilize Cu and Sn by 65% and Ni, Al, Pb and Zn by more than a percentage as high as 95, respectively.

3.6 CONCLUSION

The chapter discusses thoroughly the mechanisms of biomining by bacterial strains that can tolerate the extreme conditions of acid-mine drainage. The presence of different kinds of metallophiles opens up new avenues of biomining in varied forms of AMD from different types of mines. Biomining doesn't only help in metal recovery but also plays a vital role in the detoxification of these effluents, the release of which can severely harm the flora and fauna adjacent to them. Metagenomic studies have helped us understand the flora of the AMD in detail. Further research can thus help us use this widely available information to the advantage of our community, where detoxification of the drainage can be done using its innate flora with no artificial interventions.

REFERENCES

Amyot, G., & Vézina, S. (1997, May). Flooding as a reclamation solution to an acidic tailings pond-the Solbec case. *Proceedings 4th International Conference on Acid Rock Drainage 31*, 681–696.

Baker, B. J., & Banfield, J. F. (2003). Microbial communities in acid mine drainage. *FEMS Microbiology Ecology*, *44*(2), 139–152. DOI: 10.1016/S0168-6496(03)00028-X

Chen, G., Ye, Y., Yao, N., Hu, N., Zhang, J., & Huang, Y. (2021). A critical review of prevention, treatment, reuse, and resource recovery from acid mine drainage. *Journal of Cleaner Production*, *329*, 129666. DOI: 10.1016/j.jclepro.2021.129666

Di Marco, V., Tapparo, A., Badocco, D., D'Aronco, S., Pastore, P. and Giorio, C. (2020). Metal ion release from fine particulate matter sampled in the Po valley to an aqueous solution mimicking fog water: kinetics and solubility. *Aerosol and Air Quality Research*, 20, 720–729. https://doi.org/10.4209/aaqr.2019.10.0498

Falagán, C., Grail, B. M., & Johnson, D. B. (2017). New approaches for extracting and recovering metals from mine tailings. *Minerals Engineering*, *106*, 71–78. DOI: 10.1016/j.mineng.2016.10.008.

Fashola, M. O., Ngole-Jeme, V. M., & Babalola, O. O. (2016). Heavy metal pollution from gold mines: Environmental effects and bacterial strategies for resistance. *International Journal of Environmental Research and Public Health*, *13*(11), 1047.

Hedin, B. C., Hedin, R. S., Capo, R. C., & Stewart, B. W. (2020). Critical metal recovery potential of Appalachian acid mine drainage treatment solids. *International Journal of Coal Geology*, *231*, 103610. DOI: 10.1016/j.coal.2020.103610.

Hu, X., Yang, H., Tan, K., Hou, S., Cai, J., Yuan, X., Lan, Q., Cao, J., & Yan, S. (2022). Treatment and recovery of iron from acid mine drainage: A pilot-scale study, *Journal of Environmental Chemical Engineering*, *10*(1), 106974. DOI: 10.1016/j.jece.2021.106974.

Hua, Z. S., Han, Y. J., Chen, L. X., Liu, J., Hu, M., Li, S. J., Kuang, J. L., Chain, P. S., Huang, L. N., & Shu, W. S. (2015). Ecological roles of dominant and rare prokaryotes in acid mine drainage revealed by metagenomics and metatranscriptomics. *The ISME Journal*, *9*(6), 1280–1294.

Ighalo, J. O., Kurniawan, S. B., Iwuozor, K. O., Aniagor, C. O., Ajala, O. J., Oba, S. N., Iwuchukwu, F. U., Ahmadi, S., & Igwegbe, C. A. (2022). A review of treatment technologies for the mitigation of the toxic environmental effects of acid mine drainage (AMD). *Process Safety and Environmental Protection*, *157*, 37–58. DOI: 10.1016/j.psep.2021.11.008

Jerez, C. A. (2009). Metal extraction and biomining. *Encyclopedia of Microbiology*, *3*, 407–420.

Liljeqvist, M., Ossandon, F. J., Gonzalez, C., Rajan, S., Stell, A., Valdes, J., Holmes, D. S., & Dopson, M. (2015). Metagenomic analysis reveals adaptations to a cold-adapted lifestyle in a low-temperature acid mine drainage stream. *FEMS Microbiology Ecology*, *91*(4), fiv011. DOI: 10.1093/femsec/fiv011

Masindi, V., Foteinis, S., Renforth, P., Ndiritu, J., Maree, J. P., Tekere, M., & Chatzisymeon, E. (2022). Challenges and avenues for acid mine drainage treatment, beneficiation, and valorisation in circular economy: A review. *Ecological Engineering*, *183*, 106740. DOI: 10.1016/j.ecoleng.2022.106740

Nairn, R.W., Hedin, R.S., & Watzlaf, G.R. (1991). A preliminary review of the use of anoxic limestone drains in the passive treatment of acid mine drainage. In *Proceedings, Twelfth Annual West Virginia Surface Mine Drainage Task Force Symposium*, 3–4 April 1991, West Virginia University, Morgantown, WV.

Natarajan, K. A. (2008). Microbial aspects of acid mine drainage and its bioremediation. *Transactions of Nonferrous Metals Society of China* *18*(6), 1352–1360. DOI: 10.1016/s1003-6326(09)60008-x

Navarro, C. A., von Bernath, D., & Jerez, C. A. (2013). Heavy metal resistance strategies of acidophilic bacteria and their acquisition: Importance for biomining and bioremediation. *Biological Research*, *46*(4), 363–371. DOI: 10.4067/S0716-97602013000400008

Pan, Y., Ye, H., Li, X., Yi, X., Wen, Z., Wang, H., Lu, G., & Dang, Z. (2021). Spatial distribution characteristics of the microbial community and multi-phase distribution of toxic metals in the geochemical gradients caused by acid mine drainage, South China. *Science of the Total Environment*, *774*, 145660. DOI: 10.1016/j.scitotenv.2021.145660

Park, I., Tabelin, C. B., Jeon, S., Li, X., Seno, K., Ito, M., & Hiroyoshi, N. (2019). A review of recent strategies for acid mine drainage prevention and mine tailings recycling. *Chemosphere*, *219*, 588–606. DOI: 10.1016/j.chemosphere.2018.11.053

She, Z., Pan, X., Wang, J., Shao, R., Wang, G., Wang, S., & Yue, Z. (2021). Vertical environmental gradient drives prokaryotic microbial community assembly and species coexistence in a stratified acid mine drainage lake. *Water Research*, *206*, 117739. DOI: 10.1016/j.watres.2021.117739

Shirin, S., Jamal, A., Emmanouil, C., & Yadav, A. K. (2021). Assessment of characteristics of acid mine drainage treated with fly ash. *Applied Sciences*, *11*(9), 3910. DOI: 10.3390/app11093910

Sleator, R. D., Shortall, C., & Hill, C. (2008). Metagenomics. *Letters in Applied Microbiology*, *47*(5), 361–366.

Tiwary, R. K. (2001). Environmental impact of coal mining on water regime and its management. *Water, Air, and Soil Pollution*, *132*(1), 185–199.

Wikipedia contributors. (2021). Acidophiles in acid mine drainage. In *Wikipedia, The Free Encyclopedia*. Retrieved April 18, 2022, from https://en.wikipedia.org/w/index.php?title=Acidophiles_in_acid_mine_drainage&oldid=1039431798

4 Environmental Rehabilitation of Industrial Waste Dumping Site

Ratul Bhattacharya
Raja Rammohun Roy Mahavidyalaya

Roumi Bhattacharya
Indian Institute of Engineering, Science and Technology

Muhammad Majeed
University of Gujrat

Somya Bhandari
Modern College of Arts, Science & Commerce

Robina Aziz
Government College Women University

Dwaipayan Sinha and Ayan Mondal
Government General Degree College, Mohanpur

Subhamita Sen Niyogi
Central Water Testing Laboratory

4.1 INTRODUCTION

The recent era of population explosion, massive industrialization drives, deforestation, and indiscriminate and unmanaged disposal of urban and industrial wastes has converted natural habitats into potential sinks for different hazardous and toxic pollutants. These have proportionally developed many irreversible and dangerous health hazards in living systems and pose a severe global concern for environmental pollution. Indiscriminate, unlawful, and open dumping of municipal, urban, and industrial wastes is a dangerous practice leading to solid waste generation. Compared to many conventional technologies, bioremediation is establishing its firm hold as a cost-effective,

DOI: 10.1201/9781003451457-4

environment-friendly, and potentially efficient way of cleaning and decontaminating processes day by day. Various innovative agents like bacteria, cyanobacteria, fungi, mycorrhiza, micro- and macrophytes, and nanobioparticles are employed for decontamination. Bioremediation depends on the basic principles of adsorption, biosorption, conversion of more toxic pollutants to less toxic forms, metal precipitation, and immobilization of the pollutants (Kour et al., 2021). The application of bioremediation encompasses a wide range of conditions, like the remediation of agricultural contaminants, contaminated soils, dye-contaminated sites, heavy metal contamination, xenobiotic compounds, and industrial wastewater. The compatibility of the kind of waste generated and the bioremediation approach dictates the ease and rate of the process of waste management. Depending on the nature of the contaminants, bioremediation accomplishes its role by applying processes like biostimulation, bioaugmentation, bioaccumulation, biospurging, etc. The types of waste generated can be categorized into toxic (such as heavy metals and persistent organic pollutants), ignitable (oils, paints, and organic solvents), corrosive (like acids and alkalis), and infectious (like biohazardous and pathogen-contaminated wastes). Major industries that generate solid wastes are tanneries, sugar mills, textiles, paper, and petroleum and oil refining (Nagda et al., 2021).

Contemporary advancements and innovations in bioremediation include nanobioparticles, phytochelatins, geoengineering, microbial biosensors, and genetically modified organisms (Kuppusamy et al., 2016). Recovery of value-added products and so-called trash-to-treasure conversion motivates bioremediation through different microbe-assisted approaches, which provides a potential route toward a sustainable and circular economy (Salvador et al., 2021). Though bioremediation has many positive merits, it still faces certain challenges. The method of disposal of contaminants as well as the fate of final products have to be regulated, and microorganisms used in bioremediation have a chance of contaminating other environments. To promote a greener and more sustainable environment, bioremediation technology should be considered a viable and potent option for developing a bio-based circular economy (Figure 4.1).

4.2 MICROBES AS AN EFFICIENT TOOL OF BIOREMEDIATION

Bioremediation involves detoxifying, removing, altering, degrading, and immobilizing various physio-chemical pollutants and effluents from the natural environment through the action of many microorganisms (including bacteria and fungi) and plants. The versatile metabolic ability and the presence of a unique and beneficial enzyme profile have established them as potential tools for the bioremediation process (Thakur et al., 2019). Microbe-mediated processes are gaining prominence and priority as they are economically and environmentally sustainable and successfully mitigate pollution pressure worldwide (Raghunandan et al., 2018). Their ability to adapt to and sustain themselves in various extreme environmental conditions and their flexible nutritional dependence have made them a globally accepted agent for the bioremediation process. The microbial enzymes have proven promising in remediating persistent toxic effluents like microplastics, agrochemicals, polychlorinated biphenyls, and heavy metals by transforming the toxic form to an atoxic form and

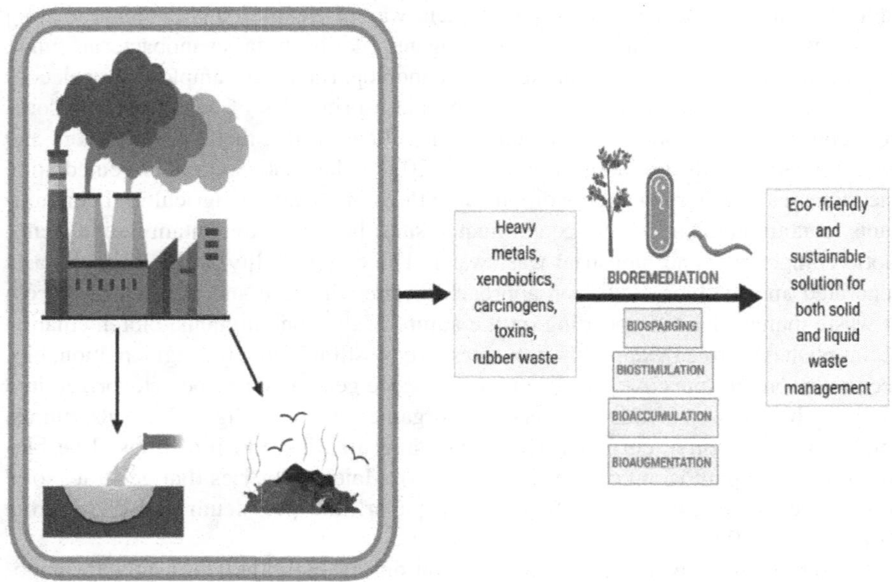

FIGURE 4.1 A pictorial representation of bioremediation of industrial wastes, its principles, and its merits over conventional techniques.

sometimes producing some novel beneficial product (Phale et al., 2019). The enzymes can be grouped under six main categories: hydrolases, isomerases, oxidoreductases, lyases, transferases, and ligases (Liu et al., 2015). The efficiency of bacteria in heavy-metal bioremediation lies in their ability to resist heavy-metal toxicity by methylation, adsorption, uptake, oxidation, and reduction (Fernández et al., 2012). Enhanced metabolic vigor in microbes at hydrocarbon-polluted sites is achieved by horizontal or vertical gene transfers and mutations, thereby acquiring genes needed in the degradation of hydrocarbons (Bidja Abena et al., 2019). According to a recent report (Lyu et al., 2021), a class of heterotrophic marine protists called Thraustochytrids are currently being associated with bioremediation and organic matter decomposition for their ability to accumulate lipids of high commercial value like carotenoids, squalene, and decahexanoic acid (Du et al., 2021; Orozco Colonia et al., 2020). These can degrade several environmental pollutants like tar balls, aromatic effluents, and animal wastes that bacteria may not completely degrade (Bongiorni et al., 2005). Many Cyanobacterial strains (including *Aphanocapsa* sp., *Oscillatoria* sp., *Phormidium* sp., and *Halobacterium* sp.) and their consortia with bacteria are also widely used in the bioremediation process, especially in heavy metal detoxification and wastewater treatment. They can reduce heavy metals and offer a potential scope for conducting genetic engineering with the parental strain to give rise to a maintenance-free remediation technology at toxic effluent dumping sites (Cuellar-Bermudez et al., 2017). They are also known as an agent of bioremediation in polluted soil environments and aquatic ecosystems as they develop symbiotic associations, further promoting the growth of microbial consortia. These also aid in the growth of plant growth-promoting rhizobacteria, which ultimately contribute to the reclamation of infertile soil and nutrient utilization (Paul and Nair, 2008).

Constant research has led bioremediation to gain a prospectus position by focusing on a sustainable approach addressing production efficiency and resource optimization balancing with limited and finite resource stock and a growing economy, thus establishing itself as an agent of the circular economy (Mccarthy et al., 2018). Microbes, and especially extremophiles are expected to drive a considerable transformation in science and biotechnology involving a sustainable bioeconomy focusing on reuse, recycling, and constant production of biomass. With microbes harboring numerous beneficial secondary metabolic pathways and biocatalysts (Ferrer et al., 2016), combined with the scope of evolved engineering technologies (Arnold, 2018), microbe-mediated remediation techniques are emerging as the backbone of future bioindustries. A holistic approach to the microbial transformation of plastics, lignocellulose, and fossil-derived polymers is a widely known topic and research avenue for many researchers and workers globally, as it aims toward a sustainable circular economy.

4.3 ROLES OF MICROBES IN THE CLEANUP OF VARIOUS CONTAMINATED SITES AND DIFFERENT APPROACHES TO BIOREMEDIATION OF INDUSTRIAL EFFLUENTS

4.3.1 Bioremediation by Phototrophic Organisms

The US Environmental Protection Agency (USEPA) recognizes bioremediation as an environmentally preferable method for restoring polluted areas and fostering long-term growth. Effluents from factories are a significant contributor to environmental degradation. Industrial hazardous waste poses a serious threat to aquatic life and, by extension, to humans. Cyanobacteria and microalgae are highly effective in removing harmful effluents because of their superior biosorption, biodegradation, and transformative powers. Increasingly, people are turning to microalgae production to detoxify waterways of harmful chemicals and metals. Algal cultivation adds value to the process by producing commercially valuable products such as fuels and various chemicals from biomass, and it can be seen as an attractive and eco-friendly means for treating waste streams because of the rapid utilization of the organic content and other nutrients present in the effluent required for biomass production. Recent advances in algae-based bioremediation of industrial pollutants and pollutants from other sources have been discussed and tabulated (Table 4.1) in this section (Ummalyma et al., 2017).

4.3.2 Phytoremediation and Phytotechnology for the Degradation and Detoxification of Wastewater

The toxic pollutants from industrial wastewater pose a potential threat to the environment and natural aquatic habitats worldwide, which should fetch the attention and awareness of environmentalists, educators, and commoners. Indiscriminate disposal of untreated contaminated water containing significant, highly toxic, recalcitrant contaminants into different water bodies creates health hazards for different water-dwelling

TABLE 4.1
Phototrophic Organism Used for Remediation of Pollutants

Type of Pollutant	Microalga Involved	Class	Removal Capacity (Max)	Reference
		Dye		
Methylene Blue	*Chlamydomonas moewusii*	Chlorophyceae	212.41 ± 4.55 mg/g	Seoane et al. (2022)
Methylene blue and Reactive blue 19	*Bifurcaria bifurcata*	Phaeophyceae	2,744.5 mg/g for methylene blue 88.7 mg/g for reactive blue 19	Bouzikri et al. (2020)
Methyl Blue, Fuchsin Acid, Rhodamine B, Methylene Blue, Bromocresol purple, Methyl Orange	*Sargassum ilicifolium*	Phaeophyceae	Methyl Blue-66.83% Fuschin Acid-34.12% Rhodamine B-18.60% Methylene blue-69.70% Bromocresol purple-16.30% Methyl Orange-13.25%	Tabaraki and Sadeghinejad (2017)
Congo red	*Chlorella vulgaris*	Chlorophyceae	200 mg/g	Hernández-Zamora et al. (2015)
Malachite green	*Haematococcus pluvialis*	Chlorophyceae	95.2%, 75.6% and 66.5% in three stages	Liu et al. (2018)
		Heavy metals		
Cadmium (Cd)	*Chlorella* sp. in beads of alginate	Chlorophyceae	59.67%	Valdez et al. (2018)
Chromium (VI) (Cr)	*Chlorella vulgaris* supported by iron sulphide	Chlorophyceae	–	Wu et al. (2021)
Cr	*Algal bloom*		155.52 mg/g	Zhang et al. (2010)
Mercury (II) (Hg)	*Chlorella vulgaris*	Chlorophyceae	52.09 mg/g	Kumar, Singh, and Sikandar (2020)

(Continued)

TABLE 4.1 (*Continued*)
Phototrophic Organism Used for Remediation of Pollutants

Type of Pollutant	Microalga Involved	Class	Removal Capacity (Max)	Reference
Uranium (U)	*Chlorella sorokiniana*	Chlorophyceae	188.7 mg/g	Embaby et al. (2022)
Copper (Cu) and Nickel (Ni)	*Chlorella vulgaris*, *Desmodesmus* sp.	Chlorophyceae	Cu: *Chlorella vulgaris*-39% *Desmodesmus* sp.-43% Ni: *Chlorella vulgaris*-32% *Desmodesmus* sp.-39%	Rugnini et al. (2017)
Lead (II) (Pb)	*Fucus spiralis*	Phaeophyceae	132 ± 14 mg/g	Filote et al. (2019)
Pesticides and insecticides				
Cypermethrin	*Fucus spiralis*	Phaeophyceae	588.24 µg/g	Năstuneac et al. (2019)
Malathion	*Nostoc muscorum*	Cyanophyceae	91% biodegradation	Ibrahim et al. (2014)
Bifenthrin	Microalga		99% through biodegradation and biosorption	Weis et al. (2020)
Pharmaceuticals and Pharmaceutical waste				
Paracetamol, Ibuprofen, Olanzapine and Simvastatin	*Nannochloropsis* sp.	Chlorophyceae	–	Encarnação et al. (2020)
Carbamazepine, Sulfamethazine, Tramadol	*Chaetoceros muelleri* and biochar	Bacillariophyceae	Carbamazepine-68.9% Sulfamethazine-64.8% Tramadol-69.3%	Mojiri et al. (2020)

(*Continued*)

TABLE 4.1 (*Continued*)
Phototrophic Organism Used for Remediation of Pollutants

Type of Pollutant	Microalga Involved	Class	Removal Capacity (Max)	Reference
17α-estradiol 17β-estradiol Estriol	*Scenedesmus dimorphus*	Cyanophyceae	17α-estradiol-85% 17β-estradiol-95% Estriol-95%	Zhang et al. (2014)
Diclofenac, Ibuprofen, Paracetamol, Metoprolol, Carbamazepine and Trimethoprim	*Chlorella sorokiniana*	Chlorophyceae	60%–100% removal Diclofenac, Ibuprofen, Paracetamol Metoprolol 30%–60% removal Carbamazepine and Trimethoprim	de Wilt et al. (2016)
Dairy wastewater	Microalgae-bacteria synergism	–	93% and 87.2% reduction of chemical oxygen demand (COD) ammonium concentration along with 100% removal of nitrates and phosphates	Biswas et al. (2021)
Tannery waste	*Scenedesmus* sp.	Cyanophyceae	Reduction of load of Cr, Cu, Pb and zinc (Zn). Decrease in nitrate and phosphate concentration	Ajayan et al. (2015)
Piggery wastewater	*Desmodesmus* sp. CHX1	Chlorophyceae	78.46% removal of ammonium nitrogen and 91.66% removal of total phosphorus	Luo et al. (2018)
Pulp and paper mill effluent	*Scenedesmus* sp.	Cyanophyceae	82% reduction of biological oxygen demand (BOD) 75% reduction of COD 65% removal of nitrate and 71.29% removal of phosphate	Usha et al. (2016)

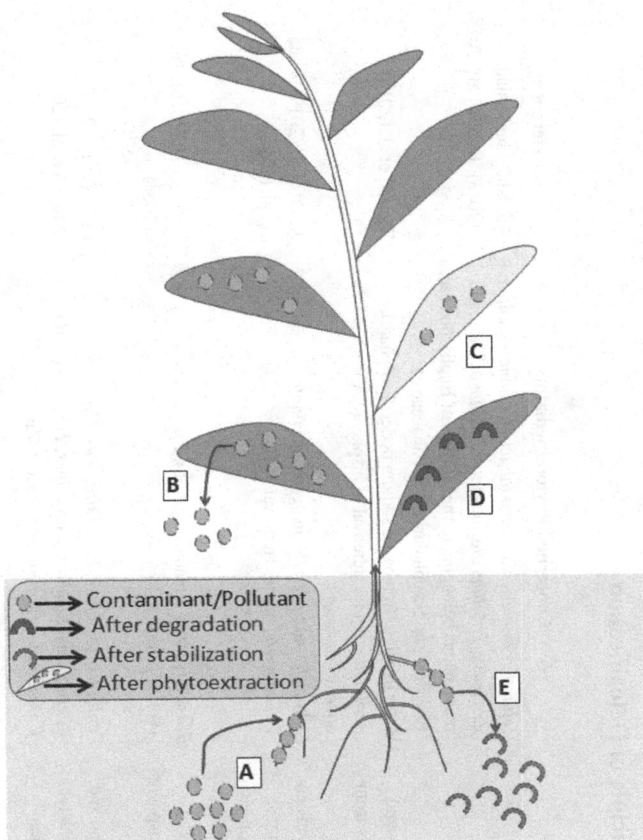

FIGURE 4.2 Fate of contaminants or pollutants (like heavy metals) after phytoremediation. A, Uptake of contaminants by root; B, Phytovolatilization; C, Phytoextraction; D, Phytodegradation; E, Phytostabilization.

micro and macroorganisms. It leads to the bioaccumulation of specific pollutants like heavy metals in living systems that consume them. Phytoremediation and other phytotechnological approaches have emerged as be a more effective decontamination procedures (Figure 4.2) for industrial wastewater than existing physio-chemical methods, as they have established themselves as stable and ecologically viable processes with comparatively fewer economic investments. Phytoremediation is a method of waste conversion, reduction, assimilation, and degradation by plant parts that involves techniques like phytoextraction, phytovolatilization, rhizofiltration, phytodegradation, and phytostabilization (Susarla et al., 2002). Phytoremediation involves plants that are tolerant to various heavy metal concentrations, are fast-growing, easily cultivable, and have extensive fibrous root structures. Different techniques involved in the process of Phytoremediation are listed below in Table 4.2, along with other attributes and examples that make them suitable for such categories.

TABLE 4.2
Different Methods of Phytoremediation and the Principle of Detoxification

Phytoremediation Methods	Principle	Concerned Phytoremediators	Reference
Phytoextraction	Uptake of contaminants especially heavy metals from wastewater and soil thereby accumulating and translocating the contaminants in their biomass.	*Helianthus annuus*, *Nicotiana tabacum*, *Cannabis sativa*, *Zea mays*, *Trifolium alexandrinum*, *Sedum alfredii*, and many plants belonging to the families of Euphorbiaceae, Brassicaceae, Scrophulariaceae, Fabaceae etc.	Ali, Khan, and Sajad (2013), Jacob et al. (2018)
Phytodegradation	Plants produce degradative enzymes, secreted around root zone removing various pollutant surfactants from wastewaters.	Water hyacinth (removing sodium dodecyl sulphate), *Azolla filiculoides* (removal of bisphenol A from aqueous surface)	Zazouli et al. (2014)
Rhizofiltration	Removal of pollutants from surface water or wastewater by plant roots as exudates from roots alter the site's pH and thereby precipitation or adsorption of pollutant metal occurs in roots.	*Callitriche stagnalis*, *Potamogeton pectinatus*, *Potamogeton natans* (removal of U)	Javed et al. (2019), Pratas et al. (2014)
Phytovolatilization	Decontamination by conversion of toxic elements to less toxic volatile form and disposed in the atmosphere in gaseous form by the plants via foliage system	*Brassica juncea* (volatalizer of Se), *Pteris vittata* (volatalizer of As)	Mahar et al. (2016), Tu (2004)
Phytostabilization	Immobilization of heavy metals from belowground and decreasing their bioavailability by using metal-tolerant plant species thereby reducing the chances of their entry into the ecosystem	*Phalaris arundinace*, *Pistia stratiotes*, *Cyperus alternifolius*, *Eichhornia crassipes*, *Thalia geniculate*, *Canna indica* (remediation of Cd^{2+}, Zn^{2+})	Burges et al. (2018), Sricoth et al. (2018)

The strategies plants adopt in phytoremediation are defense mechanisms to bear with and acclimate to severe stress, especially heavy metal stress (Thakur et al., 2016). The strategies responsible for the detoxification mechanism are tolerance and avoidance, through which a plant maintains its cellular concentration of remediated heavy metals below the tolerance limit (Hall, 2002). Implementation of these strategies by the plants involves specific innate physiological and molecular mechanisms like metal ion precipitation, root sorption, metal exclusion, chelation, heavy metal sequestration and compartmentalization in vacuoles and other plant parts, activation and overexpression of Reactive oxygen species (ROS)-scavenging genes and inducing antioxidant enzyme (for example, superoxide dismutase, peroxidase, catalase, and glutathione reductase) production (DalCorso et al., 2019; Jozefczak et al., 2012). Aquatic plants have properties that regulate the effluent metal concentration in the water medium. The submerged plant tissues harbor microorganisms and develop a microbial biofilm that releases oxygen and promotes degradation in the wastewater (Demarco et al., 2023). Water pollution by industrial wastes is a challenge ahead, and different biological agents have been analyzed for their potential for remediation of several wastewater contaminants like microplastics, pesticides, agrochemicals, polycyclic aromatic hydrocarbons (PAHs), and heavy metals (Abduro Ogo et al., 2022; Yan et al., 2022). In short, evolving and newer phytotechnologies in phytoremediation should be undertaken as a venture to decontaminate different polluted sites by exploiting their noninvasive, cost-effective, and efficient pollutant removal.

4.3.3 Nano-Bioremediation, Especially for the Treatment of Chemical Dye-Based Textile Effluents

The textile dyeing industries are among the biggest polluters, contributing to a significant adverse environmental impact (Lellis et al., 2019), generating approximately 200,000 t of dyestuff effluents annually (Ogugbue and Sawidis, 2011). They release oxides of sulfur and nitrogen, volatile organic compounds, particulate matter, and dust as air pollutants. The waste sludge discharged during the textile dyeing process contains heavy metals (such as Pb, Cr, arsenic, Ni, Cd, and Hg), micronutrients, and organic dyes. These are severely toxic to humans as well as the environment and are highly carcinogenic and mutagenic (Sharma et al., 2018). They also act as persistent environmental pollutants, ultimately impairing the ecological balance through interference and destabilization of the food chain, leading to bioaccumulation and biomagnification (Sandhya, 2010). The organic and nonbiodegradable compounds in synthetic, reactive, azo, acid, and basic dyes (Orts et al., 2018) exhibit excellent water solubility, making the effluents challenging to separate by known conventional techniques. Nanobioremediation or nanoparticles using microorganisms have opened a new and efficient avenue to mitigate and manage textile dye-contaminated wastewater pollution by virtue of its property of high reactivity with pollutants or contaminants, easing the process of immobilization of cells and enzymes, catalytic degradation of organic compounds, and disinfection of microbial flora (Anjum et al., 2019; Parvin et al., 2019). Nanomaterials can act as crucial agents for biosorption and element adsorption and can be used to decontaminate industrial effluents, including textile wastewater (Darwesh et al., 2019). The metal-based nanoparticles bond with heavy metals discharged as textile waste sludge, resulting in contaminant metal precipitation (Stensberg et al., 2011). Producing highly active nanomaterials using biological methods or microorganisms like actinomyces, bacteria, fungi, and microalgae has been reported as a greener, safer, and more environmentally friendly bioremediation process (Velsankar et al., 2021). The biosynthesis of Cu nanoparticles (Cu-NPs) by *Streptomyces* spp. and *Trichoderma* species has been shown to act as heavy metal adsorbents and decolorizing agents for textile dyes (Fouda et al., 2021; Saravanakumar et al., 2019). The photocatalytic ability of microorganism-synthesized metal-based nanoparticles (for example, ZnO-NPs, and CuO-NPs) is exploited to treat textile wastewater with high concentrations of nonbiodegradable dyes (Nwanya et al., 2019). Some of the potent biologically synthesized Metal-based nanoparticles are listed in Table 4.3, along with their functions in the bioremediation of textile effluent-based contaminants.

The immobilization of microbes and microbial enzymes on nanoparticles has considerably increased the pollutant cleaning-up ability of the nanoparticles, providing a

TABLE 4.3
List of Potent Biologically Synthesized Metal-Based Nanoparticles (MBNs) with Their Function in Bioremediationof Textile Effluent Based Contaminants

Metal-Based Nanoparticles (MBNs)	Microorganisms/ Macroorganisms Used for the Synthesis of MBNs	Function as Bioremediators	Reference
Citrofortunella macrocarpa-synthesized CuO-NP.	*Citrofortunella macrocarpa*	Photocatalytic reduction of Rhodamin B dye, capping agent.	Rafique et al. (2020)
Algae synthesized-Ag-NPs	*Caulerpa racemosa*	Reduction of Mythelene Blue	Shi et al. (2015)
MgO-NPs	*Rhizopus oryzae*	Decolourization of tannin effluents, reduction of total suspended solids (TSS), total dissolved solids (TDS), BOD, COD and conductivity.	Hassan et al. (2021)
Tio_2-NPs, Fe_3O_4-MNPs	*Phanerochaete chrysosporium*—a fungi	Pb removal from textile wastewaters.	Chen et al. (2012)
ZnO-NPs	Catharanthus roseus	Phenol red dye degradation from effluent wastewaters.	Kalaiselvi et al. (2016)
Au-MNPs	*Acinetobacter* sp. SW30 together with sodium borohydride ($NaHB_4$).	Biodegradation of the Reactive Yellow 186 dyes and Direct Black 22.	Wadhwani et al. (2018)

chance of regeneration (Xu et al., 2014). Biogenic nanoparticles are also gaining considerable importance as their inherent properties facilitate the effective separation of different organic dyes (David and Moldovan, 2020) and herbicides (Shad et al., 2021). Research involving Enzyme-assisted nanobiotechnology and bioremediation through metal-based nanoparticles still needs to flourish as a potential bioremediation strategy for textile industry effluents, considering their properties, cost-effectiveness, stability, and productivity.

4.3.4 Biological Decontamination of Heavy Metals from Dumping Sites

Heavy metals, characterized by atomic numbers higher than 20 and a density of more than 5 g/cm^3 (Ali et al., 2019), are one of the predominant parameters in landfills that are persistent, cumulative, and considered pollutants of priority concern (Briffa et al., 2020). Concerning levels of several heavy metals have been detected in landfill soil leachates as well as in the vegetation growing in the region (Beinabaj et al., 2023). Heavy metals, including arsenic (As), Cd, Mn, Pb, Hg, Cu, Cr, Ni, and Zn, are carcinogenic and mutagenic, induce phytotoxicity, and are introduced into the food chain (Pratush et al., 2018). Constant exposure to these metals generates oxidative stress, resulting in adverse health effects leading to neurological and developmental abnormalities, organ failures, and death, depending on the exposure time and dosage (Kurup et al., 2016; Mahboob et al., 2014). Among the existing treatment technologies, biological species offer an environmentally viable and cost-efficient solution for fixing high heavy metal concentrations in dumpsites via biosorption, bioaugmentation, biostimulation, and bioaccumulation (Raimondo et al., 2020). Some macro-fungal species have also been reported to accumulate heavy metals in their bodies (Damodaran et al., 2014). Recent works identifying the species related to heavy metal remediation at landfills and dumpsites are highlighted in Table 4.4. Several parameters influence the bioremediation capacity and efficiency of heavy metals in dumpsites, including initial metal concentration, specific bio-accumulating microbial dose or concentration and growth rate, presence of competing and inhibitory ions, sufficient nutrients, oxygen, light, pH, and temperature (Pant et al., 2019).

4.3.5 Management of Petroleum Industry Sludge through Bio-Surfactant-Producing Bacteria

One of the significant wastes generated by the petroleum industry during oil exploration, production, refining, and transportation is oily sludge, primarily comprising fractions of petroleum hydrocarbons (PHCs), water, asphaltenes, and heavy metals (Hu et al., 2013). According to the Resource Conservation and Recovery Act (RCRA), sludge from the petroleum industry is considered hazardous due to its high concentration of PHCs. Its contamination causes inhibition of soil microorganism activity (Suleimanov et al., 2005). Several components of oily sludge are carcinogenic, mutagenic, genotoxic, and immuno-toxicants (Mishra et al., 2001). Treatment with chemical surfactants, though fast, can cause environmental toxicity. Biosurfactants help in oil dispersion, increase bioavailability by emulsifying and solubilizing the PHCs, and offer an environmentally sustainable alternative to

TABLE 4.4
Heavy Metal Removal by Various Biological Entities

Microbial Species	Used as	Substrate Concentration	Removal								Reference
			Cd	As	Zn	Ni	Pb	Fe[a]	Cu	Cr	
Pseudomonas aeruginosa	Individual bacterial species	Cd: 50 mg/L Pb: 300 mg/L	58.80%	-	-	-	33.67%	-	-	-	Oziegbe et al. (2021)
Klebsiella edwardsii		Glucose: 5 mg/L	79.87%	-	-	-	92.41%	-	-	-	
Ankistrodesmus	Algal and cyano-bacterial consortia	5 mg/L	-	-	2.53 ± 0.7 mg/g	-	3.25 ± 1.1 mg/g	2.46 ± 1.4 mg/g	2.81 ± 0.7 mg/g	2.08 ± 2.0 mg/g	Iqbal, Javed, and Baig (2021)
Nostoc				-							
Anabaena		10 mg/L	-	-	5.73 ± 0.4 mg/g	-	7.03 ± 0.6 mg/g	5.09 ± 0.3 mg/g	5.85 ± 1.2 mg/g	3.34 ± 1.3 mg/g	
Ascomycota sp.	Fungal consortia	10–40 mg/L	-	-	0.0035/d	0.0046/d	0.0026/d	-	-	-	Hassan et al. (2019b)
Basidiomycota sp.			-	-	0.0045/d	0.0048/d	0.0029/d	-	-	-	
Ganoderma lucidum	Individual species	-	43.1%–50%	-	19.7%–27.3%	13.5%–38%	3.1%–3.7%	-	18.6–36.1%	-	Ipeaiyeda et al. (2020)
Highly tolerant fungal species[b]	Consortia	-	-	62%	-	-	-	38%	49%	42%	Hassan et al. (2019a)
Proteus mirabilis	Individual bacterial species	225 ppm each	27.68%	-	-	~24%	32.75%	-	-	~30%	Sanuth and Adekanmbi (2016)
Pseudomonas Aeruginosa		350 ppm each	~28%	-	-	~34%	~29%	-	-	~31%	
Paenalcaligenes hominis		274 ppm each	~30%	-	-	~30%	35.77%	-	-	36.89%	
		325 ppm each	32.81%	-	-	34.91%	~31%	-	-	~27%	
Bacterial species[c]	Consortia	-	~30%	-	-	~28%	~20%	-	-	32.54%	

[a] Iron.

[b] *Perenniporia subtephropora, Daldinia starbaeckii, Phanerochaete concrescens, Cerrena aurantiopora, Fusarium equiseti, Polyporales sp., Aspergillus niger, Aspergillus fumigatus,* and *Trametes versicolor.*

[c] *Proteus mirabilis, Pseudomonas aeruginosa, Paenalcaligenes hominis* and *Bordetella petrii.*

treating this hazardous waste with simultaneous oil recovery (de Sousa and Bhosle, 2012). These amphiphilic substances with hydrophobic and hydrophilic moieties interact with the aromatic hydrocarbons to reduce the surface tension of the solution (Santos et al., 2016).

Biosurfactant production has been reported in several microbial species, including *Pseudomonas* spp., *Acinetobacter* spp., *Bacillus* sp., *Streptomyces* spp., *Rhodococcus* spp., *Achromobacter* spp., *Brevibacterium* spp., and *Arthobacter* spp. These species can easily be isolated from the petroleum industry sludge, are already acclimatized to the waste characteristics, and thus have a higher efficiency in treating the oily solids. A strain of *Serratia* sp. ZS6, isolated from petroleum sludge, secretes a nonionic serrawettin-type biosurfactant and a lipase that can hydrolyze fatty acyl ester bonds (Hu et al., 2018). Another species of the same genus, an isolated oil spillage site in Assam, shows excellent remediating potential (Borah et al., 2019). *Pseudomonas aeruginosa* produces a surfactant of rhamnolipid species with an emulsifying index of 82% for diesel (Bharali and Konwar, 2011). But only a few of these strains have been applied experimentally to investigate their potential for the remediation of oily sludge and solid wastes from petroleum industries. Studies undertaken for this purpose during the last few years are summarized in Table 4.5, highlighting the efficiency of the biosurfactant and its properties. Instead of using single bacterial species, a mixed consortium has higher biodegradation rates (Wanapaisan et al., 2018). However, studies have noted the adverse effects of biosurfactant-mediated bioremediation, including cell lysis, substrate toxicity, and undesirable bacterial cell-surfactant reactions (Avramova et al., 2008), that can be avoided by optimizing the process conditions.

4.3.6 Rubber Waste: Mitigation and Management

High global production of rubber products is accompanied by a high discarded rubber waste burden, most of which is generated from the tyres of various automobiles, followed by other leather and rubber products, including furniture, clothing, footwear, toys, and sports goods. Natural rubber latex is made from poly (cis1,4-isoprene), and synthetic rubber is made from polymers of styrene, butadiene, chloroprene, and isobutylene (Myhre et al., 2012), and both take decades to degrade naturally. Commonly used rubber waste management techniques include open dumping, burning, landfilling, and grinding; alarmingly, only 1.5% of the total rubber waste is recycled and reused (Liu et al., 2020). Dumping rubber waste creates a breeding habitat for insects and induces soil ecotoxicity due to stabilizers, plasticizers, and flame retardants added during the production of the rubber product. Open burning generates toxic fumes, whereas low biodegradability limits the application of biological processes. Selected bacterial species of the genera *Gordonia*, *Xanthomonas*, *Streptomyces*, *Nocardia*, and *Mycobacterium* and strains of *Penicillium* and *Aspergillus* are observed to degrade rubber waste under optimum conditions (Tsuchii and Tokiwa, 2001). Table 4.6 shows the different rubber waste management techniques and their respective advantages and limitations. Reusing old vehicle tyres into crumb products can considerably reduce carbon emissions (Tushar et al., 2022).

TABLE 4.5
Degradation of Petroleum Industry Sludge by Bio-Surfactant-Producing Bacteria

Microbial Species	Biosurfactant Properties	Remarks from the Investigations	Reference
Ochrobactrum intermedium	* Glycolipopeptidal * Surface tension 36 ± 1 mN/m * Emulsification index: 86% for Cyclo-Hexane, 79% for hexane and 65% with waste motor oil	* 40% hydrophobic long chain aliphatic and PAHs * Up to 70% degradation of the most hydrophobic components of the petroleum sludge in 3 weeks.	Bezza, Beukes, and Chirwa (2015)
Pseudomonas balearica Z8	* Surface tension 41 mN/m * Critical micelle concentration: 90 mg/L * Emulsification index: 44% for crude oil	Maximum total petroleum hydrocarbon (TPH) removal of 35% with sludge/water ratio of 1:7 and temperature of 40°C.	Nejad et al. (2020)
Lysinibacillus sp. isolated from rhizosphere	* Surface tension 33.2 mN/m	14.4%–39.7% TPH removal	Sharuddin et al. (2021)
Enterobacter cloacae	* Rhamnolipid * Surface tension: 26.4 mN/m * Emulsification index: 100% for crude oil	* TPH degradation: 86.9% after 5 days * Removal of heavy metals like Ni, Zn, Pb, Fe, Cr, Cu	Patowary et al. (2022)
Bacillus nealsonii	* Surface tension: 34.15 ± 0.6 mN/m * Emulsification index: 55% ± 0.3% for kerosene	* Remediation of heavy engine-oil sludge: 43.6% ± 0.08% to 46.7% ± 0.01%	Phulpoto et al. (2020)
Pseudomonas aeruginosa	* Rhamnolipid * Surface tension: 26.8 mN/m * Critical micelle concentration: 0.123 g/L * Emulsification index: 66%–99% for oily sludge	* 0.060 g/L solution of the isolated biosurfactant can recover oily sludge by up to 100% Can be used for treating sticky oily sludge	Ghorbani et al. (2022)
Kocuria rosea	* Lipopeptide	* 7% crude oil recovery from petroleum sludge * 22% degradation of crude oil	Akbari et al. (2021)
Serratia marcescens	* Lipopeptide * Surface tension: 29.5 mN/m * Critical micelle concentration: 220 mg/L * Emulsification index: 56.6% ± 0.8% for n-hexane and 55.6% ± 3.0% for mineral oil	* COD removal: 48.9% * TPH removal: 64% * n-alkane removal: 65.57%	Huang et al. (2020)

PAH, polycyclic aromatic hydrocarbon

TABLE 4.6
Merits and Limitations of Using Different Waste Rubber Management Techniques at a Glance

Management Methods		Process	Advantages	Disadvantages	End Products	Reference
Thermal	Thermolysis	*By action of heat *High temperature	* Energy recovery * Material recovery	* High cost of kilns and temperature maintenance * Toxic gas generation	Oil, methane, ethyne, ethylene, ethane, propane	Li et al. (2018)
	Pyrolysis	* In absence of O_2 * 350°C–600°C	* Energy recovery	* Toxic gas generation	Alkanes, alkenes, aromatic compounds	
Chemical	Organic solvents	0.5%–10% xylene, toluene, or benzene, diphenyl disulphide	* Can be reused into other product formation	* High temperature may be required under certain cases	Hydrogen sulphide, sulphur dioxide and thiols	Chittella et al. (2021)
	Peroxides	Hydrogen peroxide or benzoyl peroxide	* Lower temperature around 80°C	* May require the presence of UV		
	Green solvents	Ionic liquids at 45°C under N	* Low temperature * Higher product recovery	* Solvents are highly expensive		
Mechanical	Ambient grinding	Passed through shredders	* Particles with high surface area and 200 µm size is produced	* Requires high amount of water for cooling	Crumb rubber, shredded rubber and ground rubber (used as modifiers to asphalt mixtures, shock and vibration absorbers and as filler material and polymer blends)	Lapkovskis et al. (2020)
	Wet grinding	Mixing water with crumbs to make slurry	* Particles with high surface area	* High energy consumption * Extra drying step		
	Cryogenic grinding	Cooled in liquid nitrogen and crushed	* Clean rubber crumb * High production rate	* High cost of liquid N * Cost of pre-grinding and drying		
	Grinding by ozone cracking	Exposed to high ozone concentration and grinding	* Less energy consumption	* End product has low surface activity		

(Continued)

TABLE 4.6 (*Continued*)
Merits and Limitations of Using Different Waste Rubber Management Techniques at a Glance

Management	Methods	Process	Advantages	Disadvantages	End Products	Reference
Physical	Microwaves	* 200°C–250°C	* No toxic compounds involved	* High energy requirement * High operation cost	Degraded rubber polymers	De Sousa et al. (2017)
	Ultrasonic frequency	*20–25 kHz, aided by high temperature				Sathiskumar and Madras (2012)
Biological	Bacterial and fungal activity	* 20°C–35°C * Optimum growth conditions for different species	* Low cost * Eco-friendly approach * No toxic end products	* Very specific species can degrade rubber * Slow rate of degradation	Nontoxic compounds	Cui et al. (2016)

4.3.7 Microplastics as Waste in Landfills and Their Bioremediation

Microplastics are synthetic organic polymers having a diameter of less than 5 mm that can either be categorized as primary microplastics (produced as beads or pellets) or secondary microplastics (from fragmentation from larger plastics) (Tang and Hadibarata, 2021). Municipal and industrial landfills and open dumping sites are considered significant sources of microplastics (Monkul and Özhan, 2021), which are leached out, contaminating soil, groundwater, nearby surface water, or ending up in the oceans (Su et al., 2019). Studies often identify the presence of microplastics in landfill leachate at concerning concentrations (He et al., 2019; Sun et al., 2021). Microplastics are potentially toxic, and when they enter into the food chain, they cause adverse ecological and health effects. The degradation of microplastics is often hypothesized to be more challenging than that of plastics (Lambert and Wagner, 2017). Plastic polymers that constitute the microplastics in landfills are observed to vary in different regions in various proportions. In a study analyzing soil from 54 dumpsites in several Asian countries for microplastic contamination, polyethylene (PE), polypropylene (PP), and polyethylene tetraphthalate (PET) were observed to be the dominant polymers (Tun et al., 2022). Another study analyzing soil from an abandoned dumpsite showed the dominance of polystyrene (PS), PET, polyamide (PA), and PP microplastics (Kazour et al., 2019). An extensive investigation involving 11 landfills in three different Nordic countries showed the presence of PE, polyurethane (PU), PET, PS, PP, polymethylmethacrylate (PMMA), polyvinyl chloride (PVC), and PA in descending order of percentages (Praagh et al., 2019).

Bioremediation, including phytoextraction, composting, and enzyme-mediated biodegradation, is considered a sustainable approach to address the microplastic contamination in dumping sites, where the substrate is finally targeted to be mineralized (Miri et al., 2022). The presence of potential plastic-degrading microorganisms with appropriate enzymes, target-specific polymers, and favorable pH, salinity, temperature, moisture content, and incubation time is essential for optimized biodegradation (Gong et al., 2012). Biodegradable additives like starch enhance the efficiency of microbial activity on PE, PVA, and PS microplastics by making the polymer hydrophilic (Zadjelovic et al., 2020). Microbial degradation of microplastics occurs in two stages, initially adhering to the polymer surface, followed by enzymatic dissimilation (Auta et al., 2018). The efficiency of bacteria toward microplastic degradation is higher in comparison to fungi and other higher species (Tang and Kristanti, 2022). Several species of bacteria belonging to *Vibrio* (Weigert et al., 2022), *Thermobifida* (Barth et al., 2015), *Alteromonadaceae* (Morohoshi et al., 2018), *Burkholderiales* (Rüthi et al., 2020), *Rhodococcus* (Rose et al., 2020), *Bacillus* (Dang et al., 2018), and *Pseudomonas* (Wilkes and Aristilde, 2017), and fungal species belonging to *Penicillium*, *Aspergillus*, *Zalerion*, and *Fusarium* are reported to efficiently degrade microplastics (Zeghal et al., 2021). Higher organisms like *Tenebrio molitor*, *Galleria mellonella*, and *Achatina fulica* decompose PS and PE microplastics using their gut microbiome (Yang et al., 2015). However, their excretion contained microplastic residues (Reichert et al., 2018). Mineralization of specific microplastic polymers occurs due to the action of enzymes, such as hydroxylases, laccases, peroxidases, and reductases, which are responsible for PE degradation, cutinases, laccases, lipases, and ureases for PU decomposition and cutinases, esterases, as well as PETase and MHETase for PET degradation (Jeyavani et al., 2021).

4.3.8 Bioremediation of Sugar-Mill Effluents

The sugar mill industry forms the backbone of agriculture-based industries worldwide and is considered a primary source of water and soil pollution through its discharge. Sugar mill effluents comprise increased amounts of total dissolved solid and total suspended solid, bagasse, chlorides, sulfates, and nitrates. These effluents lead to higher biological oxygen demand (BOD) and chemical oxygen demand (COD) ratios. These effluents significantly threaten aquatic life and agricultural lands containing paddy crops (Afrad et al., 2020). Standards laid down by the pollution control body in India have set the BOD limit of wastewater to be not more than 30 and 100 mg/L for surface water and irrigation land disposal, respectively, but total effluent generated from the sugar industry had a BOD value of 1,500 mg/L (Poddar and Sahu, 2015). Sugar industry wastewater also entails a substantial concentration of toxic metals, mainly Zn, Pb, Cu, and Hg (Fareed et al., 2022). Sugar wastewater released into water bodies such as rivers makes it hazardous and nonconsumable to the rural population for household and agricultural uses. Properties of soil affected by sugar mill effluents, such as electrical conduction, water retaining capacity, microbiological agents, and nitrogen concentrations, turned out to be higher when compared with other wastewater-affected soil (Nagaraju et al., 2009).

Various beneficial fungal species have been isolated from sugar mill effluents, such as *Aspergillus niger*, *Fusarium monoliforme*, *Penicillium pinophilum*, *Aspergillus flavus*, and *Alternaria gisen* (Pant and Adholeya, 2006). Researchers have highlighted the presence of approximately 23 genera of arbuscular mycorrhizae (AM fungi), including *Acaulospora*, *Glomus*, and *Scutellospora*, in soil contaminated with sugar industry wastewater (Bama and Ramakrishnan, 2010). Among the many fungal species found in the polluted soil, *Aspergillus niger*, *Aspergillus nidulans*, and *Aspergillus candidus* are the most abundant and common among various soil types (Awasthi et al., 2011). Therefore, bioremediation plays a considerable role in treating different types of sugar mill effluents. Bioremediation refers to using biological entities such as microbes (bacterial and fungal) and plant species for biodegradation and the removal of contaminants in the soil. Microbes, including bacterial and fungal isolates, have great potential to treat sugar-effluent-contaminated soil. The underlying processes utilized by microorganisms to potentiate degradation mainly entail: biosorption, intracellular aggregation, and reactions catalyzed by enzymes (Puyol et al., 2017).

Basidiomycetes are a fungal class that acts as an active decolorizer and biodegrades by releasing extracellular enzymes such as peroxidases, superoxide, and laccase to degrade lignocellulose (Echezonachi, 2022). There has been evidence illustrating various properties associated with bacterial extracellular polymeric substances as potential candidates for adsorbing different toxic metals, mainly Cu and Pb. These immobilized extracellular polymeric substances were reused and regenerated for up to 5 cycles in heavy metal adsorption (Ajao et al., 2020). Experiments using anaerobic sludge reactors have successfully reduced COD levels by 76% (Nacheva et al., 2009). Immobilized bacterial consortium beads of various species such as *Bacillus subtilis*, *Serratia marcescens*, and *Enterobacter asburiae* have been used as an inoculum to treat sugar mill effluent for reducing levels of physico-chemical properties and quality parameters such as BOD, COD, total suspended solid, and total dissolved solid (Fito et al., 2019).

4.3.9 Microbial Remediation for Lignocellulosic Waste Bioconversion

Lignocellulose is one of the principal components of the total renewable biomass in the environment, including residuals from agriculture, forests, and industry (Kumar et al., 2008). Lignocellulose, as the name suggests, comprises three principal constituents, i.e., cellulose, lignin, and hemicellulose, which form the structural support system of vascular plants (Betts et al., 1991). Cellulose consists of over 10,000 units of d-glucose residues arranged in unbranched linear polymeric chains linked by β-1,4-glycosidic bond (Baldrian and Valášková, 2008). Hemicellulose is a biopolymer that consists of heterogenous polysaccharides such as pentose, hexose, and acidic sugar; these monomer units are bounded by β-1,4-glycosyl bonds (Chen et al., 2022). Lignin is a highly cross-linked aromatic heteropolymer comprising 4-hydroxyphenylpropanoid monomers that are linked by different covalent and ether linkages. Different phenolic monomers, including hydroxyphenyl (H), guaiacyl (G), and syringyl (S), are in control of lignin structure in plant species. Lignin increases cell wall integrity and resistance to attack by pathogens and tends to be recalcitrant due to complex chemical bonds (Houfani et al. 2020).

People across the globe are projected to be more than 9 billion by the year 2050 (De Gonzalo et al., 2016), with an expected solid waste of approximately 90 billion tons predicted to be generated per year by 2025 (Bugg et al., 2011). Lignocellulose waste material produced after the harvest is usually burned or dumped in an unorganized manner, causing various negative impacts on humans and the environment, such as pollution and global warming (Koul et al., 2022). The increasing use of microbial populations to degrade lignocellulosic waste is credited to their ability to be cost-effective and produce bioenergy such as bioethanol and bioproducts of high nutritional value (Singh, 2019).

Cellulose is hydrolyzed into glucose monomers by various enzymes such as exo-1,4-Beta glucanases (cellobiohydrolase), endo-1,4-Beta gluconases, and cellobioses (Schwarz, 2001). The use of various enzymes, which include laccases, manganese peroxidase, and lignin peroxidase, is known to be used for the delignification process of lignocellulosic waste (Leonowicz et al., 1999). Another set of enzymes, which includes degrading enzymes such as oxidoreductase, chitinase, and laccases that possess lignocellulosic degradation potential, was found in transcriptome sequencing of the lignocellulytic fungus *Ganoderma lucidium* (Jain et al., 2020). *Trametes versicolor* has been studied in the solid-state fermentation (SSF) of corn stover and has resulted in the hydrolysis of lignocellulose material, the production of enzymes, and protein enrichment (YouShuang et al., 2011). Fungal species such as white rot fungi can be used to decompose crude oil-contaminated soil by adding lignocellulose substrates such as sawdust and corn cobs within the agricultural soil (Kumar, 2021). Fungal groups belonging to basidiomycetes (*S. coelicolor*, *S. griseus*, and *S. psammoticus*) are primarily involved in the degradation of lignin due to the occurrence of peroxidases, laccases, and manganese peroxidases (MnP), resulting in the depolymerization of lignin (Dashora et al., 2023). Researchers have established the use of *Scedosporium* species as a potential source of lignocellulolytic enzymes via varying cultural and biochemical analyses (Poirier et al., 2023). Volarization of lignin involves other biocatalysts such as cellobiose: quinone oxidoreductase, polyphenol, glyoxal oxidase, and lytic polysaccharide monooxygenase (Lee et al., 2019).

4.4 TECHNIQUES ASSOCIATED WITH THE BIOREMEDIATION PROCESS

4.4.1 Bioreactor

In the past few decades, there has been a rise in environmental contamination caused by human activity (Valiela et al., 2001). Both human health and the ecosystem are negatively impacted by pollutants, and thus, it is vital to remediate these contaminants (Singh et al., 2015). It is not possible for physical or chemical oil spill cleanup solutions to completely remove the PHCs that have been released into the ocean after an oil spill, eliminating oil residues that can be cleaned up using biotechnology (Ivshina et al., 2015). Bioremediation is a process that removes toxins from the environment in a way that is kind to the ecosystem (Mani and Kumar, 2013). During this process, microbes convert pollutants into compounds that are either less detrimental to the environment or more environmentally beneficial. Microbial remediation is an approach that can be used to remediate either in-situ or ex-situ circumstances. Though in situ bioremediation is a lengthy procedure that can be hard to control and optimize. Bioreactors emerging as a solution to these concerns as we speak. In order to accomplish the necessary remediation goals, the bioremediation methods make use of the biodegradation mechanisms that are associated with microorganisms (Kumar et al., 2018). When it comes to the growth of microorganisms, bioreactors provide the optimal environment. Bioreactors can operate in different modes, including batch, fed-batch, and continuous (Lim and Shin, 2013), thanks to the wide variety of available configurations. These include partitioning, slurry-phase, stirred-tank, bioscrubbers, biofilters, fluidized-bed, trickle-bed, airlift, packed-bed, and membrane bioreactors (Srivastava et al., 2021) and additional information regarding the primary types of bioreactors, as well as their designs, advantages, and disadvantages.

4.4.2 Land Farming

Pollutants in the environment have increased dramatically during the last few decades due to excessive human activity on energy reserves, harmful farming practices, and rapid industrialization (Bose, 2010). All of these variables have played a role in fostering this pattern. Bioremediation, the technique of cleaning up polluted regions using microbial activities, has been proven effective, reliable, and environmentally friendly (Bhatnagar and Kumari, 2013). Farming on the contaminated ground has been widely implemented as a commercial strategy for the biological removal of petroleum pollutants, and it has had some success. The technology has seen extensive use due to its ease of use and low cost. While this technology certainly has its uses, it also has physical, chemical, and biological aspects that could hinder the recovery procedure. The three principal pollutant removal methods in land farming are volatilization of low-molecular-weight volatile chemicals, biodegradation, and adsorption, and they all occur early in the contamination or treatment process (Kehrein et al., 2020). The process continues with the volatilization of volatile chemicals with a high molecular weight. Health and environmental issues are raised by the volatilization and leaching of petroleum products and the "recalcitrant" hydrocarbon residues left over when land farming is implemented (Lukić et al., 2017). However, bioreactors can help alleviate some of these problems. The bioremediation techniques of bioaugmentation and biostimulation, which involve land farming, will be useful (Lyon and Vogel, 2012). However, biostimulation should be favored in polluted areas with native pollutant-degrading bacteria because of the intrinsic difficulties associated with bioaugmentation, such as the short survival of boosted strains. While landfarming's key advantages lie in its simplicity and low cost, additional issues commonly seen as drawbacks to the technology's deployment can be overcome by applying different strategies. Large tracts of land are needed for treatment, pollutant-degrading bacteria must be readily available, and the technology must be effective at high constituent concentrations (greater than 50,000 ppm). For such high concentrations, reductions must be greater than 95%, and the technology must be adaptable enough to remove PHCs while simultaneously dealing with other contaminants that may coexist with the petroleum products (Shimp et al., 1993). Any one of these may make or break the efficacy of a treatment strategy.

4.4.3 Bio-Cell Treatment

Because of the quick pace of industrialization, components of the Earth's surface, including its soil, water, and other geomaterials, have been drastically altered (Omeje et al., 2023). Infrastructure development often induces ground contamination, which also fosters the growth of unsuitable property zones, especially within the city borders. Microorganisms such as bacteria, fungi, algae, and other similar organisms are utilized in the process of bioremediation in order to degrade and eliminate contaminants (Bala et al., 2022). The necessity of maintaining the appropriate conditions for microbial development throughout in-situ remediation presents several obstacles. These requirements include an abundance of moisture, oxygen, and nutrients. In-situ remediation techniques such as bioventing, natural attenuation, biosparging, and bioslurping are just a few strategies that encourage the growth of microorganisms as part of the cleanup process (Tyagi and Kumar, 2021). The poisoning of the soil and water supplies of the Earth is currently the most urgent concern that can be found anywhere in the world. Multiple studies have demonstrated the long-lasting detrimental effects of pollution on the robustness of ecosystems and the health of humans (Moore et al., 2004). It has been demonstrated that different approaches are effective in reducing pollution levels in both water and soil. Bioremediation (Karimi et al., 2021), often known as land-vetting or bio-cell, is a technology frequently utilized in remediating both water and soil. Adsorption, Pretreatment (Ultrasonic), Microwave, and Electrokinetic Disintegration; Bio Cell; Phytoremediation; Land Venting; Composite; Bio Venting; Bio Slurry; Bio Sludge; Composite; Bio Venting; and Bio Sludge. Because of the relatively low cost involved, bioremediation is an attractive choice to consider when treating environmental degradation. The ability of soil to be bioremediated after having been contaminated with heavy motor oil due to years of dumping waste oils from an auto repair shop was evaluated (Ololade, 2014). This particular soil was selected as a test bed for evaluating the efficacy of various process modifications because of the unique difficulties it posed to bioremediation. In addition, this soil was selected because of the unusual problems it offered to the bioremediation process. Based on the results of biocell studies, a treatment strategy combining the addition of nutrients and surfactants with bioaugmentation (Zappi et al., 1999) that uses activated sludge may remove over 80% of the TPH.

4.4.4 Composting

Poor waste management practices for crude oil sludge provide a public health risk due to the presence of toxic chemicals such as PAHs (Ahmad, 2021). Bioremediation, the process of cleaning polluted places, may include composting as a sustainable option for removing toxic waste (Thassitou and Arvanitoyannis, 2001). Fungal and bacterial species were isolated from the compost using standard culture techniques. The sludge's PAH remnants at multiple composting stages using gas chromatography and mass spectrometry (GC–MS) (Giungato et al., 2016). Composting samples showed lower levels of PAHs after decomposition, with the highest degradation efficiency seen for PAHs with the highest molecular weight. However, not all of the lower-molecular weight PAHs were broken down. Using an experimental approach, the optimum conditions for composting as a bioremediation strategy for contaminated soil can be formulated. A cheap product like compost derived from the organic portion of municipal solid waste was used as an amendment, and the model pollutant pyrene was investigated (Sayara et al., 2009). Composting is a greener waste management option that can be used for the bioremediation of hazardous wastes like crude oil sludge. Surfactants added in varied quantities did not always promote fungal growth in the treatments, but they did increase the bioavailability of PAH compounds in crude oil sludge (Mohan et al., 2006). There has been a call for more research into the potential loss of PAHs during the composting process through leaching or volatilization, the biodegradative capability of both fungi and bacteria to generate a supportive atmosphere for PAH degradation, and the diversity of PAHs that might be lost through the composting practice (Loick et al., 2009).

4.4.5 Biostimulation

The hydrocarbon-rich waste sludge (aromatic, aliphatic, and complex hydrocarbons and heavy metals) generated by petroleum industries and refineries is highly precarious, and their unconcerned disposal in huge amounts leads to environmental pollution that persistently stays in the ecosystem for a long period of time (Das and Chandran, 2011; Pal et al., 2016). Though the abundance of hydrocarbon-degrading organisms (like bacteria and archaebacteria) is naturally high in those sites, inadequate nutrient availability and inappropriate physical and chemical conditions (moisture, temperature, and pH) restrain the bioremediation process by those native microbes (Alonso-Gutiérrez et al., 2009; An et al., 2013; Hazen et al., 2013). To combat this situation and with increasing public concern, in-situ bioremediation techniques have been deployed by many scientists and workers, which, by conscious monitoring of the inhabitant microbial population, remove the toxic effluents (Megharaj et al., 2011). Biostimulation is a highly efficient, nature-friendly, and cost-effective weapon of bioremediation, especially for petrochemical products and their derivatives rich in hydrocarbons (Delille et al., 2004; Tyagi et al., 2010). Stimulation of the metabolically less efficient microbial flora of a polluted site by the addition of rate-limiting nutrients like carbon, nitrogen, and phosphorus to alleviate the remediation process is the basic principle on which biostimulation works. The efficacy and success of biostimulation depend on some factors, like the choice of microbes and the adequacy of rate-limiting nutrients (Sarkar et al., 2016). The most commonly found hydrocarbon-degrading microbes at sites contaminated with petroleum products so far reported are Fermicutes members, Gammaproteobacteria, Epsilonproteobacteria, Deltaproteobacteria, Betaproteobacteria, and Bacteroidetes (Head et al., 2014) and some methanogenic members (including Methanobacterium, Methanosarcina, Methanocella, and Methanoculleus). Groundwater contaminated with sulfate can also be detoxified using this process through the enhancement of electron donors, thereby increasing sulfate reduction (Miao et al., 2012). Despite having several prominent advantages, biostimulation has several demerits, including dependence on environmental factors that dictate its potentiality, its site-specific nature, and the requirement of constant scientific observations (Dunbar et al., 1999).

4.4.6 Bioaugmentation

The most effective way to clean up soil polluted with pesticides is bioaugmentation, a green technology described as improving the biodegradation capability of contaminated regions by adding certain microorganisms (Cycoń et al., 2017). The choice of pesticide-using microbes from diverse sources and their capacity to break down pesticides from various chemical classes in liquid media, as observed in case studies including soil in a laboratory, a greenhouse (Mohapatra et al., 2021), and outdoor settings, define the efficiency of the process. A particular focus is placed on bacterial varieties that might be used in the bioaugmentation-based environmental remediation of pesticide-contaminated soils. Numerous biotic and abiotic factors, including pH, temperature, soil type, pesticide intensity, moisture and organic material composition, supplemental carbon and nitrogen sources, inoculum, connections between created strains and native microbes, and inoculant survival, were presented because they all have a major impact on the success of microbial inoculants (Chakraborty and Akhtar, 2021).

4.4.7 Biological Treatment

The excessive amounts of minerals, organics, nitrogenous compounds, hydrogen sulfide, total dissolved solid, heavy metals like Cr and deferred particles were physiognomies of tannery effluents, which are also known for their complexity (Shukla et al., 2019). Wastewater from industries is generally treated using biochemical, physical, microbial, or a blend of these practices. The properties of the effluents produced by several tanneries and different biological treatment preferences influence the effluent sludge produced (Sreedevi et al., 2022). It was found that variations in biological loading rates, the existence of sulfides, Cr, and COD had a considerable effect on nutrient extraction efficiency and process capabilities (Singh and Chakraborty, 2021). Compared to traditional reactors, the Upflow Anaerobic Sludge Blanket Reactors performed well at successfully handling high-strength tannery effluent. The effluent from the leather industry is treated using both anaerobic and aerobic methods. The kinetic equation includes a refractory parameter to take into consideration the non-biodegradable component of the organic substances in the digester (Li et al., 2016). The data for the aerobic digestion of organic material and the denitrification of dairy wastes are applied to the kinetic equation. They all demonstrate an excellent match between the data and the kinetic equation. Moreover, it is demonstrated that the refractory factors and reaction mechanisms are unaffected by the amount of permeate organic substance (Mopper et al., 2007). This investigation supports earlier findings that permeate green waste quantity affects the effluent of biological treatment processes for organic wastes. The best choice for the cleaning of tannery effluent is physical/chemical processes mixed with biological processes (Bhat and Gogate, 2021), it may be said.

4.5 MECHANISMS OF BIOREMEDIATION

Because of their widespread use in activities, leftover pesticides and metals in land are becoming an increasingly important environmental concern (Baweja et al., 2020). Diverse defense mechanisms exist in polluted soil-dwelling organisms to counteract the toxicity of contaminants. Co-contaminated soil treatment employing the bioremediation method, which uses microbes solely, or with vegetation to remediate pollutants (Singh and Cameotra, 2004), is effective and encouraging.

4.5.1 Microbial Remediation

Microbial treatment involves the use of microbes to remove or immobilize pollutants. Many organic contaminants (Boudh et al., 2019) are degraded by microbes into end products or enzymatic intermediates that serve as nutrients for cell development. Metals can be remedied by microbes in a variety of ways, including binding, immobilization, and oxidative transformation. Metal and pesticide pollution microbial remediation may be divided into five categories: biosorption, biomagnification, bioremediation, bioaccumulation, and biodegradation (Filote et al., 2021).

4.5.1.1 Biosorption

Biosorption refers to a group of metabolism-independent mechanisms by which contaminants adhere to organisms. Because each creature has an attraction to organic and inorganic contaminants, there is considerable biosorption potential (Aksu, 2005) in a wide range of species. Complexation, electrostatic forces, electrodialysis, chemical and physical absorption, surface modification, transport, and chelation are all mechanisms involved in biosorption (Ali Redha, 2020). This happens primarily within the cell wall. The first element that prevents contaminants from entering cells is the microbe wall, which can be deposited just on the exterior or inside the retaining wall. Conversely, cell walls, mostly made of carbohydrates, lipids, and proteins, include numerous functional groups such as amino groups, carbon, thiol, hydroxyl, phosphate, and sulfate (Fomina and Gadd, 2014). To different degrees, the biosorption processes may operate concurrently. Hydrophobic chemicals, for example, can flow across membranes and be incorporated into matrix materials due to their lipophilic properties (Weiss et al., 2006). Penetration might be an essential process for organic contaminant biosorbents.

4.5.1.2 Bioaccumulation

Pollutants reach the cytoplasm through the cellular membranes and proceed through the cellular metabolic process, referred to as bioaccumulation (Elizabeth George and

Wan, 2019). The integrated physical, biochemical, and biological factors are responsible for bioaccumulation. Adsorption occurs whenever the quantity in the biome is substantially higher than in the local environment. Contaminants can affect biota by bioaccumulating in the habitat of species (Tourinho et al., 2019).

4.5.1.3 Biotransformation

Metals cannot be destroyed by bioconversion; they are transformed into yet another state. Because of changes in chemical and physical characteristics after efficient bioremediation, organisms may convert them either into a cell component or to a less hazardous form (Quagraine et al., 2005). The mechanism is primarily dependent on enzymes or chemicals produced by organisms. In numerous situations, the efficiency of xenobiotic bioconversion has been demonstrated in fungi and bacteria within microbial populations (Kumar et al., 2021). Microbial (fungal and bacterial) cleanup could be aided by bioremediation, wherein plants make excess fluid that shields the bioentities against toxic metals while also boosting the co-metabolic activities of the toxic compounds.

4.5.1.4 Biomineralization

The method by which cationic metals form mineral reserves within cells and particular organs under the direction or effect of organic biomass is referred to as biomineralization (Gadd, 2007). Based on how much the result is physiologically regulated, the process can be split into physiologically triggered mineralization and physiologically regulated mineralization.

4.5.1.5 Biodegradation

Microbial insecticide breakdown refers to employing efficient microorganism catalysts to breakdown insecticides into different minimal molecules. Pesticides can be degraded by a large range of soil bacterial species. Pesticide decomposition enzymes are encoded by the pesticide degradation (Verma et al., 2014) genes, which are found inside plasmids.

4.6 POLICIES RELATED TO THE REHABILITATION OF WASTE DUMPING SITES

Many countries across the world are concerned about the issue of soil degradation. Throughout England exclusively, 0.4% of the total land area is classified as abandoned and must be treated before it may be reused. Furthermore, there is a significant parcel of land on which land combination has deteriorated as a consequence of civil design methods (such as open-cast mining) or intensive farming. Never before has there been a greater need for services targeted at sustainable repair of construction, content, and functioning (Hossain et al., 2020). As in the United Kingdom, there is growing awareness of the potential of ecosystems to address many issues at once, including rising seas, watershed preservation, and agriculture transformation. Measurements of the ecological state are required to make immediate and adequate treatments to remediate this deterioration (Khin et al., 2012). We can measure the condition of the microbial ecosystem and thus the overall quality of soil by assessing the features of the indigenous bacterial community, in addition to the capacity for this and the process of recovery following deterioration (Gómez-Sagasti et al., 2012). Pollution is an unavoidable byproduct of human activity.

The larger the substance rotation and the more complicated and diverse the various substances generated, the more difficult it is for solid waste to fulfill the aims of "protection of individuals and ecosystem" and "conserve natural resources". Waste combustion, which was first implemented for shrinkage and sanitary reasons, produces toxic and greenhouse gases (Cui et al., 2020). Waste-to-energy operations, combined with waste avoidance and recycling strategies, directly contribute to achieving waste disposal targets. Its results also showed that these bacterial marker numbers were significantly higher in the summer compared to the winter, indicating a probable seasonal change. Ground runoff specimens from far more heavily populated sewage disposal regions had higher microbial activity on average as outflow than those from less dense regions using sewer pipes (Nolde, 2007). Temperatures, sunshine, saltiness, and the availability of nutrients are among the external conditions that influence the activity and death of groundwater microbes. Microbes found in sand live longer than all those discovered in the water stream.

Most microorganisms get first degradation estimations with k samples ranging from 0.7 to 1.5, predicting 90% death in salt and fresh water in less than 5 days. Viruses or protozoa perish once at moderate speed, leading to a longer entire lifespan. Increasing the actual structure of end-up dying studies by analyzing numerous microbes under a wider variety of circumstances might result in more precise estimations. Waste disposal has risen to the top of the agenda in several countries today. Sewage was discovered to be a possible nutrition supply for microbial culture, enabling the treatment method to be more affordable and environmentally benign (Kishor et al., 2021). As a result, the National Risks Management Research Laboratory's (NRMRL) watershed pathogen investigation aims to identify optimal methods for controlling the difficult issue of microbial pollution, assisting concerned authorities to make decisions. It contributes to the Company's Waterborne Infectious Disease Control Method by targeting sources of pollutants identified within the policy.

4.7 PROSPECTS AND CONCLUSION

Environmental pollution is continuously increasing and emerging as a matter of global concern due to its negative impact on human health and biodiversity. Industrialization, accompanied by unmanaged population growth, is responsible for the uncontrolled waste generation in every ecological sphere of the planet. Over time, indiscriminate disposal of these contaminants increases and renders the soil and water unfavorable for the sound sustenance of life processes in general. Thus, remediation of the pollutants is of extreme and prior relevance. Conventional chemical and physical processes for the remediation of pollutants are expensive, time-consuming, and nonsustainable. Thus, scientists and naturalists are in constant search of an alternate, greener remediation process to address the concern. Igniting the scope for their relentless experiments, microbes stand foremost as the model organism for this sustainable remediation process. The primary purpose of such a remediation process mediated by microorganisms or biological organisms is to rehabilitate the contaminated sites so that regular life-sustaining activities can be restored and a healthy equilibrium with the ecosystem is maintained. This chapter provided a precise and systematic overview of how microbes and living systems can be beneficially exploited to treat pollutants, along with the techniques involved in the process. To conclude, the following take-home messages are of immense importance and would also focus on future avenues to fortify the strategy of environmental rehabilitation of the contaminated sites.

- Stringent actions and rules from policymakers should be framed and imposed so that constructive guidance and direction can be obtained to dispose of the pollutants. Social awareness and mass concern for environmental issues should be the first approach, which can significantly minimize the burden of rehabilitation through the biological system.
- Regulated and responsible steps regarding the disposal of waste can significantly reduce the load of environmental pollution.
- A detailed screening of biological organisms with respect to their tolerance to pollution needs to be framed to narrow down those plants or microbes that can withstand high doses of pollutants.
- An extension of the above concept, i.e., molecular and genetic intervention in the tolerant/resistant varieties of plants and microbes and engineering them with other biological sources to check their efficacy in sequestration, degradation, and removal of pollutants, can be regarded as the prerequisites of the rehabilitation system.
- Though many experimental investigations have been done to solve this issue, field-based surveys still seem to be deficient, which should be mandated to ascertain the actual scenario of the bioremediation process.

REFERENCES

Abduro Ogo, Hirpa, Na Tang, Xiaowei Li, Xueyuan Gao, and Wei Xing. 2022. "Combined Toxicity of Microplastic and Lead on Submerged Macrophytes." *Chemosphere* 295: 133956. doi:10.1016/j.chemosphere.2022.133956.

Afrad, Md Safiul Islam, Mostakima Binta Monir, Md Enamul Haque, Aliyu Akilu Barau, and Md Manjurul Haque. 2020. "Impact of Industrial Effluent on Water, Soil and Rice Production in Bangladesh: A Case of Turag River Bank." *Journal of Environmental Health Science & Engineering* 18 (2): 825–34. doi:10.1007/s40201-020-00506-8.

Ahmad, Irshad. 2021. "Microalgae-Bacteria Consortia: A Review on the Degradation of Polycyclic Aromatic Hydrocarbons (PAHs)." *Arabian Journal for Science and Engineering* 47 (1): 19–43. doi:10.1007/s13369-021-06236-9.

Ajao, Victor, Kang Nam, Paraschos Chatzopoulos, Evan Spruijt, Harry Bruning, Huub Rijnaarts, and Hardy Temmink. 2020. "Regeneration and Reuse of Microbial Extracellular Polymers Immobilised on a Bed Column for Heavy Metal Recovery." *Water Research* 171: 115472. doi:10.1016/j.watres.2020.115472.

Ajayan, Kayil Veedu, Muthusamy Selvaraju, Pachikaran Unnikannan, and Palliyath Sruthi. 2015. "Phycoremediation of Tannery Wastewater Using Microalgae *Scenedesmus* Species." *International Journal of Phytoremediation* 17 (10): 907–16. doi:10.1080/15226514.2014.989313.

Akbari, Elham, Behnam Rasekh, Keivan Beheshti Maal, Farahnaz Karbasiun, Fatemeh Yazdian, Zarrindokht Emami-Karvani, and Reza Peighami. 2021. "A Novel Biosurfactant Producing *Kocuria rosea* ABR6 as Potential Strain in Oil Sludge Recovery and Lubrication." *AMB Express* 11 (1): 131. doi:10.1186/s13568-021-01283-9.

Aksu, Zümriye. 2005. "Application of Biosorption for the Removal of Organic Pollutants: A Review." *Process Biochemistry* 40 (3–4): 997–1026. doi:10.1016/j.procbio.2004.04.008.

Ali Redha, Ali. 2020. "Removal of Heavy Metals from Aqueous Media by Biosorption." *Arab Journal of Basic and Applied Sciences* 27 (1): 183–93. doi:10.1080/25765299.2020.1756177.

Ali, Hazrat, Ezzat Khan, and Ikram Ilahi. 2019. "Environmental Chemistry and Ecotoxicology of Hazardous Heavy Metals: Environmental Persistence, Toxicity, and Bioaccumulation." *Journal of Chemistry* 2019: 6730305. doi:10.1155/2019/6730305.

Ali, Hazrat, Ezzat Khan, and Muhammad Anwar Sajad. 2013. "Phytoremediation of Heavy Metals-Concepts and Applications." *Chemosphere* 91 (7): 869–81. doi:10.1016/j.chemosphere.2013.01.075.

Alonso-Gutiérrez, Jorge, Antonio Figueras, Joan Albaigés, Núria Jiménez, Marc Viñas, Anna M Solanas, and Beatriz Novoa. 2009. "Bacterial Communities from Shoreline Environments (Costa Da Morte, Northwestern Spain) Affected by the Prestige Oil Spill." *Applied and Environmental Microbiology* 75 (11): 3407–18. doi:10.1128/AEM.01776-08.

An, Dongshan, Sean M Caffrey, Jung Soh, Akhil Agrawal, Damon Brown, Karen Budwill, Xiaoli Dong, et al. 2013. "Metagenomics of Hydrocarbon Resource Environments Indicates Aerobic Taxa and Genes to Be Unexpectedly Common." *Environmental Science & Technology* 47 (18): 10708–17. doi:10.1021/es4020184.

Anjum, Muzammil, R Miandad, Muhammad Waqas, F Gehany, and M A Barakat. 2019. "Remediation of Wastewater Using Various Nano-Materials." *Arabian Journal of Chemistry* 12 (8): 4897–4919. doi:10.1016/j.arabjc.2016.10.004.

Arnold, Frances H 2018. "Directed Evolution: Bringing New Chemistry to Life." *Angewandte Chemie (International Ed. in English)* 57 (16): 4143–48. doi:10.1002/anie.201708408.

Auta, Helen Shnada, Chijioke Uche Emenike, B. Jayanthi, and Shahul Hamid Fauziah. 2018. "Growth Kinetics and Biodeterioration of Polypropylene Microplastics by *Bacillus* Sp. and *Rhodococcus* Sp. Isolated from Mangrove Sediment." *Marine Pollution Bulletin* 127: 15–21. doi:10.1016/j.marpolbul.2017.11.036.

Avramova, Tatyana, Anna Sotirova, Danka Galabova, and Elena Karpenko. 2008. "Effect of Triton X-100 and Rhamnolipid PS-17 on the Mineralization of Phenanthrene by *Pseudomonas* Sp. Cells." *International Biodeterioration & Biodegradation* 62 (4): 415–20. doi:10.1016/j.ibiod.2008.03.008.

Awasthi, A K, A Pandey, and Rashmi Dubey. 2011. "Diversity of Fungi in Effluents of Sugar Industries of Madhya Pradesh." *International Journal of Environmental Sciences* 1 (5): 834–38.

Bala, Saroj, Diksha Garg, Banjagere Veerabhadrappa Thirumalesh, Minaxi Sharma, Kandi Sridhar, Baskaran Stephen Inbaraj, and Manikant Tripathi. 2022. "Recent Strategies for Bioremediation of Emerging Pollutants: A Review for a Green and Sustainable Environment." *Toxics* 10 (8): 484. doi:10.3390/toxics10080484.

Baldrian, Petr, and Vendula Valášková. 2008. "Degradation of Cellulose by Basidiomycetous Fungi." *FEMS Microbiology Reviews* 32 (3): 501–21. doi:10.1111/j.1574-6976.2008.00106.x.

Bama, M Ezhil, and K Ramakrishnan. 2010. "Studies on the Occurrence and Distribution of AM Fungi in Sugar Mill Effluent Polluted Soils in Nellikuppam Region of Cuddalore District." *Journal of Phytology* 2 (1): 91–95.

Barth, Markus, Thorsten Oeser, Ren Wei, Johannes Then, Juliane Schmidt, and Wolfgang Zimmermann. 2015. "Effect of Hydrolysis Products on the Enzymatic Degradation of Polyethylene Terephthalate Nanoparticles by a Polyester Hydrolase from *Thermobifida fusca*." *Biochemical Engineering Journal* 93 (January): 222–28. doi:10.1016/j.bej.2014.10.012.

Baweja, Pooja, Savindra Kumar, and Gaurav Kumar. 2020. "Fertilizers and Pesticides: Their Impact on Soil Health and Environment." *Soil Biology*. doi:10.1007/978-3-030-44364-1_15.

Beinabaj, Seyyed Mahdi Hosseini, Hossein Heidarian, Hamed Mohammad Aleii, and Ali Hosseinzadeh. 2023. "Concentration of Heavy Metals in Leachate, Soil, and Plants in Tehran's Landfill: Investigation of the Effect of Landfill Age on the Intensity of Pollution." *Heliyon* 9: e13017. doi:10.1016/j.heliyon.2023.e13017.

Betts, W B, R K Dart, A S Ball, and S L Pedlar. 1991. "Biosynthesis and Structure of Lignocellulose." In *Biodegradation: Natural and Synthetic Materials*, pp. 139–155. London: Springer. doi:10.1007/978-1-4471-3470-1_7.

Bezza, Fisseha Andualem, Mervyn Beukes, and Evans M Nkhalambayausi Chirwa. 2015. "Application of Biosurfactant Produced by *Ochrobactrum intermedium* CN3 for Enhancing Petroleum Sludge Bioremediation." *Process Biochemistry* 50 (11): 1911–22. doi:10.1016/j.procbio.2015.07.002.

Bharali, Pranjal, and Bolin Kumar Konwar. 2011. "Production and Physico-Chemical Characterization of a Biosurfactant Produced by *Pseudomonas aeruginosa* OBP_1 Isolated from Petroleum Sludge." *Applied Biochemistry and Biotechnology* 164 (8): 1444–60. doi:10.1007/s12010-011-9225-z.

Bhat, Akash P, and Parag R Gogate. 2021. "Cavitation-Based Pre-Treatment of Wastewater and Waste Sludge for Improvement in the Performance of Biological Processes: A Review." *Journal of Environmental Chemical Engineering* 9 (2): 104743. doi:10.1016/j.jece.2020.104743.

Bhatnagar, Sonal, and Reeta Kumari. 2013. "Bioremediation: A Sustainable Tool for Environmental Management-a Review." *Annual Research & Review in Biology* 3 (4): 974–93. doi:10.1007/978-981-13-0053-0_6.

Bidja Abena, Marie Thérèse, Tongtong Li, Muhammad Naeem Shah, and Weihong Zhong. 2019. "Biodegradation of Total Petroleum Hydrocarbons (TPH) in Highly Contaminated Soils by Natural Attenuation and Bioaugmentation." *Chemosphere* 234: 864–74. doi:10.1016/j.chemosphere.2019.06.111.

Biswas, Tethi, Shashi Bhushan, Sanjeev Kumar Prajapati, and Shaon Ray Chaudhuri. 2021. "An Eco-Friendly Strategy for Dairy Wastewater Remediation with High Lipid Microalgae-Bacterial Biomass Production." *Journal of Environmental Management* 286: 112196. doi:10.1016/j.jenvman.2021.112196.

Bongiorni, L, A Pusceddu, and R Danovaro. 2005. "Enzymatic Activities of Epiphytic and Benthic Thraustochytrids Involved in Organic Matter Degradation." *Aquatic Microbial Ecology* 41: 299–305. doi:10.3354/ame041299.

Borah, Debajit, Kanika Agarwal, Ankita Khataniar, Debasish Konwar, Subrata Borgohain Gogoi, and Monem Kallel. 2019. "A Newly Isolated Strain of *Serratia* Sp. from an Oil Spillage Site of Assam Shows Excellent Bioremediation Potential." *3 Biotech* 9 (7): 283. doi:10.1007/s13205-019-1820-7.

Bose, Bimal K 2010. "Global Warming: Energy, Environmental Pollution, and the Impact of Power Electronics." *IEEE Industrial Electronics Magazine* 4 (1): 6–17. doi:10.1109/mie.2010.935860.

Boudh, Siddharth, Jay Shankar Singh, and Preeti Chaturvedi. 2019. "Microbial Resources Mediated Bioremediation of Persistent Organic Pollutants." *New and Future Developments in Microbial Biotechnology and Bioengineering*. doi:10.1016/b978-0-12-818258-1.00019-4.

Bouzikri, Said, Nadia Ouasfi, Naoual Benzidia, Anas Salhi, Salem Bakkas, and Layachi Khamliche. 2020. "Marine Alga '*Bifurcaria bifurcata*': Biosorption of Reactive Blue 19 and Methylene Blue from Aqueous Solutions." *Environmental Science and Pollution Research* 27 (27): 33636–48. doi:10.1007/s11356-020-07846-w.

Briffa, Jessica, Emmanuel Sinagra, and Renald Blundell. 2020. "Heavy Metal Pollution in the Environment and Their Toxicological Effects on Humans." *Heliyon* 6 (9): e04691. doi:10.1016/j.heliyon.2020.e04691.

Bugg, Timothy D H, Mark Ahmad, Elizabeth M Hardiman, and Rahul Singh. 2011. "The Emerging Role for Bacteria in Lignin Degradation and Bio-Product Formation." *Current Opinion in Biotechnology* 22 (3): 394–400. doi:10.1016/j.copbio.2010.10.009.

Burges, Aritz, Itziar Alkorta, Lur Epelde, and Carlos Garbisu. 2018. "From Phytoremediation of Soil Contaminants to Phytomanagement of Ecosystem Services in Metal Contaminated Sites." *International Journal of Phytoremediation* 20 (4): 384–97. doi:10.1080/15226514.2017.1365340.

Chakraborty, Tanushree, and Nasim Akhtar. 2021. "Biofertilizers: Characteristic Features and Applications." *Biofertilizers*. doi:10.1002/9781119724995.ch15.

Chen, Guiqiu, Song Guan, Guangming Zeng, Xiaodong Li, Anwei Chen, Cui Shang, Ying Zhou, Huanke Li, and Jianmin He. 2012. "Cadmium Removal and 2,4-Dichlorophenol Degradation by Immobilized *Phanerochaete chrysosporium* Loaded with Nitrogen-Doped TiO_2 Nanoparticles." *Applied Microbiology and Biotechnology* 97 (7): 3149–57. doi:10.1007/s00253-012-4121-1.

Chen, Zhu, Yuguang Wang, Haina Cheng, and Hongbo Zhou. 2022. "Hemicellulose Degradation: An Overlooked Issue in Acidic Deep Eutectic Solvents Pretreatment of Lignocellulosic Biomass." *Industrial Crops and Products* 187: 115335. doi:10.1016/j.indcrop.2022.115335.

Chittella, Harika, Li Wan Yoon, Suganti Ramarad, and Zee-Wei Lai. 2021. "Rubber Waste Management: A Review on Methods, Mechanism, and Prospects." *Polymer Degradation and Stability* 194: 109761. doi:10.1016/j.polymdegradstab.2021.109761.

Cuellar-Bermudez, Sara P, Gibran S Aleman-Nava, Rashmi Chandra, J Saul Garcia-Perez, Jose R Contreras-Angulo, Giorgos Markou, Koenraad Muylaert, Bruce E Rittmann, and Roberto Parra-Saldivar. 2017. "Nutrients Utilization and Contaminants Removal. A Review of Two Approaches of Algae and Cyanobacteria in Wastewater." *Algal Research* 24: 438–49. doi:10.1016/j.algal.2016.08.018.

Cui, Caiyun, Yong Liu, Bo Xia, Xiaoyan Jiang, and Martin Skitmore. 2020. "Overview of Public-Private Partnerships in the Waste-to-Energy Incineration Industry in China: Status, Opportunities, and Challenges." *Energy Strategy Reviews* 32: 100584. doi:10.1016/j.esr.2020.100584.

Cui, Xiaoxiao, Suhe Zhao, and Bingwu Wang. 2016. "Microbial Desulfurization for Ground Tire Rubber by Mixed Consortium-*Sphingomonas* Sp. and *Gordonia* Sp." *Polymer Degradation and Stability* 128: 165–71. doi:10.1016/j.polymdegradstab.2016.03.011.

Cycoń, Mariusz, Agnieszka Mrozik, and Zofia Piotrowska-Seget. 2017. "Bioaugmentation as a Strategy for the Remediation of Pesticide-Polluted Soil: A Review." *Chemosphere* 172: 52–71. doi:10.1016/j.chemosphere.2016.12.129.

DalCorso, Giovanni, Elisa Fasani, Anna Manara, Giovanna Visioli, and Antonella Furini. 2019. "Heavy Metal Pollutions: State of the Art and Innovation in Phytoremediation." *International Journal of Molecular Sciences* 20 (14): 3412. doi:10.3390/ijms20143412.

Damodaran, Dilna, K Vidya Shetty, and B Raj Mohan. 2014. "Uptake of Certain Heavy Metals from Contaminated Soil by Mushroom-Galerina Vittiformis." *Ecotoxicology and Environmental Safety* 104: 414–22. doi:10.1016/j.ecoenv.2013.10.033.

Dang, Thi Cam Ha, Dang Thang Nguyen, Hoang Thai, Thuy Chinh Nguyen, Thi Thu Hien Tran, Viet Hung Le, Van Huynh Nguyen, et al. 2018. "Plastic Degradation by Thermophilic *Bacillus* Sp. BCBT21 Isolated from Composting Agricultural Residual in Vietnam." *Advances in Natural Sciences: Nanoscience and Nanotechnology* 9 (1): 015014. doi:10.1088/2043-6254/aaabaf.

Darwesh, Osama, Ibrahim Matter, Mohamed Eida, Hassan Moawad, and You-Kwan Oh. 2019. "Influence of Nitrogen Source and Growth Phase on Extracellular Biosynthesis of Silver Nanoparticles Using Cultural Filtrates of *Scenedesmus Obliquus*." *Applied Sciences* 9 (7): 1465. doi:10.3390/app9071465.

Das, Nilanjana, and Preethy Chandran. 2011. "Microbial Degradation of Petroleum Hydrocarbon Contaminants: An Overview." *Biotechnology Research International* 2011: 941810. doi:10.4061/2011/941810.

Dashora, Kavya, Meghana Gattupalli, Gyan Datta Tripathi, Zoya Javed, Shweta Singh, Maria Tuohy, Prakash Kumar Sarangi, Deepti Diwan, Harikesh B Singh, and Vijai Kumar Gupta. 2023. "Fungal Assisted Valorisation of Polymeric Lignin: Mechanism, Enzymes and Perspectives." *Catalysts* 13 (1): 149. doi:10.3390/catal13010149.

David, Luminita, and Bianca Moldovan. 2020. "Green Synthesis of Biogenic Silver Nanoparticles for Efficient Catalytic Removal of Harmful Organic Dyes." *Nanomaterials* 10 (2): 202. doi:10.3390/nano10020202.

De Gonzalo, Gonzalo, Dana I Colpa, Mohamed H M Habib, and Marco W Fraaije. 2016. "Bacterial Enzymes Involved in Lignin Degradation." *Journal of Biotechnology* 236: 110–19. doi:10.1016/j.jbiotec.2016.08.011.

De Sousa, Fabiula DB, Carlos H Scuracchio, Guo-Hua Hu, and Sandrine Hoppe. 2017. "Devulcanization of Waste Tire Rubber by Microwaves." *Polymer Degradation and Stability* 138: 169–81. doi:10.1016/j.polymdegradstab.2017.03.008.

de Sousa, Trelita, and Saroj Bhosle. 2012. "Isolation and Characterization of a Lipopeptide Bioemulsifier Produced by *Pseudomonas nitroreducens* TSB.MJ10 Isolated from a Mangrove Ecosystem." *Bioresource Technology* 123 (November): 256–62. doi:10.1016/j.biortech.2012.07.056.

de Wilt, Arnoud, Andrii Butkovskyi, Kanjana Tuantet, Lucia Hernandez Leal, Tânia V Fernandes, Alette Langenhoff, and Grietje Zeeman. 2016. "Micropollutant Removal in an Algal Treatment System Fed with Source Separated Wastewater Streams." *Journal of Hazardous Materials* 304: 84–92. doi:10.1016/j.jhazmat.2015.10.033.

Delille, Daniel, Frederic Coulon, and Emilien Pelletier. 2004. "Effects of Temperature Warming during a Bioremediation Study of Natural and Nutrient-Amended Hydrocarbon-Contaminated Sub-Antarctic Soils." *Cold Regions Science and Technology* 40 (1–2): 61–70. doi:10.1016/j.coldregions.2004.05.005.

Demarco, Carolina Faccio, Maurízio Silveira Quadro, Filipe Selau Carlos, Simone Pieniz, Luiza Beatriz Gamboa Araújo Morselli, and Robson Andreazza. 2023. "Bioremediation of Aquatic Environments Contaminated with Heavy Metals: A Review of Mechanisms, Solutions and Perspectives." *Sustainability* 15 (2): 1411. doi:10.3390/su15021411.

Du, Fei, Yu-Zhou Wang, Ying-Shuang Xu, Tian-Qiong Shi, Wen-Zheng Liu, Xiao-Man Sun, and He Huang. 2021. "Biotechnological Production of Lipid and Terpenoid from Thraustochytrids." *Biotechnology Advances* 48: 107725. doi:10.1016/j.biotechadv.2021.107725.

Dunbar, John, Shannon Takala, Susan M Barns, Jody A Davis, and Cheryl R Kuske. 1999. "Levels of Bacterial Community Diversity in Four Arid Soils Compared by Cultivation and 16S RRNA Gene Cloning." *Applied and Environmental Microbiology* 65 (4): 1662–69. doi:10.1128/AEM.65.4.1662-1669.1999.

Echezonachi, Samuel Okere. 2022. "The Role of White Rot Fungi in Bioremediation." *Microbes and Microbial Biotechnology for Green Remediation*. doi:10.1016/b978-0-323-90452-0.00034-7.

Elizabeth George, S, and Yongshan Wan. 2019. "Advances in Characterizing Microbial Community Change and Resistance upon Exposure to Lead Contamination: Implications for Ecological Risk Assessment." *Critical Reviews in Environmental Science and Technology* 50 (21): 2223–70. doi:10.1080/10643389.2019.1698260.

Embaby, Mohamed A, El-Sayed A Haggag, Ahemd S El-Sheikh, and Diaa A Marrez. 2022. "Biosorption of Uranium from Aqueous Solution by Green Microalga *Chlorella sorokiniana*." *Environmental Science and Pollution Research International* 29 (38): 58388–404. doi:10.1007/s11356-022-19827-2.

Encarnação, Telma, Cátia Palito, Alberto A C C Pais, Artur J M Valente, and Hugh D Burrows. 2020. "Removal of Pharmaceuticals from Water by Free and Imobilised Microalgae." *Molecules* 25 (16): 3639. doi:10.3390/molecules25163639.

Fareed, Ghulam, Jameel Ahmed Baig, Tasneem Gul Kazi, Hassan Imran Afridi, Khalil Akhtar, and Imam Bakhsh Solangi. 2022. "Heavy Metals Contamination Levels in the Products of Sugar Industry along with Their Impact from Sugar to the End Users." *International Journal of Environmental Analytical Chemistry*, 1–10. doi:10.1080/03067319.2022.2062238.

Fernández, Pablo M, María M Martorell, Julia I Fariña, and Lucia I C Figueroa. 2012. "Removal Efficiency of Cr6+ by Indigenous *Pichia* Sp. Isolated from Textile Factory Effluent." *The ScientificWorld Journal* 2012: 708213. doi:10.1100/2012/708213.

Ferrer, Manuel, Mónica Martínez-Martínez, Rafael Bargiela, Wolfgang R Streit, Olga V Golyshina, and Peter N Golyshin. 2016. "Estimating the Success of Enzyme Bioprospecting through Metagenomics: Current Status and Future Trends." *Microbial Biotechnology* 9 (1): 22–34. doi:10.1111/1751-7915.12309.

Filote, Catalina, Irina Volf, Sîlvia C R Santos, and Cidália M S Botelho. 2019. "Bioadsorptive Removal of Pb(II) from Aqueous Solution by the Biorefinery Waste of *Fucus spiralis*." *Science of the Total Environment* 648: 1201–9. doi:10.1016/j.scitotenv.2018.08.210.

Filote, Cătălina, Mihaela Roşca, Raluca Hlihor, Petronela Cozma, Isabela Simion, Maria Apostol, and Maria Gavrilescu. 2021. "Sustainable Application of Biosorption and Bioaccumulation of Persistent Pollutants in Wastewater Treatment: Current Practice." *Processes* 9 (10): 1696. doi:10.3390/pr9101696.

Fito, Jemal, Nurelegne Tefera, and Stijn W H Van Hulle. 2019. "Sugarcane Biorefineries Wastewater: Bioremediation Technologies for Environmental Sustainability." *Chemical and Biological Technologies in Agriculture* 6 (1). doi:10.1186/s40538-019-0144-5.

Fomina, Marina, and Geoffrey Michael Gadd. 2014. "Biosorption: Current Perspectives on Concept, Definition and Application." *Bioresource Technology* 160: 3–14. doi:10.1016/j.biortech.2013.12.102.

Fouda, Amr, Saad El-Din Hassan, Mohamed Ali Abdel-Rahman, Mohamed M S Farag, Amr Shehal-deen, Asem A Mohamed, Sultan M Alsharif, Ebrahim Saied, Saad A Moghanim, and Mohamed Salah Azab. 2021. "Catalytic Degradation of Wastewater from the Textile and Tannery Industries by Green Synthesized Hematite (α-Fe_2O_3) and Magnesium Oxide (MgO) Nanoparticles." *Current Research in Biotechnology* 3: 29–41. doi:10.1016/j.crbiot.2021.01.004.

Gadd, Geoffrey M 2007. "Geomycology: Biogeochemical Transformations of Rocks, Minerals, Metals and Radionuclides by Fungi, Bioweathering and Bioremediation." *Mycological Research* 111 (1): 3–49. doi:10.1016/j.mycres.2006.12.001.

Ghorbani, Monire, Morteza Hosseini, Ghasem Najafpour, and Reza Hajimohammadi. 2022. "Synthesis and Characterization of Rhamnolipid Biosurfactant Produced by *Pseudomonas aeruginosa* PTCC 1340 for Emulsification of Oil Sludge in Oil Storage Tank." *Arabian Journal for Science and Engineering* 47 (1): 219–26. doi:10.1007/s13369-021-05872-5.

Giungato, Pasquale, Gianluigi de Gennaro, Pierluigi Barbieri, Sara Briguglio, Martino Amodio, Lucrezia de Gennaro, and Francesco Lasigna. 2016. "Improving Recognition of Odors in a Waste Management Plant by Using Electronic Noses with Different Technologies, Gas Chromatography-Mass Spectrometry/Olfactometry and Dynamic Olfactometry." *Journal of Cleaner Production* 133: 1395–1402. doi:10.1016/j.jclepro.2016.05.148.

Gómez-Sagasti, María T, Itziar Alkorta, José M Becerril, Lur Epelde, Mikel Anza, and Carlos Garbisu. 2012. "Microbial Monitoring of the Recovery of Soil Quality during Heavy Metal Phytoremediation." *Water, Air, & Soil Pollution* 223 (6): 3249–62. doi:10.1007/s11270-012-1106-8.

Gong, Jixian, Nan Duan, and Xueming Zhao. 2012. "Evolutionary Engineering of *Phaffia rhodozyma* for Astaxanthin-Overproducing Strain." *Frontiers of Chemical Science and Engineering* 6 (2): 174–78. doi:10.1007/s11705-012-1276-3.

Hall, J L 2002. "Cellular Mechanisms for Heavy Metal Detoxification and Tolerance." *Journal of Experimental Botany* 53 (366): 1–11. doi:10.1093/jxb/53.366.1.

Hassan, Auwalu, Agamuthu Pariatamby, Aziz Ahmed, Helen Shnada Auta, and Fauziah Shahul Hamid. 2019a. "Enhanced Bioremediation of Heavy Metal Contaminated Landfill Soil Using Filamentous Fungi Consortia: A Demonstration of Bioaugmentation Potential." *Water, Air, & Soil Pollution* 230 (9). doi:10.1007/s11270-019-4227-5.

Hassan, Auwalu, Agamuthu Periathamby, Aziz Ahmed, Ossai Innocent, and Fauziah Shahul Hamid. 2019b. "Effective Bioremediation of Heavy Metal-Contaminated Landfill Soil through Bioaugmentation Using Consortia of Fungi." *Journal of Soils and Sediments* 20 (1): 66–80. doi:10.1007/s11368-019-02394-4.

Hassan, Saad El-Din, Amr Fouda, Ebrahim Saied, Mohamed M S Farag, Ahmed M Eid, Mohammed G Barghoth, Mohamed A Awad, Mohammed F Hamza, and Mohamed F Awad. 2021. "Rhizopus Oryzae-Mediated Green Synthesis of Magnesium Oxide Nanoparticles (MgO-NPs): A Promising Tool for Antimicrobial, Mosquitocidal Action, and Tanning Effluent Treatment." *Journal of Fungi* 7 (5): 372. doi:10.3390/jof7050372.

Hazen, Terry C, Andrea M Rocha, and Stephen M Techtmann. 2013. "Advances in Monitoring Environmental Microbes." *Current Opinion in Biotechnology* 24 (3): 526–33. doi:10.1016/j.copbio.2012.10.020.

He, Pinjing, Liyao Chen, Liming Shao, Hua Zhang, and Fan Lü. 2019. "Municipal Solid Waste (MSW) Landfill: A Source of Microplastics? -Evidence of Microplastics in Landfill Leachate." *Water Research* 159: 38–45. doi:10.1016/j.watres.2019.04.060.

Head, Ian M, Neil D Gray, and Stephen R Larter. 2014. "Life in the Slow Lane; Biogeochemistry of Biodegraded Petroleum Containing Reservoirs and Implications for Energy Recovery and Carbon Management." *Frontiers in Microbiology* 5: 566. doi:10.3389/fmicb.2014.00566.

Hernández-Zamora, Miriam, Eliseo Cristiani-Urbina, Fernando Martínez-Jerónimo, Hugo Virgilio Perales-Vela, Teresa Ponce-Noyola, María del Carmen Montes-Horcasitas, and Rosa Olivia Cañizares-Villanueva. 2015. "Bioremoval of the Azo Dye Congo Red by the Microalga *Chlorella vulgaris*." *Environmental Science and Pollution Research* 22 (14): 10811–23. doi:10.1007/s11356-015-4277-1.

Hossain, Md Uzzal, S Thomas Ng, Prince Antwi-Afari, and Ben Amor. 2020. "Circular Economy and the Construction Industry: Existing Trends, Challenges and Prospective Framework for Sustainable Construction." *Renewable and Sustainable Energy Reviews* 130: 109948. doi:10.1016/j.rser.2020.109948.

Houfani, Aicha Asma, Nico Anders, Antje C. Spiess, Petr Baldrian, and Said Benallaoua. 2020. "Insights from Enzymatic Degradation of Cellulose and Hemicellulose to Fermentable Sugars- A Review." *Biomass and Bioenergy* 134: 105481. doi:10.1016/j.biombioe.2020.105481.

Hu, Guangji, Jianbing Li, and Guangming Zeng. 2013. "Recent Development in the Treatment of Oily Sludge from Petroleum Industry: A Review." *Journal of Hazardous Materials* 261: 470–90. doi:10.1016/j.jhazmat.2013.07.069.

Hu, Xingcui, Tao Cheng, and Jianhua Liu. 2018. "A Novel *Serratia* Sp. ZS6 Isolate Derived from Petroleum Sludge Secretes Biosurfactant and Lipase in Medium with Olive Oil as Sole Carbon Source." *AMB Express* 8 (1): 165. doi:10.1186/s13568-018-0698-9.

Huang, Yi, Hanghai Zhou, Gang Zheng, Yanhong Li, Qinglin Xie, Shaohong You, and Chunfang Zhang. 2020. "Isolation and Characterization of Biosurfactant-Producing Serratia Marcescens ZCF25 from Oil Sludge and Application to Bioremediation." *Environmental Science and Pollution Research* 27 (22): 27762–72. doi:10.1007/s11356-020-09006-6.

Ibrahim, Wael M, Mohamed A Karam, Reda M El-Shahat, and Asmaa A Adway. 2014. "Biodegradation and Utilization of Organophosphorus Pesticide Malathion by Cyanobacteria." *BioMed Research International* 2014: 392682. doi:10.1155/2014/392682.

Ipeaiyeda, Ayodele Rotimi, Clementina Oyinkansola Adenipekun, and Oluwatola Oluwole. 2020. "Bioremediation Potential of *Ganoderma lucidum* (Curt:Fr) P. Karsten to Remove Toxic Metals from Abandoned Battery Slag Dumpsite Soil and Immobilisation of Metal Absorbed Fungi in Bricks." *Cogent Environmental Science* 6 (1): 1847400. doi:10.1080/23311843.2020.1847400.

Iqbal, Jamshaid, Atif Javed, and Muhammad Anwar Baig. 2021. "Heavy Metals Removal from Dumpsite Leachate by Algae and Cyanobacteria." *Bioremediation Journal* 26 (1): 31–40. doi:10.1080/10889868.2021.1884530.

Ivshina, Irena B, Maria S Kuyukina, Anastasiya V Krivoruchko, Andrey A Elkin, Sergey O Makarov, Colin J Cunningham, Tatyana A Peshkur, Ronald M Atlas, and James C Philp. 2015. "Oil Spill Problems and Sustainable Response Strategies through New Technologies." *Environmental Science: Processes & Impacts* 17 (7): 1201–19. doi:10.1039/c5em00070j.

Jacob, Jaya Mary, Chinnannan Karthik, Rijuta Ganesh Saratale, Smita S Kumar, Desika Prabakar, K Kadirvelu, and Arivalagan Pugazhendhi. 2018. "Biological Approaches to Tackle Heavy Metal Pollution: A Survey of Literature." *Journal of Environmental Management* 217: 56–70. doi:10.1016/j.jenvman.2018.03.077.

Jain, Kavish Kumar, Amit Kumar, Akshay Shankar, Dhananjay Pandey, Bhupendra Chaudhary, and Krishna Kant Sharma. 2020. "*De Novo* Transcriptome Assembly and Protein Profiling of Copper-Induced Lignocellulolytic Fungus *Ganoderma lucidum* MDU-7 Reveals Genes Involved in Lignocellulose Degradation and Terpenoid Biosynthetic Pathways." *Genomics* 112 (1): 184–98. doi:10.1016/j.ygeno.2019.01.012.

Javed, Muhammad Tariq, Kashif Tanwir, Muhammad Sohail Akram, Muhammad Shahid, Nabeel Khan Niazi, and Sylvia Lindberg. 2019. "Phytoremediation of Cadmium-Polluted Water/Sediment by Aquatic Macrophytes: Role of Plant-Induced PH Changes." *Cadmium Toxicity and Tolerance in Plants*. doi:10.1016/b978-0-12-814864-8.00020-6.

Jeyavani, Jeyaraj, Ashokkumar Sibiya, Sivakumar Shanthini, Cyril Ravi, Sekar Vijayakumar, Durairaj Karthick Rajan, and Baskaralingam Vaseeharan. 2021. "A Review on Aquatic Impacts of Microplastics and Its Bioremediation Aspects." *Current Pollution Reports* 7 (3): 286–99. doi:10.1007/s40726-021-00188-2.

Jozefczak, Marijke, Tony Remans, Jaco Vangronsveld, and Ann Cuypers. 2012. "Glutathione is a Key Player in Metal-Induced Oxidative Stress Defenses." *International Journal of Molecular Sciences* 13 (3): 3145–75. doi:10.3390/ijms13033145.

Kalaiselvi, Aasaithambi, Selvaraj Mohana Roopan, Gunabalan Madhumitha, C Ramalingam, Al-Dhabi Naif Abdullah, and Mariadhas Valan Arasu. 2016. "*Catharanthus roseus* -Mediated Zinc Oxide Nanoparticles against Photocatalytic Application of Phenol Red under UV@365 Nm." *Current Science* 111 (11): 1811. doi:10.18520/cs/v111/i11/1811-1815.

Karimi, Hoda, Shahriar Mahdavi, Behnam Asgari Lajayer, Ebrahim Moghiseh, Vishnu D Rajput, Tatiana Minkina, and Tess Astatkie. 2021. "Insights on the Bioremediation Technologies for Pesticide-Contaminated Soils." *Environmental Geochemistry and Health* 44 (4): 1329–54. doi:10.1007/s10653-021-01081-z.

Kazour, Maria, Sarah Terki, Khalef Rabhi, Sharif Jemaa, Gaby Khalaf, and Rachid Amara. 2019. "Sources of Microplastics Pollution in the Marine Environment: Importance of Wastewater Treatment Plant and Coastal Landfill." *Marine Pollution Bulletin* 146: 608–18. doi:10.1016/j.marpolbul.2019.06.066.

Kehrein, Philipp, Mark van Loosdrecht, Patricia Osseweijer, Marianna Garfí, Jo Dewulf, and John Posada. 2020. "A Critical Review of Resource Recovery from Municipal Wastewater Treatment Plants - Market Supply Potentials, Technologies and Bottlenecks." *Environmental Science: Water Research & Technology* 6 (4): 877–910. doi:10.1039/c9ew00905a.

Khin, Mya Mya, A Sreekumaran Nair, V Jagadeesh Babu, Rajendiran Murugan, and Seeram Ramakrishna. 2012. "A Review on Nanomaterials for Environmental Remediation." *Energy & Environmental Science* 5 (8): 8075. doi:10.1039/c2ee21818f.

Kishor, Roop, Diane Purchase, Ganesh Dattatraya Saratale, Rijuta Ganesh Saratale, Luiz Fernando Romanholo Ferreira, Muhammad Bilal, Ram Chandra, and Ram Naresh Bharagava. 2021. "Ecotoxicological and Health Concerns of Persistent Coloring Pollutants of Textile Industry Wastewater and Treatment Approaches for Environmental Safety." *Journal of Environmental Chemical Engineering* 9 (2): 105012. doi:10.1016/j.jece.2020.105012.

Koul, Bhupendra, Mohammad Yakoob, and Maulin P. Shah. 2022. "Agricultural Waste Management Strategies for Environmental Sustainability." *Environmental Research* 206 (April): 112285. doi:10.1016/j.envres.2021.112285.

Kour, Divjot, Tanvir Kaur, Rubee Devi, Ashok Yadav, Manali Singh, Divya Joshi, Jyoti Singh, et al. 2021. "Beneficial Microbiomes for Bioremediation of Diverse Contaminated Environments for Environmental Sustainability: Present Status and Future Challenges." *Environmental Science and Pollution Research* 28 (20): 24917–39. doi:10.1007/s11356-021-13252-7.

Kumar, Amit, Ajar Nath Yadav, Raju Mondal, Divjot Kour, Gangavarapu Subrahmanyam, Aftab A Shabnam, Shakeel A Khan, et al. 2021. "Myco-Remediation: A Mechanistic Understanding of Contaminants Alleviation from Natural Environment and Future Prospect." *Chemosphere* 284: 131325. doi:10.1016/j.chemosphere.2021.131325.

Kumar, Mahendra, Alak Kumar Singh, and Mohammad Sikandar. 2020. "Biosorption of Hg (II) from Aqueous Solution Using Algal Biomass: Kinetics and Isotherm Studies." *Heliyon* 6 (1): e03321. doi:10.1016/j.heliyon.2020.e03321.

Kumar, Raj, Sompal Singh, and Om V Singh. 2008. "Bioconversion of Lignocellulosic Biomass: Biochemical and Molecular Perspectives." *Journal of Industrial Microbiology & Biotechnology* 35 (5): 377–91. doi:10.1007/s10295-008-0327-8.

Kumar, Vankayalapati Vijaya. 2021. "Microbial Remediation: A Natural Approach for Environmental Pollution Management." *Fungal Biology*. doi:10.1007/978-3-030-54422-5_7.

Kumar, Vineet, S K Shahi, and Simranjeet Singh. 2018. "Bioremediation: An Eco-Sustainable Approach for Restoration of Contaminated Sites." *Microbial Bioprospecting for Sustainable Development*. doi:10.1007/978-981-13-0053-0_6.

Kuppusamy, Saranya, Thavamani Palanisami, Mallavarapu Megharaj, Kadiyala Venkateswarlu, and Ravi Naidu. 2016. "In-Situ Remediation Approaches for the Management of Contaminated Sites: A Comprehensive Overview." *Reviews of Environmental Contamination and Toxicology*. doi:10.1007/978-3-319-20013-2_1.

Kurup, Pradeep, Connor Sullivan, Ryan Hannagan, Shiran Yu, Hesameddin Azimi, Seth Robertson, David Ryan, Ramaswamy Nagarajan, Timothy Ponrathnam, and Gary Howe. 2016. "A Review of Technologies for Characterization of Heavy Metal Contaminants." *Indian Geotechnical Journal* 47 (4): 421–36. doi:10.1007/s40098-016-0214-6.

Lambert, Scott, and Martin Wagner. 2017. "Microplastics are Contaminants of Emerging Concern in Freshwater Environments: An Overview." In *The Handbook of Environmental Chemistry*. Springer International Publishing. doi:10.1007/978-3-319-61615-5_1.

Lapkovskis, Vjaceslavs, Viktors Mironovs, Andrei Kasperovich, Vadim Myadelets, and Dmitri Goljandin. 2020. "Crumb Rubber as a Secondary Raw Material from Waste Rubber: A Short Review of End-of-Life Mechanical Processing Methods." *Recycling* 5 (4): 32. doi:10.3390/recycling5040032.

Lee, Siseon, Minsik Kang, Jung-Hoon Bae, Jung-Hoon Sohn, and Bong Hyun Sung. 2019. "Bacterial Valorization of Lignin: Strains, Enzymes, Conversion Pathways, Biosensors, and Perspectives." *Frontiers in Bioengineering and Biotechnology* 7: 209. doi:10.3389/fbioe.2019.00209.

Lellis, Bruno, Cíntia Zani Fávaro-Polonio, João Alencar Pamphile, and Julio Cesar Polonio. 2019. "Effects of Textile Dyes on Health and the Environment and Bioremediation Potential of Living Organisms." *Biotechnology Research and Innovation* 3 (2): 275–90. doi:10.1016/j.biori.2019.09.001.

Leonowicz, Andrzej, Anna Matuszewska, Jolanta Luterek, Dirk Ziegenhagen, Maria Wojtaś-Wasilewska, Nam-Seok Cho, Martin Hofrichter, and Jerzy Rogalski. 1999. "Biodegradation of Lignin by White Rot Fungi." *Fungal Genetics and Biology* 27 (2–3): 175–85. doi:10.1006/fgbi.1999.1150.

Li, Dong, Xianbo Huang, Qingjing Wang, Yuexiang Yuan, Zhiying Yan, Zhidong Li, Yajun Huang, and Xiaofeng Liu. 2016. "Kinetics of Methane Production and Hydrolysis in Anaerobic Digestion of Corn Stover." *Energy* 102: 1–9. doi:10.1016/j.energy.2016.02.074.

Li, Qinghai, Fuxin Li, Aihong Meng, Zhongchao Tan, and Yanguo Zhang. 2018. "Thermolysis of Scrap Tire and Rubber in Sub/Super-Critical Water." *Waste Management* 71: 311–19. doi:10.1016/j.wasman.2017.10.017.

Lim, Henry C, and Hwa Sung Shin. 2013. *Fed-Batch Cultures*. Cambridge University Press. doi:10.1017/cbo9781139018777.

Liu, HuiLin, XiaoPing Wang, and DeMin Jia. 2020. "Recycling of Waste Rubber Powder by Mechano-Chemical Modification." *Journal of Cleaner Production* 245: 118716. doi:10.1016/j.jclepro.2019.118716.

Liu, J H, L Zhang, D C Zha, L Q Chen, X X Chen, and Z M Qi. 2018. "Biosorption of Malachite Green onto *Haematococcus pluvialis* Observed through Synchrotron Fourier-Transform Infrared Microspectroscopy." *Letters in Applied Microbiology* 67 (4): 348–53. doi:10.1111/lam.13043.

Liu, Lei, Xiuying Zhang, and Taiyang Zhong. 2015. "Pollution and Health Risk Assessment of Heavy Metals in Urban Soil in China." *Human and Ecological Risk Assessment: An International Journal* 22 (2): 424–34. doi:10.1080/10807039.2015.1078226.

Loick, Nadine, Phil J Hobbs, Mike D C Hale, and Davey L Jones. 2009. "Bioremediation of Poly-Aromatic Hydrocarbon (PAH)-Contaminated Soil by Composting." *Critical Reviews in Environmental Science and Technology* 39 (4): 271–332. doi:10.1080/10643380701413682.

Lukić, Borislava, Antonio Panico, David Huguenot, Massimiliano Fabbricino, Eric D van Hullebusch, and Giovanni Esposito. 2017. "A Review on the Efficiency of Landfarming Integrated with Composting as a Soil Remediation Treatment." *Environmental Technology Reviews* 6 (1): 94–116. doi:10.1080/21622515.2017.1310310.

Luo, Long-zao, Yu Shao, Shuang Luo, Fan-jian Zeng, and Guang-ming Tian. 2018. "Nutrient Removal from Piggery Wastewater by *Desmodesmus* Sp.CHX1 and Its Cultivation Conditions Optimization." *Environmental Technology* 40 (21): 2739–46. doi:10.1080/09593330.2018.1449903.

Lyon, Delina Y, and Timothy M Vogel. 2012. "Bioaugmentation for Groundwater Remediation: An Overview." In *Bioaugmentation for Groundwater Remediation*. New York: Springer. doi:10.1007/978-1-4614-4115-1_1.

Lyu, Lu, Qiuzhen Wang, and Guangyi Wang. 2021. "Cultivation and Diversity Analysis of Novel Marine Thraustochytrids." *Marine Life Science & Technology* 3 (2): 263–75. doi:10.1007/s42995-020-00069-5.

Mahar, Amanullah, Ping Wang, Amjad Ali, Mukesh Kumar Awasthi, Altaf Hussain Lahori, Quan Wang, Ronghua Li, and Zengqiang Zhang. 2016. "Challenges and Opportunities in the Phytoremediation of Heavy Metals Contaminated Soils: A Review." *Ecotoxicology and Environmental Safety* 126: 111–21. doi:10.1016/j.ecoenv.2015.12.023.

Mahboob, Shahid, H F Alkkahem Al-Balwai, F Al-Misned, K A Al-Ghanim, and Z Ahmad. 2014. "A Study on the Accumulation of Nine Heavy Metals in Some Important Fish Species from a Natural Reservoir in Riyadh, Saudi Arabia." *Toxicological & Environmental Chemistry* 96 (5): 783–98. doi:10.1080/02772248.2014.957485.

Mani, D, and Chitranjan Kumar. 2013. "Biotechnological Advances in Bioremediation of Heavy Metals Contaminated Ecosystems: An Overview with Special Reference to Phytoremediation." *International Journal of Environmental Science and Technology* 11 (3): 843–72. doi:10.1007/s13762-013-0299-8.

McCarthy, Andrew, Rob Dellink, and Ruben Bibas. 2018. "The Macroeconomics of the Circular Economy Transition: A Critical Review of Modelling Approaches." *OECD Environment Working Papers*. Organisation for Economic Co-Operation and Development (OECD). doi:10.1787/af983f9a-en.

Megharaj, Mallavarapu, Balasubramanian Ramakrishnan, Kadiyala Venkateswarlu, Nambrattil Sethunathan, and Ravi Naidu. 2011. "Bioremediation Approaches for Organic Pollutants: A Critical Perspective." *Environment International* 37 (8): 1362–75. doi:10.1016/j.envint.2011.06.003.

Miao, Z, M L Brusseau, K C Carroll, C Carreón-Diazconti, and B Johnson. 2012. "Sulfate Reduction in Groundwater: Characterization and Applications for Remediation." *Environmental Geochemistry and Health* 34 (4): 539–50. doi:10.1007/s10653-011-9423-1.

Miri, Saba, Rahul Saini, Seyyed Mohammadreza Davoodi, Rama Pulicharla, Satinder Kaur Brar, and Sara Magdouli. 2022. "Biodegradation of Microplastics: Better Late than Never." *Chemosphere* 286: 131670. doi:10.1016/j.chemosphere.2021.131670.

Mishra, S, J Jyot, R C Kuhad, and B Lal. 2001. "Evaluation of Inoculum Addition to Stimulate in Situ Bioremediation of Oily-Sludge-Contaminated Soil." *Applied and Environmental Microbiology* 67 (4): 1675–81. doi:10.1128/AEM.67.4.1675-1681.2001.

Mohan, S Venkata, Takuro Kisa, Takeru Ohkuma, Robert A Kanaly, and Yoshihisa Shimizu. 2006. "Bioremediation Technologies for Treatment of PAH-Contaminated Soil and Strategies to Enhance Process Efficiency." *Reviews in Environmental Science and Bio/Technology* 5 (4): 347–74. doi:10.1007/s11157-006-0004-1.

Mohapatra, D, S K Rath, and P K Mohapatra. 2021. "Soil Fungi for Bioremediation of Pesticide Toxicants: A Perspective." *Geomicrobiology Journal* 39 (3–5): 352–72. doi:10.1080/01490451.2021.2019855.

Mojiri, Amin, Maedeh Baharlooeian, Reza Andasht Kazeroon, Hossein Farraji, and Ziyang Lou. 2020. "Removal of Pharmaceutical Micropollutants with Integrated Biochar and Marine Microalgae." *Microorganisms* 9 (1): 4. doi:10.3390/microorganisms9010004.

Monkul, Mehmet Murat, and Hakkı O Özhan. 2021. "Microplastic Contamination in Soils: A Review from Geotechnical Engineering View." *Polymers* 13 (23): 4129. doi:10.3390/polym13234129.

Moore, Michael N, Michael H Depledge, James W Readman, and D R Paul Leonard. 2004. "An Integrated Biomarker-Based Strategy for Ecotoxicological Evaluation of Risk in Environmental Management." *Mutation Research/Fundamental and Molecular Mechanisms of Mutagenesis* 552 (1–2): 247–68. doi:10.1016/j.mrfmmm.2004.06.028.

Mopper, Kenneth, Aron Stubbins, Jason D Ritchie, Heidi M Bialk, and Patrick G Hatcher. 2007. "Advanced Instrumental Approaches for Characterization of Marine Dissolved Organic Matter: Extraction Techniques, Mass Spectrometry, and Nuclear Magnetic Resonance Spectroscopy." *Chemical Reviews* 107 (2): 419–42. doi:10.1021/cr050359b.

Morohoshi, Tomohiro, Kento Ogata, Tetsuo Okura, and Shunsuke Sato. 2018. "Molecular Characterization of the Bacterial Community in Biofilms for Degradation of Poly(3-Hydroxybutyrate-Co-3-Hydroxyhexanoate) Films in Seawater." *Microbes and Environments* 33 (1): 19–25. doi:10.1264/jsme2.ME17052.

Myhre, Marvin, Sitisaiyidah Saiwari, Wilma Dierkes, and Jacques Noordermeer. 2012. "Rubber Recycling: Chemistry, Processing, and Applications." *Rubber Chemistry and Technology* 85 (3): 408–49. doi:10.5254/rct.12.87973.

Nacheva, P Mijaylova, G Moeller Chávez, J Matías Chacón, and A Canul Chuil. 2009. "Treatment of Cane Sugar Mill Wastewater in an Upflow Anaerobic Sludge Bed Reactor." *Water Science and Technology* 60 (5): 1347–52. doi:10.2166/wst.2009.402.

Nagaraju, M, G Narasimha, and V Rangaswamy. 2009. "Impact of Sugar Industry Effluents on Soil Cellulase Activity." *International Biodeterioration & Biodegradation* 63 (8): 1088–92. doi:10.1016/j.ibiod.2009.09.006.

Nagda, Adhishree, Mukesh Meena, and Maulin P Shah. 2021. "Bioremediation of Industrial Effluents: A Synergistic Approach." *Journal of Basic Microbiology* 62 (3–4): 395–414. doi:10.1002/jobm.202100225.

Năstuneac, Violeta, Mirela Panainte-Lehăduş, Emilian Florin Moşneguţu, Simona Gavrilaş, Gabriela Cioca, and Florentina-Daniela Munteanu. 2019. "Removal of Cypermethrin from Water by Using *Fucus spiralis* Marine Alga." *International Journal of Environmental Research and Public Health* 16 (19): 3663. doi:10.3390/ijerph16193663.

Nejad, Yaser Soltani, Neematollah Jaafarzadeh, Mehdi Ahmadi, Mehrnoosh Abtahi, Shokouh Ghafari, and Sahand Jorfi. 2020. "Remediation of Oily Sludge Wastes Using Biosurfactant Produced by Bacterial Isolate *Pseudomonas balearica strain* Z8." *Journal of Environmental Health Science & Engineering* 18 (2): 531–39. doi:10.1007/s40201-020-00480-1.

Nolde, Erwin. 2007. "Possibilities of Rainwater Utilisation in Densely Populated Areas Including Precipitation Runoffs from Traffic Surfaces." *Desalination* 215 (1–3): 1–11. doi:10.1016/j.desal.2006.10.033.

Nwanya, Assumpta Chinwe, Lovasoa Christine Razanamahandry, A K H Bashir, Chinwe O Ikpo, Stephen C Nwanya, Subelia Botha, S K O Ntwampe, Fabian I Ezema, Emmanuel I Iwuoha, and Malik Maaza. 2019. "Industrial Textile Effluent Treatment and Antibacterial Effectiveness of *Zea mays* L. Dry Husk Mediated Bio-Synthesized Copper Oxide Nanoparticles." *Journal of Hazardous Materials* 375: 281–89. doi:10.1016/j.jhazmat.2019.05.004.

Ogugbue, Chimezie Jason, and Thomas Sawidis. 2011. "Bioremediation and Detoxification of Synthetic Wastewater Containing Triarylmethane Dyes by *Aeromonas hydrophila* Isolated from Industrial Effluent." *Biotechnology Research International* 2011: 967925. doi:10.4061/2011/967925.

Ololade, Isaac A. 2014. "An Assessment of Heavy-Metal Contamination in Soils within Auto-Mechanic Workshops Using Enrichment and Contamination Factors with Geoaccumulation Indexes." *Journal of Environmental Protection* 05 (11): 970–82. doi:10.4236/jep.2014.511098.

Omeje, Emmanuel T, Daniel N Obiora, Francisca N Okeke, Johnson C Ibuot, Desmond O Ugbor, and Victor D Omeje. 2023. "Investigation of Aquifer Vulnerability and Sensitivity Analysis of Modified Drastic and Sintacs Models: A Case Study of Ovogovo Area, Eastern Nigeria." *Acta Geophysica*. doi:10.1007/s11600-022-00992-4.

Orozco Colonia, Brigitte Sthepani, Gilberto Vinícius de Melo Pereira, and Carlos Ricardo Soccol. 2020. "Omega-3 Microbial Oils from Marine Thraustochytrids as a Sustainable and Technological Solution: A Review and Patent Landscape." *Trends in Food Science & Technology* 99: 244–56. doi:10.1016/j.tifs.2020.03.007.

Orts, F, A I del Río, J Molina, J Bonastre, and F Cases. 2018. "Electrochemical Treatment of Real Textile Wastewater: Trichromy Procion HEXL(r)." *Journal of Electroanalytical Chemistry* 808: 387–94. doi:10.1016/j.jelechem.2017.06.051.

Oziegbe, O, A O Oluduro, E J Oziegbe, E F Ahuekwe, and S J Olorunsola. 2021. "Assessment of Heavy Metal Bioremediation Potential of Bacterial Isolates from Landfill Soils." *Saudi Journal of Biological Sciences* 28 (7): 3948–56. doi:10.1016/j.sjbs.2021.03.072.

Pal, Sreela, Fawzi Banat, Ali Almansoori, and Mohammad Abu Haija. 2016. "Review of Technologies for Biotreatment of Refinery Wastewaters: Progress, Challenges and Future Opportunities." *Environmental Technology Reviews* 5 (1): 12–38. doi:10.1080/21622515.2016.1164252.

Pant, Deepak, and Alok Adholeya. 2006. "Enhanced Production of Ligninolytic Enzymes and Decolorization of Molasses Distillery Wastewater by Fungi under Solid State Fermentation." *Biodegradation* 18 (5): 647–59. doi:10.1007/s10532-006-9097-z.

Pant, Gaurav, Alka Singh, Mitali Panchpuri, Ravi Gyana Prasuna, Kaizar Hossain, Syed Zaghum Abbas, Akil Ahmad, Norli Ismail, and Mohd Rafatullah. 2019. "Enhancement of Biosorption Capacity of Cyanobacterial Strain to Remediate Heavy Metals." *Desalination and Water Treatment* 165: 244–52. doi:10.5004/dwt.2019.24509.

Parvin, Fahmida, Sharmin Yousuf Rikta, and Shafi M Tareq. 2019. "Application of Nanomaterials for the Removal of Heavy Metal from Wastewater." *Nanotechnology in Water and Wastewater Treatment*. doi:10.1016/b978-0-12-813902-8.00008-3.

Patowary, Rupshikha, Bhagyalakshmi Rajbongshi, Arundhuti Devi, and Manisha Goswami. 2022. "Concurrent Degradation of Petroleum Sludge and Simultaneous Rhamnolipid Biosurfactant Production: An Aesthetic Bioremediation Approach." *Research Square Platform LLC*. doi:10.21203/rs.3.rs-2281581/v1.

Paul, Diby, and Sudha Nair. 2008. "Stress Adaptations in a Plant Growth Promoting Rhizobacterium (PGPR) with Increasing Salinity in the Coastal Agricultural Soils." *Journal of Basic Microbiology* 48 (5): 378–84. doi:10.1002/jobm.200700365.

Phale, Prashant S, Amrita Sharma, and Kamini Gautam. 2019. "Microbial Degradation of Xenobiotics like Aromatic Pollutants from the Terrestrial Environments." In *Pharmaceuticals and Personal Care Products: Waste Management and Treatment Technology*. Elsevier. doi:10.1016/b978-0-12-816189-0.00011-1.

Phulpoto, Irfan Ali, Zhisheng Yu, Bowen Hu, Yanfen Wang, Fabrice Ndayisenga, Jinmei Li, Hongxia Liang, and Muneer Ahmed Qazi. 2020. "Production and Characterization of Surfactin-like Biosurfactant Produced by Novel Strain *Bacillus nealsonii* S2MT and It's Potential for Oil Contaminated Soil Remediation." *Microbial Cell Factories* 19 (1): 145. doi:10.1186/s12934-020-01402-4.

Poddar, Pradeep Kumar, and Omprakash Sahu. 2015. "Quality and Management of Wastewater in Sugar Industry." *Applied Water Science* 7 (1): 461–68. doi:10.1007/s13201-015-0264-4.

Poirier, Wilfried, Jean-Philippe Bouchara, and Sandrine Giraud. 2023. "Lignin-Modifying Enzymes in *Scedosporium* Species." *Journal of Fungi* 9 (1): 105. doi:10.3390/jof9010105.

Praagh, Martijn van, Cornelia Hartman, and Emma Brandmyr. 2019. "Microplastics in Landfill Leachates in the Nordic Countries." *TemaNord*. doi:10.6027/tn2018-557.

Pratas, João, Carlos Paulo, Paulo J C Favas, and Perumal Venkatachalam. 2014. "Potential of Aquatic Plants for Phytofiltration of Uranium-Contaminated Waters in Laboratory Conditions." *Ecological Engineering* 69: 170–76. doi:10.1016/j.ecoleng.2014.03.046.

Pratush, Amit, Ajay Kumar, and Zhong Hu. 2018. "Adverse Effect of Heavy Metals (As, Pb, Hg, and Cr) on Health and Their Bioremediation Strategies: A Review." *International Microbiology* 21 (3): 97–106. doi:10.1007/s10123-018-0012-3.

Puyol, Daniel, Damien J Batstone, Tim Hülsen, Sergi Astals, Miriam Peces, and Jens O Krömer. 2017. "Resource Recovery from Wastewater by Biological Technologies: Opportunities, Challenges, and Prospects." *Frontiers in Microbiology* 7: 2106. doi:10.3389/fmicb.2016.02106.

Quagraine, E K, H G Peterson, and J V Headley. 2005. "In Situ Bioremediation of Naphthenic Acids Contaminated Tailing Pond Waters in the Athabasca Oil Sands Region-Demonstrated Field Studies and Plausible Options: A Review." *Journal of Environmental Science and Health, Part A* 40 (3): 685–722. doi:10.1081/ese-200046649.

Rafique, Muhammad, Falak Shafiq, Syed Sajid Ali Gillani, Muhammad Shakil, Muhammad Bilal Tahir, and Iqra Sadaf. 2020. "Eco-Friendly Green and Biosynthesis of Copper Oxide Nanoparticles Using *Citrofortunella microcarpa* Leaves Extract for Efficient Photocatalytic Degradation of Rhodamin B Dye Form Textile Wastewater." *Optik* 208: 164053. doi:10.1016/j.ijleo.2019.164053.

Raghunandan, Kerisha, Ashwani Kumar, Santhosh Kumar, Kugenthiren Permaul, and Suren Singh. 2018. "Production of Gellan Gum, an Exopolysaccharide, from Biodiesel-Derived Waste Glycerol by *Sphingomonas* spp." *3 Biotech* 8 (1): 71. doi:10.1007/s13205-018-1096-3.

Raimondo, Enzo E, Juliana M Saez, Juan D Aparicio, María S Fuentes, and Claudia S Benimeli. 2020. "Bioremediation of Lindane-Contaminated Soils by Combining of Bioaugmentation and Biostimulation: Effective Scaling-up from Microcosms to Mesocosms." *Journal of Environmental Management* 276: 111309. doi:10.1016/j.jenvman.2020.111309.

Reichert, Jessica, Johannes Schellenberg, Patrick Schubert, and Thomas Wilke. 2018. "Responses of Reef Building Corals to Microplastic Exposure." *Environmental Pollution* 237: 955–60. doi:10.1016/j.envpol.2017.11.006.

Rose, Ruth-Sarah, Katherine H. Richardson, Elmeri Johannes Latvanen, China A. Hanson, Marina Resmini, and Ian A. Sanders. 2020. "Microbial Degradation of Plastic in Aqueous Solutions Demonstrated by CO_2 Evolution and Quantification." *International Journal of Molecular Sciences* 21 (4): 1176. doi:10.3390/ijms21041176.

Rugnini, L, G Costa, R Congestri, and L Bruno. 2017. "Testing of Two Different Strains of Green Microalgae for Cu and Ni Removal from Aqueous Media." *Science of the Total Environment* 601–602: 959–67. doi:10.1016/j.scitotenv.2017.05.222.

Rüthi, Joel, Damian Bölsterli, Lucrezia Pardi-Comensoli, Ivano Brunner, and Beat Frey. 2020. "The 'Plastisphere' of Biodegradable Plastics is Characterized by Specific Microbial Taxa of Alpine and Arctic Soils." *Frontiers in Environmental Science* 8 (September). doi:10.3389/fenvs.2020.562263.

Salvador, Rodrigo, Fabio N Puglieri, Anthony Halog, Fernanda G de Andrade, Cassiano M Piekarski, and Antonio C De Francisco. 2021. "Key Aspects for Designing Business Models for a Circular Bioeconomy." *Journal of Cleaner Production* 278: 124341. doi:10.1016/j.jclepro.2020.124341.

Sandhya, S. 2010. "Biodegradation of Azo Dyes under Anaerobic Condition: Role of Azoreductase." In *The Handbook of Environmental Chemistry*. Springer, Berlin, Heidelberg. doi:10.1007/698_2009_43.

Santos, Danyelle, Raquel Rufino, Juliana Luna, Valdemir Santos, and Leonie Sarubbo. 2016. "Biosurfactants: Multifunctional Biomolecules of the 21st Century." *International Journal of Molecular Sciences* 17 (3): 401. doi:10.3390/ijms17030401.

Sanuth, Hassan, and Abimbola Adekanmbi. 2016. "Biosorption of Heavy Metals in Dumpsite Leachate by Metal-Resistant Bacteria Isolated from Abule-Egba Dumpsite, Lagos State, Nigeria." *British Microbiology Research Journal* 17 (3): 1–8. doi:10.9734/bmrj/2016/28011.

Saravanakumar, Kandasamy, Sabarathinam Shanmugam, Nipun Babu Varukattu, Davoodbasha MubarakAli, Kandasamy Kathiresan, and Myeong-Hyeon Wang. 2019. "Biosynthesis and Characterization of Copper Oxide Nanoparticles from Indigenous Fungi and Its Effect of Photothermolysis on Human Lung Carcinoma." *Journal of Photochemistry and Photobiology B: Biology* 190: 103–9. doi:10.1016/j.jphotobiol.2018.11.017.

Sarkar, Jayeeta, Sufia K Kazy, Abhishek Gupta, Avishek Dutta, Balaram Mohapatra, Ajoy Roy, Paramita Bera, Adinpunya Mitra, and Pinaki Sar. 2016. "Biostimulation of Indigenous Microbial Community for Bioremediation of Petroleum Refinery Sludge." *Frontiers in Microbiology* 7: 1407. doi:10.3389/fmicb.2016.01407.

Sathiskumar, P S, and Giridhar Madras. 2012. "Ultrasonic Degradation of Butadiene, Styrene and Their Copolymers." *Ultrasonics Sonochemistry* 19 (3): 503–8. doi:10.1016/j.ultsonch.2011.09.003.

Sayara, Tahseen, Montserrat Sarrà, and Antoni Sánchez. 2009. "Optimization and Enhancement of Soil Bioremediation by Composting Using the Experimental Design Technique." *Biodegradation* 21 (3): 345–56. doi:10.1007/s10532-009-9305-8.

Schwarz, W H. 2001. "The Cellulosome and Cellulose Degradation by Anaerobic Bacteria." *Applied Microbiology and Biotechnology* 56 (5–6): 634–49. doi:10.1007/s002530100710.

Seoane, Raquel, Sergio Santaeufemia, Julio Abalde, and Enrique Torres. 2022. "Efficient Removal of Methylene Blue Using Living Biomass of the Microalga *Chlamydomonas moewusii*: Kinetics and Equilibrium Studies." *International Journal of Environmental Research and Public Health* 19 (5): 2653. doi:10.3390/ijerph19052653.

Shad, Salma, Nadia Bashir, Marie-France Belinga-Desaunay Nault, and Iseult Lynch. 2021. "Incorporation of Biogenic Zinc Nanoparticles into a Polymeric Membrane: Impact on the Capture of Organic Herbicides." *Cleaner Engineering and Technology* 5 (December): 100339. doi:10.1016/j.clet.2021.100339.

Sharma, Babita, Arun Kumar Dangi, and Pratyoosh Shukla. 2018. "Contemporary Enzyme Based Technologies for Bioremediation: A Review." *Journal of Environmental Management* 210: 10–22. doi:10.1016/j.jenvman.2017.12.075.

Sharuddin, Siti Shilatul Najwa, Siti Rozaimah Sheikh Abdullah, Hassimi Abu Hasan, Ahmad Razi Othman, and Nur 'Izzati Ismail. 2021. "Potential Bifunctional Rhizobacteria from Crude Oil Sludge for Hydrocarbon Degradation and Biosurfactant Production." *Process Safety and Environmental Protection* 155: 108–21. doi:10.1016/j.psep.2021.09.013.

Shi, Chaohong, Nengwu Zhu, Yanlan Cao, and Pingxiao Wu. 2015. "Biosynthesis of Gold Nanoparticles Assisted by the Intracellular Protein Extract of *Pycnoporus sanguineus* and Its Catalysis in Degradation of 4-Nitroaniline." *Nanoscale Research Letters* 10: 147. doi:10.1186/s11671-015-0856-9.

Shimp, J F, J C Tracy, L C Davis, E Lee, W Huang, L E Erickson, and J L Schnoor. 1993. "Beneficial Effects of Plants in the Remediation of Soil and Groundwater Contaminated with Organic Materials." *Critical Reviews in Environmental Science and Technology* 23 (1): 41–77. doi:10.1080/10643389309388441.

Shukla, Pradeep K, Pragati Misra, Navodita Maurice, and Pramod W Ramteke. 2019. "Heavy Metal Toxicity and Possible Functional Aspects of Microbial Diversity in Heavy Metal-Contaminated Sites." In *Microbial Genomics in Sustainable Agroecosystems*. Springer, Singapore. doi:10.1007/978-981-32-9860-6_15.

Singh, Ajay. 2019. "Environmental Problems of Salinization and Poor Drainage in Irrigated Areas: Management through the Mathematical Models." *Journal of Cleaner Production* 206: 572–79. doi:10.1016/j.jclepro.2018.09.211.

Singh, Pooja, and Swaranjit Singh Cameotra. 2004. "Enhancement of Metal Bioremediation by Use of Microbial Surfactants." *Biochemical and Biophysical Research Communications* 319 (2): 291–97. doi:10.1016/j.bbrc.2004.04.155.

Singh, Rachana, Samiksha Singh, Parul Parihar, Vijay Pratap Singh, and Sheo Mohan Prasad. 2015. "Arsenic Contamination, Consequences and Remediation Techniques: A Review." *Ecotoxicology and Environmental Safety* 112: 247–70. doi:10.1016/j.ecoenv.2014.10.009.

Singh, Shweta, and Saswati Chakraborty. 2021. "Bioremediation of Acid Mine Drainage in Constructed Wetlands: Aspect of Vegetation (*Typha latifolia*), Loading Rate and Metal Recovery." *Minerals Engineering* 171: 107083. doi:10.1016/j.mineng.2021.107083.

Sreedevi, P R, K Suresh, and Guangming Jiang. 2022. "Bacterial Bioremediation of Heavy Metals in Wastewater: A Review of Processes and Applications." *Journal of Water Process Engineering* 48: 102884. doi:10.1016/j.jwpe.2022.102884.

Sricoth, Theeta, Weeradej Meeinkuirt, Patompong Saengwilai, John Pichtel, and Puntaree Taeprayoon. 2018. "Aquatic Plants for Phytostabilization of Cadmium and Zinc in Hydroponic Experiments." *Environmental Science and Pollution Research* 25 (15): 14964–76. doi:10.1007/s11356-018-1714-y.

Srivastava, Akhileshwar Kumar, Rajesh Kumar Singh, and Divya Singh. 2021. "Microbe-Based Bioreactor System for Bioremediation of Organic Contaminants: Present and Future Perspective." *Microbe Mediated Remediation of Environmental Contaminants*. Elsevier. doi:10.1016/b978-0-12-821199-1.00020-1.

Stensberg, Matthew Charles, Qingshan Wei, Eric Scott McLamore, David Marshall Porterfield, Alexander Wei, and María Soledad Sepúlveda. 2011. "Toxicological Studies on Silver Nanoparticles: Challenges and Opportunities in Assessment, Monitoring and Imaging." *Nanomedicine* 6 (5): 879–98. doi:10.2217/nnm.11.78.

Su, Yinglong, Zhongjian Zhang, Dong Wu, Lu Zhan, Huahong Shi, and Bing Xie. 2019. "Occurrence of Microplastics in Landfill Systems and Their Fate with Landfill Age." *Water Research* 164: 114968. doi:10.1016/j.watres.2019.114968.

Suleimanov, R R, I M Gabbasova, and R N Sitdikov. 2005. "Changes in the Properties of Oily Gray Forest Soil during Biological Reclamation." *Biology Bulletin* 32 (1): 93–99. doi:10.1007/s10525-005-0014-5.

Sun, Jing, Zhuo-Ran Zhu, Wei-Hua Li, Xiaofang Yan, Li-Kun Wang, Lu Zhang, Jianbin Jin, Xiaohu Dai, and Bing-Jie Ni. 2021. "Revisiting Microplastics in Landfill Leachate: Unnoticed Tiny Microplastics and Their Fate in Treatment Works." *Water Research* 190: 116784. doi:10.1016/j.watres.2020.116784.

Susarla, Sridhar, Victor F Medina, and Steven C McCutcheon. 2002. "Phytoremediation: An Ecological Solution to Organic Chemical Contamination." *Ecological Engineering* 18 (5): 647–58. doi:10.1016/s0925-8574(02)00026-5.

Tabaraki, Reza, and Negar Sadeghinejad. 2017. "Biosorption of Six Basic and Acidic Dyes on Brown Alga *Sargassum ilicifolium*: Optimization, Kinetic and Isotherm Studies." *Water Science and Technology* 75 (11): 2631–38. doi:10.2166/wst.2017.136.

Tang, Kuok Ho Daniel, and Risky Ayu Kristanti. 2022. "Bioremediation of Perfluorochemicals: Current State and the Way Forward." *Bioprocess and Biosystems Engineering* 45 (7): 1093–1109. doi:10.1007/s00449-022-02694-z.

Tang, Kuok Ho Daniel, and Tony Hadibarata. 2021. "Microplastics Removal through Water Treatment Plants: Its Feasibility, Efficiency, Future Prospects and Enhancement by Proper Waste Management." *Environmental Challenges* 5: 100264. doi:10.1016/j.envc.2021.100264.

Thakur, Meghna, Igor L Medintz, and Scott A Walper. 2019. "Enzymatic Bioremediation of Organophosphate Compounds-Progress and Remaining Challenges." *Frontiers in Bioengineering and Biotechnology* 7: 289. doi:10.3389/fbioe.2019.00289.

Thakur, Sveta, Lakhveer Singh, Zularisam Ab Wahid, Muhammad Faisal Siddiqui, Samson Mekbib Atnaw, and Mohd Fadhil Md Din. 2016. "Plant-Driven Removal of Heavy Metals from Soil: Uptake, Translocation, Tolerance Mechanism, Challenges, and Future Perspectives." *Environmental Monitoring and Assessment* 188 (4). doi:10.1007/s10661-016-5211-9.

Thassitou, P K, and I S Arvanitoyannis. 2001. "Bioremediation: A Novel Approach to Food Waste Management." *Trends in Food Science & Technology* 12 (5–6): 185–96. doi:10.1016/s0924-2244(01)00081-4.

Tourinho, Paula S, Vladimír Kočí, Susana Loureiro, and Cornelis A M van Gestel. 2019. "Partitioning of Chemical Contaminants to Microplastics: Sorption Mechanisms, Environmental Distribution and Effects on Toxicity and Bioaccumulation." *Environmental Pollution* 252: 1246–56. doi:10.1016/j.envpol.2019.06.030.

Tsuchii, Akio, and Yutaka Tokiwa. 2001. "Microbial Degradation of Tyre Rubber Particles." *Biotechnology Letters* 23 (12): 963–69. doi:10.1023/a:1010593807416.

Tu, S. 2004. "Effects of Arsenic Species and Phosphorus on Arsenic Absorption, Arsenate Reduction and Thiol Formation in Excised Parts of *Pteris vittata* L." *Environmental and Experimental Botany* 51 (2): 121–31. doi:10.1016/j.envexpbot.2003.08.003.

Tun, Thant Zin, Tatsuya Kunisue, Shinsuke Tanabe, Maricar Prudente, Annamalai Subramanian, Agus Sudaryanto, Pham Hung Viet, and Haruhiko Nakata. 2022. "Microplastics in Dumping Site Soils from Six Asian Countries as a Source of Plastic Additives." *Science of the Total Environment* 806: 150912. doi:10.1016/j.scitotenv.2021.150912.

Tushar, Quddus, Joao Santos, Guomin Zhang, Muhammed A. Bhuiyan, and Filippo Giustozzi. 2022. "Recycling Waste Vehicle Tyres into Crumb Rubber and the Transition to Renewable Energy Sources: A Comprehensive Life Cycle Assessment." *Journal of Environmental Management* 323 (December): 116289. doi:10.1016/j.jenvman.2022.116289.

Tyagi, Bhawna, and Naveen Kumar. 2021. "Bioremediation: Principles and Applications in Environmental Management." *Bioremediation for Environmental Sustainability*. doi:10.1016/b978-0-12-820524-2.00001-8.

Tyagi, Meenu, M Manuela R da Fonseca, and Carla C C R de Carvalho. 2010. "Bioaugmentation and Biostimulation Strategies to Improve the Effectiveness of Bioremediation Processes." *Biodegradation* 22 (2): 231–41. doi:10.1007/s10532-010-9394-4.

Ummalyma, Sabeela Beevi, Ashok Pandey, Rajeev K Sukumaran, and Dinabandhu Sahoo. 2017. "Bioremediation by Microalgae: Current and Emerging Trends for Effluents Treatments for Value Addition of Waste Streams." In *Biosynthetic Technology and Environmental Challenges*. Springer, Singapore. doi:10.1007/978-981-10-7434-9_19.

Usha, M T, T Sarat Chandra, R Sarada, and V S Chauhan. 2016. "Removal of Nutrients and Organic Pollution Load from Pulp and Paper Mill Effluent by Microalgae in Outdoor Open Pond." *Bioresource Technology* 214: 856–60. doi:10.1016/j.biortech.2016.04.060.

Valdez, Christian, Yomaira Perengüez, Bence Mátyás, and María Fernanda Guevara. 2018. "Analysis of Removal of Cadmium by Action of Immobilized *Chlorella* Sp. Micro-Algae in Alginate Beads." *F1000Research* 7: 54. doi:10.12688/f1000research.13527.1.

Valiela, Ivan, Jennifer L Bowen, and Joanna K York. 2001. "Mangrove Forests: One of the World's Threatened Major Tropical Environments." *Bioscience* 51 (10): 807. doi:10.1641/0006-3568(2001)051[0807:mfootw]2.0.co;2.

Velsankar, K, S Suganya, P Muthumari, S Mohandoss, and S Sudhahar. 2021. "Ecofriendly Green Synthesis, Characterization and Biomedical Applications of CuO Nanoparticles Synthesized Using Leaf Extract of Capsicum Frutescens." *Journal of Environmental Chemical Engineering* 9 (5): 106299. doi:10.1016/j.jece.2021.106299.

Verma, Jay Prakash, Durgesh Kumar Jaiswal, and R Sagar. 2014. "Pesticide Relevance and Their Microbial Degradation: A-State-of-Art." *Reviews in Environmental Science and Bio/Technology* 13 (4): 429–66. doi:10.1007/s11157-014-9341-7.

Wadhwani, Sweety A, Utkarsha U Shedbalkar, Shradhda Nadhe, Richa Singh, and Balu A. Chopade. 2018. "Decolorization of Textile Dyes by Combination of Gold Nanocatalysts Obtained from *Acinetobacter* Sp. SW30 and NaBH4." *Environmental Technology & Innovation* 9 (February): 186–97. doi:10.1016/j.eti.2017.12.001.

Wanapaisan, Pagakrong, Natthariga Laothamteep, Felipe Vejarano, Joydeep Chakraborty, Masaki Shintani, Chanokporn Muangchinda, Tomomi Morita, et al. 2018. "Synergistic Degradation of Pyrene by Five Culturable Bacteria in a Mangrove Sediment-Derived Bacterial Consortium." *Journal of Hazardous Materials* 342 (January): 561–70. doi:10.1016/j.jhazmat.2017.08.062.

Weigert, Sebastian, Pablo Perez-Garcia, Florian J. Gisdon, Andreas Gagsteiger, Kristine Schweinshaut, G. Matthias Ullmann, Jennifer Chow, Wolfgang R. Streit, and Birte Höcker. 2022. "Investigation of the Halophilic PET Hydrolase PET6 from *Vibrio gazogenes*." *Protein Science* 31 (12). doi:10.1002/pro.4500.

Weis, Leticia, Rosana de Cassia de Souza Schneider, Michele Hoeltz, Alexandre Rieger, Schirley Tostes, and Eduardo A Lobo. 2020. "Potential for Bifenthrin Removal Using Microalgae from a Natural Source." *Water Science and Technology* 82 (6): 1131–41. doi:10.2166/wst.2020.160.

Weiss, Jochen, Paul Takhistov, and D Julian McClements. 2006. "Functional Materials in Food Nanotechnology." *Journal of Food Science* 71 (9). Wiley: R107–16. doi:10.1111/j.1750-3841.2006.00195.x.

Wilkes, R.A., and L. Aristilde. 2017. "Degradation and Metabolism of Synthetic Plastics and Associated Products by *Pseudomonas* Sp.: Capabilities and Challenges." *Journal of Applied Microbiology* 123 (3): 582–93. doi:10.1111/jam.13472.

Wu, Jun, Hao Zheng, Jun Hou, Lingzhan Miao, Fang Zhang, Raymond Jianxiong Zeng, and Baoshan Xing. 2021. "In Situ Prepared Algae-Supported Iron Sulfide to Remove Hexavalent Chromium." *Environmental Pollution* 274: 115831. doi:10.1016/j.envpol.2020.115831.

Xu, Jiakun, Jingjing Sun, Yuejun Wang, Jun Sheng, Fang Wang, and Mi Sun. 2014. "Application of Iron Magnetic Nanoparticles in Protein Immobilization." *Molecules* 19 (8): 11465–86. doi:10.3390/molecules190811465.

Yan, Haifeng, Zaisheng Yan, Luming Wang, Zheng Hao, and Juan Huang. 2022. "Toward Understanding Submersed Macrophyte Vallisneria Natans-Microbe Partnerships to Improve Remediation Potential for PAH-Contaminated Sediment." *Journal of Hazardous Materials* 425: 127767. doi:10.1016/j.jhazmat.2021.127767.

Yang, Yu, Jun Yang, Wei-Min Wu, Jiao Zhao, Yiling Song, Longcheng Gao, Ruifu Yang, and Lei Jiang. 2015. "Biodegradation and Mineralization of Polystyrene by Plastic-Eating Mealworms: Part 2. Role of Gut Microorganisms." *Environmental Science & Technology* 49 (20): 12087–93. doi:10.1021/acs.est.5b02663.

YouShuang, Zhu, Zhang HaiBo, Zhang YingLong, and Huang Feng. 2011. "Lignocellulose Degradation, Enzyme Production and Protein Enrichment by Trametes Versicolor during Solid-State Fermentation of Corn Stover." *African Journal of Biotechnology* 10 (45): 9182–92. doi:10.5897/ajb11.810.

Zadjelovic, Vinko, Audam Chhun, Mussa Quareshy, Eleonora Silvano, Juan R Hernandez-Fernaud, María M Aguilo-Ferretjans, Rafael Bosch, Cristina Dorador, Matthew I Gibson, and Joseph A Christie-Oleza. 2020. "Beyond Oil Degradation: Enzymatic Potential of Alcanivorax to Degrade Natural and Synthetic Polyesters." *Environmental Microbiology* 22 (4): 1356–69. doi:10.1111/1462-2920.14947.

Zappi, Mark, Yue Jing, Gayle Albritton, Al Crawley, Judy Singletary, Ronald Tarbutton, and Neil Hall. 1999. "Use of Biocells for Enhanced Bioremediation of Soils Contaminated with Heavy Petroleum Hydrocarbons." *Transportation Research Record: Journal of the Transportation Research Board* 1675 (1): 51–60. doi:10.3141/1675-07.

Zazouli, Mohammad Ali, Yousef Mahdavi, Edris Bazrafshan, and Davoud Balarak. 2014. "Phytodegradation Potential of BisphenolA from Aqueous Solution by Azolla Filiculoides." *Journal of Environmental Health Science & Engineering* 12: 66. doi:10.1186/2052-336X-12-66.

Zeghal, Emna, Annika Vaksmaa, Hortense Vielfaure, Teun Boekhout, and Helge Niemann. 2021. "The Potential Role of Marine Fungi in Plastic Degradation - A Review." *Frontiers in Marine Science* 8 (November). doi:10.3389/fmars.2021.738877.

Zhang, Hong, Yi Tang, Dongqing Cai, Xianan Liu, Xiangqin Wang, Qing Huang, and Zengliang Yu. 2010. "Hexavalent Chromium Removal from Aqueous Solution by Algal Bloom Residue Derived Activated Carbon: Equilibrium and Kinetic Studies." *Journal of Hazardous Materials* 181 (1–3): 801–8. doi:10.1016/j.jhazmat.2010.05.084.

Zhang, Yongli, Mussie Y Habteselassie, Eleazer P Resurreccion, Vijaya Mantripragada, Shanshan Peng, Sarah Bauer, and Lisa M Colosi. 2014. "Evaluating Removal of Steroid Estrogens by a Model Alga as a Possible Sustainability Benefit of Hypothetical Integrated Algae Cultivation and Wastewater Treatment Systems." *ACS Sustainable Chemistry & Engineering* 2 (11): 2544–53. doi:10.1021/sc5004538.

5 Metagenomics-Based Characterization of Microbial Diversity across Industrial Waste Dumping Sites

Meesha Singh, Rupsha Karmakar, Sayak Ganguli, and Mahashweta Mitra Ghosh
St Xavier's College (Autonomous), Kolkata

5.1 INTRODUCTION

Environmental pollution is a serious concern that has arisen as a result of rapid industrialization, urbanization, and an increase in people's living standards. Increased industrialization has brought with it the challenge of massive waste output, both solid and liquid. Industrial wastes are defined as those residual substances that are produced during manufacturing processes in industrial operations. Elements like chemicals, metals, paper, and even radioactive wastes can all be categorized as industrial effluents. Thousands of small and large industrial facilities discharge their wastes straight onto open land areas or into local water resources with no treatment. The issues regarding disposal of industrial solid wastes are linked to the lack of infrastructure facilities and concerns about taking adequate precautions. Though large- and medium-scale industries show a basic degree of waste disposal efficiency, numerous small-scale industries have been found to dispose their wastes here and there, resulting in the intermingling of industrial wastes with domestic and clinical effluents. This expands the scope of the problem. Table 5.1 encloses a list of hazardous industrial wastes like heavy metals, hydrocarbons, insecticides, organic compounds, hazardous or radioactive materials, and other hazardous industrial wastes, along with their respective sources. Industrial wastes can broadly be classified into two major types: dry cleaning fluids and embalming fluids. Perchloroethyleneis a primary component of cleaning fluids, whereas embalming fluids contain carcinogens. Both can pollute community water supplies through contamination of groundwater (Millati et al., 2019). Heavy metals and their corresponding metalloid contamination are one of the trending environmental threats associated with the unrestricted disposal of toxic compounds from diverse origins, including industrial operations (Sharma et al., 2021a). The wastewater discharged from manufacturing operations in many sectors has high

 DOI: 10.1201/9781003451457-5

TABLE 5.1
Sources of Generation of Some of the Major Industrial Solid Wastes (Millati et al., 2019)

Industrial Solid Waste	Source
Coal Ash	Coal based thermal power plants, brickworks and cement industries
Red mud	Metal like aluminium and copper extraction industries
Phosphogypsum	Ammonium phosphate and phosphoric acid plants
Lime sludge	Paper and pulp mills
Blast Furnace slags containing silicates, aluminosilicates, and calcium-alumina-silicates.	Steel and iron plants
Coal washery dust	Coal mines
Mica scraper waste	Mica mining areas
Untanned and tanned collagen, noncollagenous protein wastes	Tannery Industry

amounts of sodium carbonate, sodium sulfide, sodium hydroxide, hydrochloric acid, and calcium oxide (Sharma et al., 2021b). As per the reports of Selvi et al. (2019), a rise in some crucial parameters such as Biological Oxygen Demand, Chemical Oxygen Demand, turbidity, etc. due to the emptying of even secondarily treated effluent from industrial waste treatment plants into the adjacent natural water resources containing various organic pollutants and heavy metals exerts negative impacts on the aquatic flora and fauna in the habitat. Eutrophication is another problem that is facilitated by the direct dumping of industrial sludge into freshwater habitats. Pollutants such as inorganic substances such as phosphorus, nitrogen, and total organic carbon promote the growth of toxic algal blooms on the water surface, which in turn cause hypoxic conditions in the water body, leading to the death of aquatic life. Overabundances of toxic algal communities have been reported to dramatically increase water acidity and decrease aquatic biodiversity. There are different ways in which industrial waste can pollute the environment. Direct and untreated discharge of sludge and associated wastewater, land leaching, dumping in landfills, and ship pollution are a few of them (Selvi et al., 2019).

Hazardous wastes that cannot be recycled or repurposed are eventually disposed of in secure landfills or incinerators. According to the latest report of the CPCB in 2019, India has a total of 3,159 dumpsites, with Uttar Pradesh, Madhya Pradesh, Maharashtra, Karnataka, Rajasthan, and Gujarat being the top six states with the most dumpsites. Figure 5.1 and Table 5.2 enlist the state-wise count of the existing solid waste dumpsites in India. There are 69,308 hazardous waste-producing plants in India with an annual capacity of 39.46 million metric tons of solid waste generation (CPCB, 2019). Whereas the data of the Central Pollution Control Board (CPCB), 2019–2020, shows that states like Gujarat, Maharashtra, and Tamil Nadu top the list of hazardous waste-generating industries contributing a huge amount of landfillable waste in India per annum (Table 5.3)

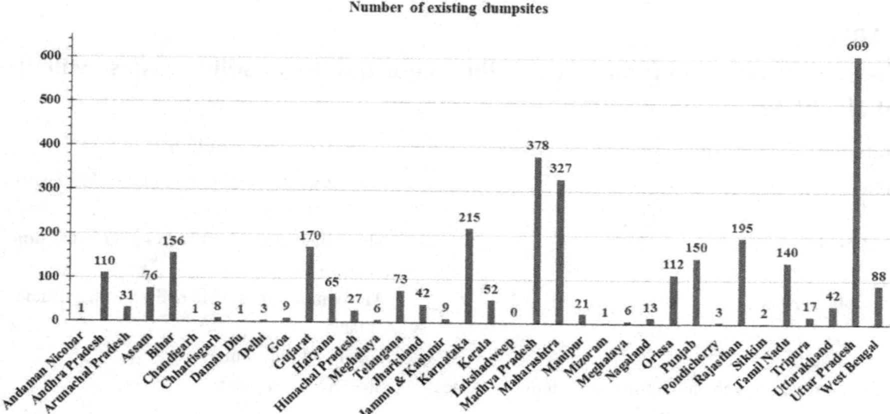

FIGURE 5.1 State-wise count of the existing solid waste dumpsites in India (CPCB, 2019.)

TABLE 5.2
State Wise Status of the Number of Solid Waste Dumpsites in India (CPCB, 2019)

State	Number of Existing Dumpsites
Andaman Nicobar	1
Andhra Pradesh	110
Arunachal Pradesh	31
Assam	76
Bihar	156
Chandigarh	1
Chhattisgarh	8
Daman Diu	1
Delhi	3
Goa	9
Gujarat	170
Haryana	65
Himachal Pradesh	27
Meghalaya	6
Telangana	73
Jharkhand	42
Jammu & Kashmir	9
Karnataka	215
Kerala	52
Lakshadweep	0
Madhya Pradesh	378
Maharashtra	327

(*Continued*)

TABLE 5.2 (*Continued*)
State Wise Status of the Number of Solid Waste Dumpsites in India (CPCB, 2019)

State	Number of Existing Dumpsites
Manipur	21
Mizoram	1
Meghalaya	6
Nagaland	13
Orissa	112
Punjab	150
Pondicherry	3
Rajasthan	195
Sikkim	2
Tamil Nadu	140
Tripura	17
Uttarakhand	42
Uttar Pradesh	609
West Bengal	88
TOTAL	3,159

TABLE 5.3
Data on the Hazardous Wastes Generating Industries and Their Contribution to the Annual Wastes Generation Capacity across Various States in India (CPCB, 2019)

State	Number of Hazardous Wastes Generating Industries	Quantity of Waste Generated as per Annual Return (MT)
Andaman & Nicobar Islands	3	59.76
Andhra Pradesh	2,648	620,952
Assam	183	52,394
Bihar	166	7,630
Chandigarh	1,363	2,125
Chhattisgarh	413	172,438
DD&DNH	410	4,631
Delhi	1,912	2,683
Goa	1,628	28,569
Gujarat	19,662	2,485,317
Haryana	4,845	200,606
Himachal Pradesh	2,436	27,725
Jammu & Kashmir	238	1,213
Jharkhand	109	209

(*Continued*)

TABLE 5.3 (*Continued*)
Data on the Hazardous Wastes Generating Industries and Their Contribution to the Annual Wastes Generation Capacity across Various States in India (CPCB, 2019)

State	Number of Hazardous Wastes Generating Industries	Quantity of Waste Generated as per Annual Return (MT)
Kerala	1,616	311,042
Madhya Pradesh	2,863	232,199
Maharashtra	7,257	999,566
Manipur	374	0
Meghalaya	38	276
Mizoram	40	20
Nagaland	38	29
Odisha	360	679,860
Puducherry	131	34,907
Punjab	3,262	122,167
Rajasthan	2,094	587,554
Sikkim	49	1,722
Tamil Nadu	3,961	1,059,387
Telangana	3,024	317,091
Uttar Pradesh	2,597	362,114
Uttarakhand	4,340	21,818
West Bengal	809	131,412

(CPCB, 2019). Similarly, in most African cities, around 20% of urban waste is dumped in landfill, contributing to one-third of the country's disease burden (Fazzo et al., 2017).

Uncontrolled and unmonitored industrial waste landfills have the potential to percolate harmful compounds into the environment, which can pose a human health impact in many ways. Hazardous compounds that migrate from the site have the potential to spread and harm environmental media, including ground and surface water, air, soil, etc. (Sharma et al., 2022).

Reviewing the topic of health hazards from waste dumping sites should begin with a study that reported that a large number of toxic materials and chemicals leaked from the landfill sites in New York State into the adjacent streams, freshwater aquifers, and indoor air of houses, posing a serious health threat to the local population. Following a similar trend, State Inspectorate for Environment Protection data has highlighted that 60% of hazardous waste sites listed in Poland pose potential or actual harm to public health and the environment (Rushton, 2003). Following are a few of the adverse effects of industrial solid waste on the environment and human health:

- Refugee dump leachate percolates into the earth, contaminating subsurface water. Spillage of certain toxic chemicals, such as polychlorinated biphenyls, cyanides, mercury, and heavy metals like Cu, Pb, Co, Cd, and Cr, into groundwater can cause serious health issues, including birth defects, cancer, kidney and liver dysfunction, immune function disorders, neurotoxic disorders, etc.

- Organic solid wastes generate an unpleasant stench as they decompose, polluting the environment.
- Scavengers and stray animals may invade roadside debris and scatter it across a vast region, inflicting significant aesthetic damage to the environment.
- Multiple vectors, for example, rats and insects, may infest garbage dumps and spread a variety of illnesses in humans, viz. diarrhea, amoebic dysentery, etc., through contaminating food and drinking water sources.
- Diseases such as plague, trichinosis, salmonellosis, endemic typhus, and others can be spread by rats living near infectious solid wastes.
- The clogging of drains and pits by solid waste causes water logging, which encourages mosquito breeding and the transmission of illnesses such as malaria and plague.
- Extensive research in the field of reproductive disorders and birth defects associated with landfill sites has proved that various reproductive impairments, spontaneous abortion, and premature birth are all highly prevalent with industrial solid waste contamination. A study including around 21 landfill sites in Europe indicated that living within 3 km of an industrial waste dump site was related to a considerably increased incidence of congenital abnormalities, cardiac malformations, neural tube abnormalities, etc. (Chadar and Chadar, 2017).

Parallelly, Lead acid battery (LAB) industries act as a principal dissemination site of hazardous metals like Pb, Cu, Cd, Cr, and Ni into the environment. Because of the potential toxicity of heavy metals on public health, academics all around the world are interested in studying human subjection to toxic heavy metals and the health hazards that come with it. Investigations of human exposure to heavy metals and the related health hazards have piqued the interest of academics throughout the world. Assessment of health risks from heavy metals in soil adjacent to an industrial landfill area in Ibadan, Nigeria, has proved that higher titers of toxic metals like Pb, Mn, Cu, Cd, Cr, and Ni pose considerable health threats to the local residents through dermal contact, ingestion, and inhalation (Ogundele et al., 2019).

Apart from the pollutants, the microbial diversity in the leachates, sludge, and soil of municipal landfill sites is equally noticeable. Members of campylobacterota, Proteobacteria, Actinobacteria, and Firmicutes are a few of them to be noted due to their higher frequencies (Parks et al., 2018). Pollution-induced changes in soil and water composition can bring about changes in the microbial profile of the habitat and even a loss of microbial diversity. On the flip side, the metabolic plasticity of a few members provides selective fitness to the pollutant-tolerant bacterial populations, leading to the establishment of a new microbial community in the niche. Members of the class Deltaproteobacteria and Gammaproteobacteria are found to be predominant in the soils and sediments contaminated with polymeric hydrocarbons and metals such as gold, uranium, etc., whereas mangrove soils receiving industrial oil spills showed higher titers of a few of the oil-pollutant-resistant genera like *Marinobacterium*, *Haliea*, and *Cycloclasticus*. Hence, detailed profiling of the microbial community in a landfill setting can be an effective biomarker (Selvarajan et al., 2022). Dumpsite effluents are also known to contribute to antimicrobial resistance, a leading concern of the 21st century, with their ability to accumulate and percolate various antibiotic compounds as well as pathogenic microbial members from community sewage. A study in Aizawl municipal landfill identified that few pathogenic as well as nonpathogenic bacteria like *Arthrobacter* sp., *Staphylococcus* sp., *Achromobacter* sp., and *Bacillus thuringiensis* showed resistance to about 84% of antibiotics tested against, including beta-lactam, quinolone, and glycopeptide classes of antibiotics (Zothanpuia et al., 2021).

Many of the microbial members. i.e., bacteria and fungi, in the leachate and corresponding soil of solid landfills bear the potential to biodegrade hazardous pollutants, producing lesser toxic byproducts and end products like water, methane, carbon dioxide, etc. A greater titer of catabolic and carbon utilization capacity was found for the members of the bacterial community that are active in the dumpsite soils. Furthermore, microorganisms present in the landfills have proven their abilities to remediate as well as lower heavy metal trace concentrations in the landfills (Zainun and Simarani, 2018). Investigations in the Okhla dumpsites in Delhi, India, for example, revealed the prevalence of members of the phylum *Saccharomonospora* playing a major role in the process of composting solid wastes (Yadav et al., 2020). The isolation of two potent plastic-degrading bacterial genera, *Arthrobacter* sp. and *Pseudomonas* sp., from the dumpsites is also noticeable (Nauendorf et al., 2016).

Hence, gaining better knowledge of the microbial architecture at solid waste disposal sites, particularly leachates, might help advance the field of bioremediation as an effective ecological management tool. Although various culture-dependent laboratory-based studies as well as a pilot study including a landfill bioreactor have traced the fundamental microbial response, the structural and functional characteristics of microbial communities in the actual landfill have yet to be uncovered. This is the place where the application of metagenomics studies is mandatory, which can overcome the limitations of cultivability.

For many years, metagenomics techniques have offered a wide range of applications in terms of environmental sustainability. Metagenomics encompasses research at three primary levels that are all interlinked: sample processing, DNA sequencing, and functional analysis, which thereby result in the determination of the functional gene composition of microbial communities. Thus, this high-throughput culture-independent investigation can provide a comprehensive picture of the microbial profile in the dumpsites. The metagenomics study of soil samples collected from municipal dumpsites accumulating domestic and industrial waste in Malaysia showed the predominance of phyla like Proteobacteria, Firmicutes, and Acidobacteria in the functional and nonfunctional landfills, respectively (Zainun and Simarani, 2018). Similarly, shotgun metagenomic techniques aided in noting the abundance of bacterial phyla, viz., Proteobacteria, Actinobacteria, Bacteroidetes, and Firmicutes, in the leachate samples collected from Pirana landfill in Gujarat, India (Kumar et al., 2021). In a study including two landfills in China, Wan et al. (2021) highlighted that of 69 pathogenic microbes—Actinobacteria, Proteobacteria, and Bacteroidetes were the most numerous (Wan et al., 2021). Furthermore, metagenome-based examination of landfill soil samples could predict possible disease threats and environmental menaces from landfill inhabitant pathogens like *Bacillus cereus, Bordetella pertussis, Klebsiella pneumoniae, and Pseudomonas aeruginosa.* Antibiotic resistance genes linked to the above-mentioned pathogens were also identified in the municipal dumpsites through KEGG pathway analysis (Wan et al., 2021). 16S rRNA metabarcoding of a water sample from a municipal landfill contaminated aquifer identified higher microbial abundances as well as diversity in comparison to the uncontaminated well water. Followed by *Proteobacteria; Acidobacteria, Actinobacteria, Verrucomicrobia,* and *Bacteroidetes* were the noticeable ones in terms of abundances (Abiriga et al., 2021). Hence, uncovering the diversity in addition to the functional profiling of the bacterial community in the soil and leachates at community landfill sites is indispensable for proposing effective waste management strategies and public health monitoring.

5.2 METHODOLOGY

The inability of conventional culture-dependent laboratory-based studies to report a large number of microbial taxa has led culture independent high-throughput sequencing technology to emerge as one of the primary techniques for deep understanding of microbial members in municipal and industrial landfill sites, including their relative abundances, functional diversities, enriched metabolic outputs, and pathogenic load. Metagenomics studies involve the application of bioinformatics tools and genomic technologies, which thereby aid in direct access to the genetic content of microbial communities. In a typical sequence-based metagenomics project, multiple steps are involved, as shown in Figure 5.2. It yields a taxonomic classification of microbial populations. It aids in the exploration of rare microbial taxa (Sogin et al., 2006).

5.2.1 Selection of Study Area

The selection of a suitable study area in proximity to industrial waste dump sites is the foremost step to be performed.

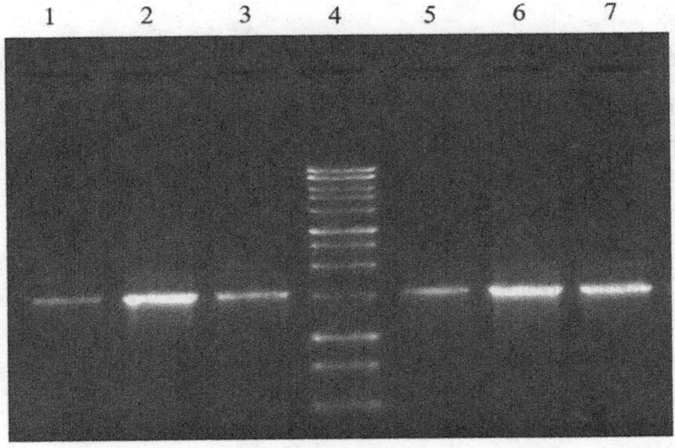

FIGURE 5.2 Proposed pipeline for the characterization of microflora abundant in industrial waste dump sites

5.2.2 Sampling and Processing

Sample collection is the first and most critical step in metagenomics-based assays. Soil, leachate, or sludge samples need to be collected from the vicinity of industrial landfills in sterile containers. Sampling should be performed at least at the replicate of three. The samples are then transferred to the laboratory at 4°C within 10 hours of the collection. The storage temperature is –20°C. The DNA extracted should represent all the cells present in the sample (Thomas et al., 2012).

5.2.3 Physico-Chemical Analyses

Followed by sample collection, temperature, pH, salinity, and turbidity of the study area and sample ought to be recorded. There are additional parameters that also should be analyzed in the collected samples, as shown in Table 5.4.

5.2.4 DNA Extraction and Quality Control

DNA can be extracted from the raw samples using various standard protocols. DNA extraction using any protocol should be initiated with bacterial cell isolation from solid waste through mixing 2 g of solid waste with 6 mL of phosphate buffer solution (PBS) (pH 7.0), vortexing, and incubation at room temperature for 5 minutes. The quality and quantity checks of extracted DNA can be carried out using Nanodrop and 1.2% Agarose gel electrophoresis.

5.2.5 Design of PCR Primers and Illumina Sequence Adapters

16S Metagenomic Sequencing Library preparation and Polymerase chain reaction (PCR) primer preparation against the conserved sequence of V3 & V4 regions of the 16S rRNA gene should be the next steps to be performed for amplification. PCR

TABLE 5.4
Various Physico-Chemicals Parameters to Be Tested in the Collected Sample

S. No.	Organics	Inorganics	Heavy Metals
1.1.	BOD (Biological Oxygen Demand)	Ammonia	Arsenic (Ar)
2.2.	COD (Chemical Oxygen Demand)	Total Kjeldahl Nitrogen	Cadmium (Cd)
3.3.	TOC (Total Organic Carbon)	Nitrite/Nitrate	Chromium (Cr)
4.4.	Oil and Grease	Total Phosphorus	Lead (Pb)
5.5.	TS (Total Solids)	Chlorides	Mercury (Hg)
6.6.	TSS (Total Suspended Solids)	Sulphides	Nickel (Ni)
7.7.	TDS (Total Dissolved Solids)		Thallium (Tl)
8.8.	TVS (Total Volatile Solids)		Copper (Cu)

primers should be designed with appropriate sample barcoding index sequences and Illumina adapter sequences. While designing these primers, attention must be paid to allowing multiplexing of samples while still preventing the production of primer dimers or secondary structures that would lower PCR efficiency.

5.2.6 PCR Amplification of the Target Region

Each of the primer sets that differs from the others merely in their barcoding indices should have its PCR conditions optimized. For purification of the amplicons originated from amplifications of the V3 & V4 regions, Agencourt XP beads can be used. The QUBIT ds DNA HS kit needs to be used for quantification of the libraries, followed by normalization to 0 nMol/L for sequencing. The quality check of the library by Agarose is the next stage to be performed. Figure 5.3 shows an image of amplified libraries on 2% agarose gel electrophoresis.

5.2.7 Sequencing

The order of DNA nucleotides, i.e. adenines, cytosines, guanines, and thymines, in the metagenome is revealed through genome sequencing. Next-generation sequencing is commonly used these days. For metagenomic samples, 16S metagenomic sequencing platforms such as 454/Roche and the Illumina Hiseq, Miseq, Novaseq, or Solexa platforms are extensively used. Ion Torrent is also used these days (Thomas et al., 2012). All the platforms mentioned above are able to generate ~150 Mb of data per sample.

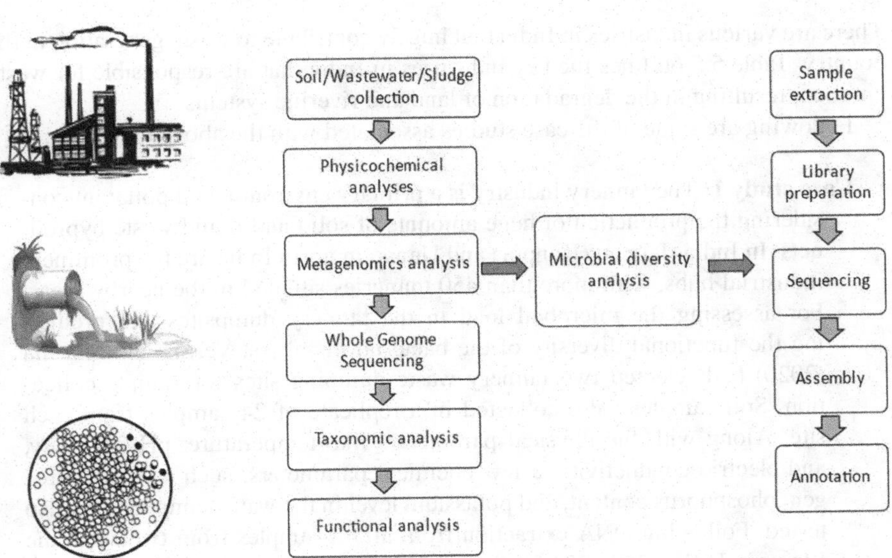

FIGURE 5.3 Representative image showing amplified libraries for six different samples on agarose gel electrophoresis. (Lane 1: Sample 1; Lane 2: Sample 2; Lane 3: Sample 3; Lane 4: PCR Ladder; Lane 5: Sample 4; Lane 6: Sample 5; and Lane 7: Sample 6.)

5.2.8 Assembly

The metagenome assemblies are performed on short read pairs, which result in longer microbial sequence contigs. There are two strategies for assembly: reference-based assembly (co-assembly) and *de novo* assembly. The adapter trimming is performed, prior to which the raw paired-end sequences are filtered. ORFs are then predicted from these assembled contigs (Thomas et al., 2012).

5.2.9 Bioinformatics Analysis

In the first step, the quality of the raw reads of Illumina sequencing should be checked for ambiguous bases, Phred score, read length, nucleotide base content, and other relevant parameters by using the FASTQC toolkit (http://www.bioinformatics.babraham.ac.uk/projects/fastqc). In the next step, quality-processed paired or unpaired reads need to be analyzed with any of the standard bioinformatic pipelines like QIIME 2.0, DADA 2, or MOTHUR for a comprehensive and comparative analysis of the metagenome data. Annotation of metagenomic sequencing data includes Taxon calling and OTU clustering of the quality-checked sequence reads, diversity analysis, functional potential prediction, gene ontology classification of biological pathways, and pathogenic load prediction through taxon set analysis. KEGG, eggNOG, TIGRFAM, COG/KOG, and PFAM are some of the reference databases that can be utilized. (Thomas et al., 2012).

5.3 FEW CASE STUDIES

There are various industries in India that hugely contribute to waste generation in the country. Table 5.5 outlines the key industries in India that are responsible for waste creation, resulting in the degradation of land and riverine systems.

Following are some of the case studies associated with the above industries:

Case study 1: The tannery industry is a primary environmental pollutant considering the production of huge amounts of solid and liquid waste byproducts. In India, Jajmau (Kanpur) and Unnao in north India are the prominent industrial hubs, with more than 450 tanneries situated in the nearby areas. For assessing the microbial load in the tannery dumpsites and predicting the functional diversity of the bacterial members, Verma and Sharma (2020) had selected two tannery waste dumping sites for sample collection. Soil samples were collected in a replicate of 24 samples from each site. Along with the physical parameters like temperature, pH, turbidity, and electric conductivity, a few chemical parameters, such as total nitrogen, phosphorus content, and potassium level in the water sample. were also tested. Following DNA extraction from all 48 samples from two sites, the Illumina HiSeq 2500 Platform was used for sequencing the samples. For diversity analysis of microbial communities and functional annotation, the NCBI-nr and KEGG databases were used. Among the 181 bacterial genera identified at both sites, about 70 were found to be common across all the

TABLE 5.5
Major Industries Responsible for Waste Generation in India

S. No.	Major Industries Generating Pollutants in India	Primary Pollution Caused	Total Discharge of Pollutants
1.	Tannery	Soil and water	*Dissolved solids = 860 tonnes/day *Chloride = 360 tonnes/day *Chromium = 11 tonnes/day *Vegetable tannins and nontannins *Nitrogen, phosphorous, sulfate, ammonium salts, lime Mohammed (2017)
2.	Pulp and paper	Water (riverine systems)	*Organic pollutants: fibres, starch, hemi-cellulose and organic acids *Organochlorine pollutants: chlorinated derivatives of acids, phenols, dibenzo-p-dioxins/furans *Chemicals: chloroform, chloro-acetones, aldehydes and acetic acid * Dioxins and furans Bajpai (2015)
3.	Pharmaceutical	Water (riverine systems)	* Antibiotic residues (AR) *Antibiotic resistant genes (ARG) *Antibiotic resistant bacteria (ARB) Thai et al. (2018)

samples. *Bacillus, Bacteroides, Burkholderia, Candidatus, Entotheonella, Candidatus, Clostridium, Desulfovibrio, Desulfuromonas, Nitrosococcus, Nocardioides, Paenibacillus, Planctomyces, Proteiniphilum, Pseudomonas, Rhizobium, Rhodopirellula, Rhodothermus, Salinibacter, Sorangium, Spirochaeta, Streptomyces,* and *Sulfurovum* were among the common members. Table 5.6 lists the unique bacterial genera identified in the two study sites (Verma & Sharma, 2020).

Figure 5.4 represents relative distribution of the bacterial community across the two study sites. 39.3% of bacterial genera were found to be common, with the prevalence of *Pseudomonas, Bacillus*, etc. On the other hand, microbial abundance in Unnao was reported to be higher than that in Jajmau.

The common phylum and genera found in the present study may be useful to identify the inherent bioremediation potential of the microflora thriving in industrial waste dumpsites. The functional diversity analysis of the metagenomes revealed the abundance of metabolic pathways: carbohydrates, amino acids, and proteins, which thereby shows the potential of microbial communities to bioremediate tannery wastes and heavy metals (Verma & Sharma, 2020).

TABLE 5.6
Microbial Diversity Analysis in Metagenomic Samples Collected from Tannery Waste Dumping Sites in Jajmau and Unnao

Site	Bacterial Genera Identified
Jajmau	*Alkalispirochaeta, Aminobacterium, Anaerophaga, Anaerosalibacter, Arhodomonas, Balneola, Brevibacterium, Chelativorans, Clostridiisalibacter, Conexibacter, Corynebacterium, Cyclobacterium, Fodinicurvata, Geobacillus, Halanaerobium, Halococcus, Halothermothrix, Halovivax, Legionella, Marinobacterium, Marinospirillum, Methanoculleus, Methylohalobius, Nocardiopsis, Oceanobacillus, Oligella, Pusillimonas, Rubinisphaera, Ruminiclostridium, Saccharomonospora, Salinicoccus, Salinisphaera, Salisaeta, Stappia, Tepidanaerobacter, Tepidicaulis, Thiomicrospira, Vibrio, Virgibacillus, Yaniella*
Unnao	*Acinetobacter, Anaeromyxobacter, Arenimonas, Azoarcus, Azospirillum, Bacteriovorax, Bdellovibrio, Burkholderiales Genera Incertae Sedis, Candidatus Accumulibacter, Candidatus Nitrosotenuis, Candidatus Solibacter, Castellaniella, Caulobacter, Chloroflexus, Cystobacter, Desulfobulbus, Desulfococcus, Desulfomicrobium, Devosia, Fulvivirga, Gemmata, Hahella, Hassallia, Herbaspirillum, Herpetosiphon, Hyphomonas, Ilumatobacter, Kangiella, Ktedonobacter, Leptonema, Leptospira, Marivirga, Methylocaldum, Micavibrio, Microvirga, Muricauda, Myxococcus, Nitrosomonas, Nitrososphaera, Nitrosospira, Nitrospira, Nocardia, Novosphingobium, Opitutus, Pedobacter, Pedosphaera, Phenylobacterium, Pirellula, Pontibacter, Porphyromonas, Pyrinomonas, Rhodanobacter, Rhodococcus, Roseiflexus, Salinimicrobium, Sedimentibacter, Sedimenticola, Singulisphaera, Solimonas, Sphingobium, Sphingomonas, Sphingopyxis, Sulfuricurvum, Sulfurimonas, Thauera, Thermomonas, Thiobacillus, Xanthomonas*

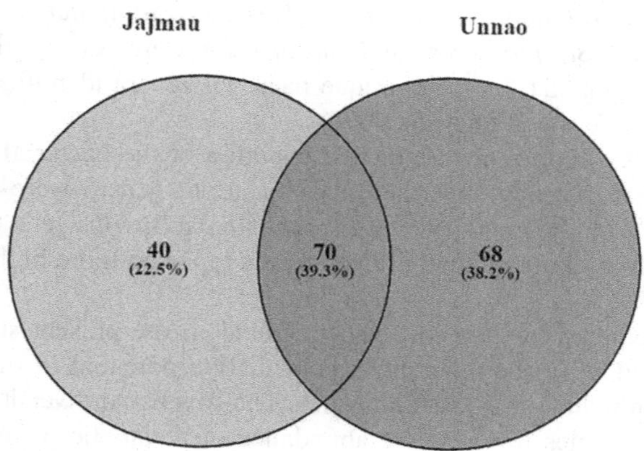

FIGURE 5.4 Relative distribution of bacterial genera across the two study sites, Jajmau and Unnao.

Case study 2: The pulp and paper mill industries are one of the largest and major sources of water pollution. In India, there are around 861 pulp paper units, including small paper industries (Annual Report of CPPRI, 2019–2020). The approximate production of hazardous pollutants from the pulp and paper industries is 100 million kg annually. K. R. Pulp and Paper Limited in Shahjahanpur, India (27° 53′ 1.4748 ″ N and 79° 54′ 44.0892″ E) is known for producing wood pulp, writing paper, and kraft paper, as well as for recycling used papers. In a study by Sharma et al. (2021a), metagenomic profiling of the collected wastewater and leachate samples originated from the paper mill using the Illumina MiSeq sequencing platform revealed bacterial genera (Bacteroidetes, Actinobacteria, Firmicutes, Acidobacter, and Proteobacteria) to be the dominant ones. Along with the biodegrading microbes, including candidates of the class Gammaproteobacteria, methanotrophic members of the class Alphaproteobacteria showed their prevalence in the sample. Table 5.7 represents microbial diversity in the collected sample at different taxon levels (Sharma et al., 2021b).

Case study 3: The Yamuna, a major tributary of the Ganga, is one of the most polluted rivers in India. The effluents from around 18 major drains in the Delhi-NCR region enter the river (Central Pollution Control Board (CPCB, 2019). The untreated effluents of urban, clinical, municipal, and industrial origins enter directly into the river, thereby making it an ideal environment to determine the threshold of pollution on the microbial communities thriving in the riverine system. In order to evaluate the diversity and functional characterization of the microbial community in the river water contaminated by industrial effluent in the embankment of the Yamuna River, Mittal et al. (2019) collected water samples twice a year. One in the pre-monsoon period and a second one in the post-monsoon season. Followed by collection, raw samples were subjected to 16S metagenomics sequencing using the Illumina HiSeq platform and whole-genome shotgun sequencing. Taxonomic assignments of the raw sequence reads were facilitated by the QIIME pipeline. Microbial diversity in the collected samples is represented in Table 5.8. Members of the phylum Proteobacteria showed the highest prevalence in

TABLE 5.7
Microbial Diversity Analysis in Metagenomic Samples Collected from Paper and Pulp Waste Dumping Sites in Shahjahanpur

Diversity Analysis	Dominant Bacterial Members
At phylum level	Actinobacteria, Proteobacteria, Spirochaetes, Acidobacteria, Verrucomicrobia, Bacteroidetes, Firmicutes, Chloroflexi
At class level	Bacteroidia, Anaerolineae, Spirochaetia, Clostridia, Deltaproteobacteria, Gammaproteobacteria, Alphaproteobacteria, Verrucomicrobiae
At genus level	*Christenesenellaceae*-R-7-group, *Leptolinea*, *Anaerolinea*, *Bdellovibrio*, *Desulfovibrio*, *Candidatus-Solibacter*

TABLE 5.8
Microbial Diversity Analysis in Metagenomic Samples Collected from Yamuna River Water in Delhi-NCR Region

Diversity Analysis	Dominant Bacterial Members
At phylum level	Proteobacteria, Actinobacteria, Firmicutes, Bacteroidetes
At genus level	*Burkholderia, Achromobacter, Thauera, Pseudomonas, Acinetobacter, Comamonas, Enterobacter, Aeromonas, Methylotenera, Klebsiella, Variovorax, Sphingobium, Dechloromonas, Ferrimonas, Shewanella, Acidovorax, Delftia, Flavobacterium, Plesiomonas, Allochromatium, Cloacibacterium, Kaistobacter, Polaromonas, Xanthomonas, Stenotrophomonas, Rhodobacter, Novosphingobium*

the sample genera, followed by members of the family Aeromonadaceae. *Burkholderia*, *Achromobacter*, *Thauera*, *Pseudomonas*, *Acinetobacter*, *Comamonas*, and *Enterobacter* were among the noticeable genera.

The most abundant pathways were related to ABC transporters, a two-component system, carbon metabolism, and amino acid biosynthesis. The CARD database was used for the identification of ARGs. The genes conferring resistance against antibiotics such as macrolides, β-lactams, chloramphenicol, aminocoumarin, tetracycline, lipopeptides, polymyxins, elfamycin, lincosamides, aminoglycosides, isoniazid, and trimethoprim were also found. The BacMet database was used to identify genes related to metal resistance. Cu-, Ni-, Zn-, Hg-, Co-, and Ar-resistant genes were found to be abundant (Mittal et al., 2019).

5.4 DISCUSSIONS AND CONCLUSIONS

Industrial waste is an agglomeration of different substrates, thereby making it an ideal environment for the growth and multiplication of numerous microorganisms. Industrial discharges are a significant contributor to water pollution and have the ability to change the microbial ecosphere. There are a lot of publications on the microbiological quality of household municipal effluents and their alleged consequences on the environment, but there is a knowledge void about the contribution of individual industrial waste disposal into the receiving water bodies. Ample evidence on the adaptability of microbes to toxic environments has necessitated the researchers to assess microbial composition in different habitats, including community solid waste dumpsites.

A comparative study of unique and common bacterial phyla reflected that members of the phylum Proteobacteria and Firmicutes were common across all sites under our study. Whereas 65.6% of bacterial genera came out to be unique in Jajmau and Unnao, which is in case study site 1. Opportunistic bacterial pathogens like *Burkholderia*, *Flavobacterium*, and *Pseudomonas* were commonly abundant in the wastewater samples from Jajmau and Unnao (Case study site 1) and the Yamuna River (Case study 3). These pathogens are known to cause multiple serious illnesses like respiratory troubles, pneumonia, different hospital-acquired infections, columnar diseases, etc. So, when considering human health, the risks posed by pathogenic microorganisms living at industrial waste dump sites cannot be understated. On the other hand, microbial communities inhabiting different effluents play a role in wastewater treatment. For example, metabolic processes conducted by denitrifying, iron-oxidizing, and sulfur-reducing bacteria help treat wastewater worldwide. Similar kinds of results have been recorded from our selected study sites. Sulfate, nitrate, and nitrite-reducing bacteria *Desulfovibrio* and *Anaerolinea*, known for bioremediation, are prevalent in the soil samples receiving dumping from the tannery industry and paper mills that are in case study sites 1 and 2, respectively (Figure 5.5).

Therefore, industrial waste dump sites can act as potential sources for the isolation, characterization, and identification of a wide spectrum of microbial communities. In this context, culture-independent techniques have been proven to score better not only in identifying the taxonomic fingerprint of microbial communities but also in other parameters including data volume, data prediction ability, data generation time, and expenses. Metagenomics can be employed to understand the identities and functional diversities of the microbial populations thriving in such waste dump sites. Tools and methods based on metagenomics give essential knowledge on the environmental microbes involved in biogeochemical cycles, which aid in the breakdown and detoxification of environmental pollutants. Agenda 21 under the Sustainable Development Goals, which deals with the handling of solid wastes and effluents, has highlighted the need for regulated waste dumping, waste recycling, and bioremediation for a healthier environment. In this context, it is very important to identify the dynamics of the various interactions among the microbial populations in response to environmental pollution. This will aid the researchers in assessing the efficiency of bioremediation techniques in light of the ongoing pollutant pressure in the environment and ecosystem.

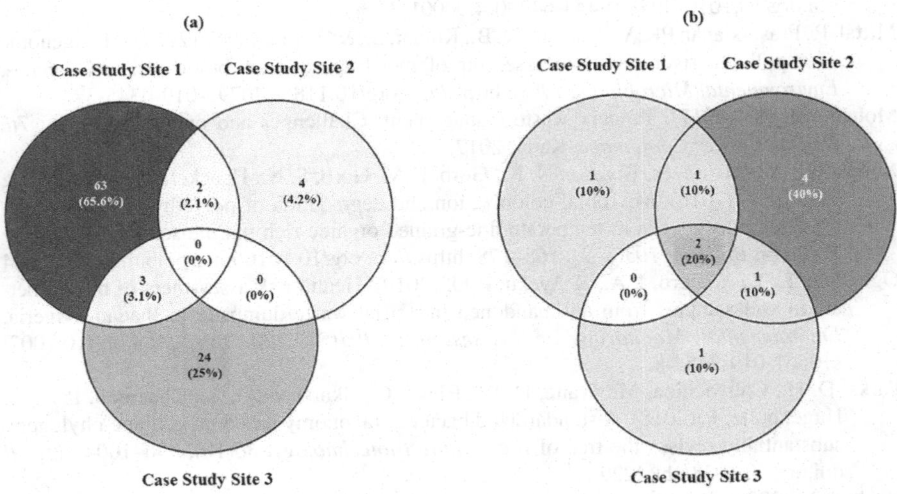

FIGURE 5.5 Comparative account of unique and common bacterial members (a) at Genus level (b) at Phylum level across the three study sites.

ACKNOWLEDGEMENT

The authors acknowledge the funding provided by the DBT Grant (BT/INF/22/SP41296/2020) from the Department of Biotechnology, Government of India and Intramural Research Grant (IMSXC2022-23/010_003) provided by St. Xavier's College (Autonomous), Kolkata.

REFERENCES

Abiriga, D., Jenkins, A., Alfsnes, K., Vestgarden, L. S., & Klempe, H. (2021). Characterisation of the bacterial microbiota of a landfill-contaminated confined aquifer undergoing intrinsic remediation. *Science of the Total Environment, 785*, 147349. https://doi.org/10.1016/j.scitotenv.2021.147349

Bajpai, P. (2015). *Green Chemistry and Sustainability in Pulp and Paper Industry*. Springer, Cham. https://doi.org/10.1007/978-3-319-18744-0.

CPCB (2019). *Latest Data on Dumpsites*. Central Pollution Control Board, Delhi.

Chadar, S. N., & Chadar, K. (2017). Solid waste pollution: A hazard to environment. *Recent Advances in Petrochemical Science, 2*(3), 555586. https://doi.org/10.19080/RAPSCI.2017.02.555586.

Fazzo, L., Minichilli, F., Santoro, M., Ceccarini, A., Della Seta, M., Bianchi, F., Comba, P., & Martuzzi, M. (2017). Hazardous waste and health impact: A systematic review of the scientific literature. *Environmental Health: A Global Access Science Source, 16*(1), 107. https://doi.org/10.1186/s12940-017-0311-8

Kumar, R., Pandit, P., Kumar, D., Patel, Z., Pandya, L., Kumar, M., Joshi, C., & Joshi, M. (2021). Landfill microbiome harbour plastic degrading genes: A metagenomic study of solid waste dumping site of Gujarat, India. *The Science of the Total Environment, 779*, 146184. https://doi.org/10.1016/j.scitotenv.2021.146184

Millati, R., Cahyono, R. B., Ariyanto, T., Azzahrani, I. N., Putri, R. U., & Taherzadeh, M. J. (2019). Agricultural, industrial, municipal, and forest wastes: an overview. In: *Sustainable Resource Recovery and Zero Waste Approaches*. Elsevier, pp: 1–22. https://doi.org/10.1016/B978-0-444-64200-4.00001-3.

Mittal, P., Prasoodanan PK, V., Dhakan, D. B., Kumar, S., & Sharma, V. K. (2019). Metagenome of a polluted river reveals a reservoir of metabolic and antibiotic resistance genes. *Environmental Microbiome, 14*(5). https://doi.org/10.1186/s40793-019-0345-3

Mohammed, K. (2017). Tannery waste management: Challenges and opportunities. In: *7th National Water Conference,* Kano, 2017.

Nauendorf, A., Krause, S., Bigalke, N. K., Gorb, E. V., Gorb, S. N., Haeckel, M., Wahl, M., & Treude, T. (2016). Microbial colonization and degradation of polyethylene and biodegradable plastic bags in temperate fine-grained organic-rich marine sediments. *Marine Pollution Bulletin, 103*(1–2), 168–178. https://doi.org/10.1016/j.marpolbul.2015.12.024

Ogundele, L. T., Adejoro, I. A., & Ayeku, P. O. (2019). Health risk assessment of heavy metals in soil samples from an abandoned industrial waste dumpsite in Ibadan, Nigeria. *Environmental Monitoring and Assessment, 191*(5), 290. https://doi.org/10.1007/s10661-019-7454-8.

Parks, D. H., Chuvochina, M., Waite, D. W., Rinke, C., Skarshewski, A., Chaumeil, P. A., & Hugenholtz, P. (2018). A standardized bacterial taxonomy based on genome phylogeny substantially revises the tree of life. *Nature Biotechnology, 36*(10), 996–1004. https://doi.org/10.1038/nbt.4229

Rushton L. (2003). Health hazards and waste management. *British Medical Bulletin, 68,* 183–197. https://doi.org/10.1093/bmb/ldg034.

Selvi, A., Rajasekar, A., Theerthagiri, J., Ananthaselvam, A., Sathishkumar, K., Madhavan, J., & Rahman, P. K. (2019). Integrated remediation processes toward heavy metal removal/recovery from various environments a review. *Frontiers in Environmental Science*, 7, 66.

Selvarajan, R., Ogola, H., Kalu, C. M., Sibanda, T., & Obize, C. (2022). Bacterial communities in informal dump sites: A rich source of unique diversity and functional potential for bioremediation applications. *Applied Sciences, 12*(24), 12862. https://doi.org/10.3390/app122412862

Sharma, P., Gaur, V. K., Gupta, S., Varjani, S., Pandey, A., Gnansounou, E., You, S., Ngo, H. H., & Wong, J. (2022). Trends in mitigation of industrial waste: Global health hazards, environmental implications and waste derived economy for environmental sustainability. *The Science of the Total Environment, 811*, 152357. https://doi.org/10.1016/j.scitotenv.2021.152357

Sharma, P., Tripathi, S., & Chandra, R. (2021a). Highly efficient phytoremediation potential of metal and metalloids from the pulp paper industry waste employing *Eclipta alba* (L) and *Alternanthera philoxeroide* (L): Biosorption and pollution reduction. *Bioresource Technology*, *319*, 124147. https://doi.org/10.1016/j.biortech.2020.124147

Sharma, P., Tripathi, S., & Chandra, R. (2021b). Metagenomic analysis for profiling of microbial communities and tolerance in metal-polluted pulp and paper industry wastewater. *Bioresource Technology*, *324*, 124681. https://doi.org/10.1016/j.biortech.2021.124681

Sogin, M. L., Morrison, H. G., Huber, J. A., Mark Welch, D., Huse, S. M., Neal, P. R., Arrieta, J. M., & Herndl, G. J. (2006). Microbial diversity in the deep sea and the underexplored "rare biosphere". *Proceedings of the National Academy of Sciences of the United States of America*, *103*(32), 12115–12120. https://doi.org/10.1073/pnas.0605127103

Thai, P. K., Ky, L. X., Binh, V. N., Nhung, P. H., Nhan, P. T., Hieu, N. Q., Dang, N., Tam, N., & Anh, N. (2018). Occurrence of antibiotic residues and antibiotic-resistant bacteria in effluents of pharmaceutical manufacturers and other sources around Hanoi, Vietnam. *The Science of the Total Environment*, *645*, 393–400. https://doi.org/10.1016/j.scitotenv.2018.07.126

Thomas, T., Gilbert, J., & Meyer, F. (2012). Metagenomics - a guide from sampling to data analysis. *Microbial Informatics and Experimentation*, 2(1), 3. https://doi.org/10.1186/2042-5783-2-3

Verma, S. K., & Sharma, P. C. (2020). NGS-based characterization of microbial diversity and functional profiling of solid tannery waste metagenomes. *Genomics*, *112*(4), 2903–2913. https://doi.org/10.1016/j.ygeno.2020.04.002

Wan S, Xia M, Tao J, Pang Y, Yu F, Wu J, & Chen S. (2021). Metagenomics analysis reveals the microbial communities, antimicrobial resistance gene diversity and potential pathogen transmission risk of two different landfills in China. *Diversity*, *13*(6), 230. https://doi.org/10.3390/d13060230

Yadav, B., Johri, A. K., & Dua, M. (2020). Metagenomic analysis of the microbial diversity in solid waste from Okhla landfill, New Delhi, India. *Microbiology Resource Announcements*, *9*(46), e00921-20. https://doi.org/10.1128/MRA.00921-20

Zainun, M. Y., & Simarani, K. (2018). Metagenomics profiling for assessing microbial diversity in both active and closed landfills. *The Science of the Total Environment, 616–617*, 269–278.

Zothanpuia, Zomuansangi, R., Leo, V. V., Passari, A. K., Yadav, M. K., & Singh, B. P. (2021). Antimicrobial sensitivity profiling of bacterial communities recovered from effluents of municipal solid waste dumping site. *3 Biotech, 11*(2), 37. https://doi.org/10.1007/s13205-020-02548-z

6 Application of Microbes in Leaching Rare Earth Elements and Radioactive Elements

Bioleaching of Rare Earth Elements: A Challenge and Success

Madhumita Majumder
Raidighi College

Anjalika Roy
Visva-Bharti, Santiniketan

6.1 INTRODUCTION

The definition of rare earth elements (REEs) has evolved throughout time. REEs are a group of 15 elements that range from lanthanum to lutetium (atomic numbers 57–71). These elements, sometimes known as lanthanoids, are frequently discussed together due to their chemical and physical similarities (Theodore, 2009). Although yttrium (Y) and scandium (Sc) are not lanthanoids, they have certain similarities. As a result, REEs are now classified as lanthanoids that include Y and Sc. The estimated average REE concentration in the Earth's crust ranges from 130 to 240 mg/g, which is substantially higher than other commonly mined elements and their corresponding chondritic abundances (Dushyantha et al., 2020). A precise classification that is routinely used in extraction is as follows (Australian Industry Commission, 1995):

- Light rare earth elements (LREEs) include lanthanum (La), cerium (Ce), praseodymium (Pr), neodymium (Nd), and promethium (Pr) (Pm).
- Medium rare earth elements (MREEs) include samarium (Sm), europium (Eu), and gadolinium (Gd).
- Heavy rare earth elements (HREEs) include terbium (Tb), dysprosium (Dy), holmium (Ho), erbium (Er), thulium (Tm), ytterbium (Yb), lutetium (Lu), scandium (Sc), and yttrium (Yt) (Y).

DOI: 10.1201/9781003451457-6

Due to its similarity in numerous characteristics, yttrium is included in the HREE. Rare-earth elements (REEs) have grown in importance over the last century as their electrochemical, magnetic, alloy strengthening, and luminescent properties have become increasingly vital in contemporary electric autos, renewable energy generation, electronics, and high-performance aircraft. This is reflected in the worldwide rare-earth metal market, which is predicted to rise at a compound annual growth rate of 13.7% between 2017 and 2021 (Ramprasad et al., 2022). They offer tremendous financial potential as well as scientific advancements in the realm of green energy. These advantages, however, come at a hefty social and environmental price, such as unlawful mining, which results in environmental devastation.

Several articles on rare earth elements appeared in the media, and the world was terrified that China was about to destroy Western countries high-tech industries due to export restrictions. At the time, China produced 95% of the world's REEs. Only a few major firms trade REEs, and they are sent to only a few countries: China, Canada, Australia, the United States, Russia, India, and Japan. The world's greatest supplier of REEs is China, whereas the leading importers are the United States, Japan, and Germany (Roskill, 2015). REEs may be found in phosphates, carbonates, fluorides, and silicates and grow one ore body at a time. REEs are present in nature as part of the chemistry of the host material rather than as native elemental metals. As a result, recovering REMs requires advanced processing processes to chemically break down the minerals that contain the REEs. Despite the fact that over 200 REE-bearing minerals have been found, bastnäsite [(Ce, La) (CO_3) F], xenotime (YPO_4), and monazite [(Ce, La) PO_4) are the most feasible REE mineral ores for REM extraction. (Tyler 2004).

When alloyed or mixed with more common metals like iron in tiny proportions, these metals exhibit exceptional luminous, conductive, and magnetic properties, making them exceedingly valuable. These metals are abundant in diverse parts of the planet, with some elements in the earth's crust occurring in about equal amounts as copper or tin. Rare earths, on the other hand, are rarely found in substantial numbers and are frequently combined with other elements or radioactive elements such as uranium and thorium. The ore was given the designations "rare" and "earth" since it had never been seen before, and "earth" was the word used in 18th-century geology for acid-soluble rocks. After the site where it was found in 1794, a scientist dubbed this hitherto unknown "earth" Yttria. Mines in the Ytterby area mined rocks throughout time, resulting in four elements named after the town (yttrium, ytterbium, terbium, and erbium) (Science History Institute, 2022). Nuclear characteristics vary widely across REEs, and REEs have a wide spectrum of nuclear, metallurgical, chemical, catalytic, electrical, magnetic, and optical properties. This has resulted in an increase in applications like lighter flints, glass polishing, catalysts in the petroleum refining sector, alloying agents (used to improve the oxidation resistance of alloys) in metallurgical processes, optics, permanent magnets, and electronics, to name a few. Rare earths are being used to absorb UV radiation in vehicle glass, corrosion prevention, and metal coatings in corrosive and salty settings. High-temperature superconductivity, safe hydrogen storage, and transport are futuristic applications of REE for a post hydrocarbon economy (Haxel, 2002).

Bastnasite (REE-FCO_3), monazite (light REE-PO_4), and xenotime (heavy REE-PO_4) are the three primary REE ores being mined for production, accounting for over 95% of all known REE minerals (Gupta and Krishnamurthy, 1992; Rosenblum and Fleischer, 1995). In addition to REE-PO_4, monazite ore typically includes Th and, on rare occasions, U, both of which are radioactive and provide a separation and disposal issue when extracted with REEs (Gupta and Krishnamurthy, 1992). Chemical leaching at high temperatures (140°C) with either concentrated sulfuric acid or concentrated NaOH, followed by acid treatment, is the traditional technique of extracting REE from ores (Gupta and Krishnamurthy, 1992). This process solubilizes metal sulfide as metal cations in the leaching medium in ore/minerals. Bioleaching has the potential to be more ecologically friendly than traditional mineral extraction. Bioleaching of REEs has been studied in a few studies, including one on *Penicillium* bioleaching in red mud (Qu and Lian, 2013) and two recent studies on bioleaching of monazite concentrate (Hassanien et al., 2014) and monazite-containing ore (Hassanien et al., 2014). The bioleaching method involves the use of microorganisms to remove metals from low-grade ores. Large amounts of low-grade ores are often thrown in trash heaps during the separation of higher-grade ores (Shin et al., 2015). Chemical processes cannot profitably treat metals from such ores. There are enormous amounts of low-grade ores, particularly copper ores, that may be economically treated using bio-leaching. Several metals, including copper, zinc, uranium, cobalt, nickel, and others, have been effectively dissolved using the bioleaching technique. Due to the scarcity of natural resources, researchers have shifted their emphasis to secondary solid waste bioleaching technology. Various secondary solid wastes created by industrial and mining operations are dangerous to biodiversity and pose a number of concerns. Ignitibility, corrosiveness, reactivity, and toxicity are all harmful traits that these solid wastes might have (Devi et al. 2018). In comparison to other processes like hydrometallurgy and pyrometallurgy, bioleaching has proven to be a cost-effective and environmentally benign method.

This review aims to provide an updated understanding of the global REE scenario, beginning with their applications in high-tech products, occurrence, various types of economic deposits on land and in the oceans, behavior in various geological systems, state-of-the-art chemical characterization techniques, and recycling. Other themes covered include their use in agriculture and medicine, as well as their environmental effects and the search for REE substitutes in diverse applications.

6.2 HISTORY OF REEs

Historical records state that ancient civilizations in Asia (100–200 years BCE) and Europe (around 200 years CE) used oxidizing bacteria to recover metals from mineral ores (Ehrlich, 2001). Nevertheless, the first formal industrial application of the technology did not appear until the late 1950s, when Zimmerley and colleagues patented the first heap-leaching process to leach metals from, among others, sulfide-bearing pyrite and oxide-sulfide copper (Cu) ores (Ehrlich, 2001). Earth's crust accounts for 169.1 ppm of the total abundance of REEs, including 137.8 ppm of light rare earth elements and 31.3 ppm of HREEs (Dushyantha et al., 2020). Therefore, in this context, the word 'rare' essentially implies that REEs are not commonly concentrated as economic ore deposits, though viable deposits are distributed across the world in relatively low concentrations (Haxel et al., 2002). Voncken (2016) mentioned that actually "Rare Earth Element" is a term used in the 19th century and is based on the discovery of the only REE deposit in the world at that time, namely in the Ytterby region of Sweden, thus this deposit is considered something rare. This country became the first to discover REE. The International Union of Pure and Applied Chemistry (IUPAC) includes yttrium (Y), scandium (Sc), and 15 metallic elements in a group of REEs. Among them, Promethium (Pm) is the only rare earth element that does not occur naturally and was discovered due to a nuclear reaction (De Lima et al. 2015). REE processing from ores into purified form is mostly concentrated in the Asia-Pacific region, especially China, which produces 90% of current world demand. Thus, REE was used as a powerful tool by China in its trade war against other countries, especially the USA, in 2019 (Dushyantha et al., 2020). New applications of REE are opening challenges for its manufacturing and global marketing. The governments curtail restrictions for the promotion of global trade and shipping all around the world.

China took control over RE exports in 2010, stating the reason to build up a stockpile of REE and reduce environmental erosion, which ultimately hiked the prices of REE as well as the anxiety of stakeholders from 2010 to 2013. Therefore, this duration is well known as the Rare Earth Crisis, and China portrays the monoplastic grip on REEs (Voncken, 2016) as global suppliers and consumers. The 2010 crisis forced the car industry to design vehicles with the minimum use of rare earths and repair or reuse consumer electronics rather than discard them. REOs production in China, for example, increased from 73 thousand tonnes to 120 thousand tonnes from 2000 to 2018 (United States Geological Survey (USGS), 2018). The Chinese ministry of industry and information technology took initiatives in October 2016 to restrict illegal RE production, transaction, and processing in China and address other internal issues, such as overproduction, less innovative measures, low efficiency, and lax environmental standards in the RE industry (Ilankoon et al., 2018; Mancheri et al., 2019). Despite their relative abundance in the Earth's crust, REE resources with minable concentrations are less common. At the present time, about 850 minable REE deposits have been identified in a few locations, which are mainly in China, Vietnam, Brazil, Russia, India, and Australia (Zhou et al., 2017).

6.3 PROSPECTIVE ADVANTAGES OF BIOLEACHING OVER CHEMICAL LEACHING

The development of bioleaching techniques plays a key role in the sustainable mining of rare earth elements, ultimately resulting in a green economy. Chemical leaching of REE employs some corrosive reagents that have harmful effects on the environment. Several stages of the RE industry, such as mining, processing, transportation, waste disposal, and decomposition, always give evidence of occupational and environmental exposures to REEs (Rim, 2017). Pollution due to the chemical reagents, such as sulfates, oxalates, and extract solvents used in the extraction, processing, and recycling of REEs, is another aspect of the environmental issue if those are discharged to the environment without any pre-treatments (De Baar et al., 1985; Ali, 2014). Bioleaching avoids the emission of poisonous sulfur dioxide, which is harmful to the environment and the health of miners. The use of fossil fuels causes high emissions of carbon by consuming a lot of energy in the course of RE mining and processing, which has serious environmental implications as well as limited stock. For example, the production of 1 kg of Sm-Eu-Gd oxide emits about 55 kg of CO_2 (Weng et al., 2013). A better future on this globe can be achieved by adopting the United Nations Sustainable Development Goals of responsible production and consumption for the rare earth industry. Bioleaching by autotrophic and heterotrophic microorganisms maximizes the outputs and minimizes residues and environmental hazards.

The extraction process of REE from ores has a definite sequence, like roasting at different temperatures and using acidic and alkaline leaching solutions depending on the mineralogy for refining. Other physico-chemical methods used for recovery and separation of REE include electrowining, ion exchange, fractional crystallization, and solvent extraction (Kronholm et al., 2013). These methods have several disadvantages, i.e., high acid/base consumption, long residence times, the requirement of high temperatures, and a high cost due to process complexity (Walawalkar, 2016). Nowadays, biotechnological approaches for the recovery or mining of REEs involve implementing phosphate-solubilizing bacteria in fusion with waste electrical and electronic equipment, making it cost-efficient and environmentally friendly. Usually, bioleaching has a lower yield rate compared to chemical leaching. However, bioleaching of the Fluid Catalytic cracking (FCC) catalyst using *Gluconobacter oxydans* spent medium containing 10–15 mM gluconic acid was more effective. Barnett et al. (2018) compared bioleaching of REEs from anion-adsorption clay using *Aspergillus* and *Bacillus* cultures to leaching with 0.5 M ammonium sulfate, demonstrating that bioleaching improved Ce extraction and maximum REE yields were observed between 10 and 60 days of bioleaching, presumably caused by the differing solubilization mechanisms. It has been reported that three microbial strains, *Candida bombicola, Phanerochaete chrysoporium,* and *Cryptococcus curvatus,* have high extracting efficiencies of REE from coal fly ash, which is much more feasible than chemical leaching (Park and Liang, 2019). In Russia, incubation of Coal ash with an acidophilic chemolithotropic microbial community recovered 52% Se, 52.6% Y, and 59.5% La (Muravyov et al., 2015). Egyptian monazite mining by two different microbes at different temperatures showed some variations in results, like *Aspergillus ficuum* at 30°C yielding 75.4% and *Pseudomonas aeruginosa* at 30°C yielding 63.5%.

6.3.1 Principal Facts of Bioleaching

1. The microbes attached themselves to the surface of the mineral by their growth mediated by extracellular polymeric substances, i.e., sugars, lipids, proteins, or nucleic acids, of surrounding cells.
2. After proper interaction of microbes with minerals, they secrete organic acids (mostly citric, gluconic, and oxalic), amino acids, or enzyme complexes.

Organic acids promote mineral dissolution by donating H^+ to the proton-donating dissolution process, forming inner-sphere surface complexes that dislodge structural metals from the mineral surface. The formation of an aqueous metal ligand reduces the relative solution saturation of minerals undergoing dissolution (Goyne et al., 2010). Thus, Bioleaching is preferred for its having promising results due to several main reasons:

1. Environment friendly
2. Low energy input
3. Low chemical cost
4. Waste minimization
5. Resource recovery

6.4 PROCESS OF BIOLEACHING

For extracting REEs, the application of the biohydrometallurgy method gives a green alternative to the conventional methods, which are complex and energy-intensive. Bastnaesite, loparite, monazite, xenotime, and the lateric ion-adsorption clays are the main economic REE-bearing minerals. Autotrophic and heterotrophic microorganisms are equally capable of solubilizing REEs, and the selection of these microbes for bioleaching processes depends on the type of mineral. Autotrophic bacteria have been applied for the extraction of scandium from ore minerals containing metal sulfides, whereas heterotrophic bacteria are mostly employed for REE carbonates and phosphates (Rasoulnia et al., 2021). Bastnaesite is a rare earth fluorocarbonate ore (REE (CO_3) F) commonly containing cerium, lanthanum, or yttrium. A different technique involves a thermal activation stage at 400°C followed by HCl leaching, which reduces fluorine but not carbonate release (Peelman et al., 2016). The bioleaching of REEs has been developed, especially on monazite, using microorganisms able to dissolve phosphorous from inorganic rocks, named phosphate solubilizing microorganisms (Figure 6.2). Several organisms, like bacteria, fungi, actinomycetes, and algae, mobilize insoluble phosphorous. Various species like *Bacillus*, *Pseudomonas*, *Burkholderia*, *Enterobacter* or *Azotobacter*, and fungi like *Aspergillus*, *Penicillium*, *Trichoderma* or *Rhizoctonia* are reported as phosphate solubilizers (Prabhu et al., 2019). Phosphate solubilizing microorganisms have been generally used as biofertilizers in agriculture to increase crop production (Alori et al., 2017), but few reports that are related to the recovery of valuable metals from phosphate minerals are available (Delvasto et al., 2008; Shin et al., 2015).

The process of bioleaching is conducted in two ways:

i. Direct bioleaching
ii. Indirect bioleaching

(i) Direct Bioleaching

Thiobacillus ferrooxidans is an autotrophic, aerobic, gram-negative rod-shaped bacterium generally used in microbial leaching. Through the process of CO_2 fixation, it synthesizes its carbon substances. It derives the required energy for CO_2 fixation

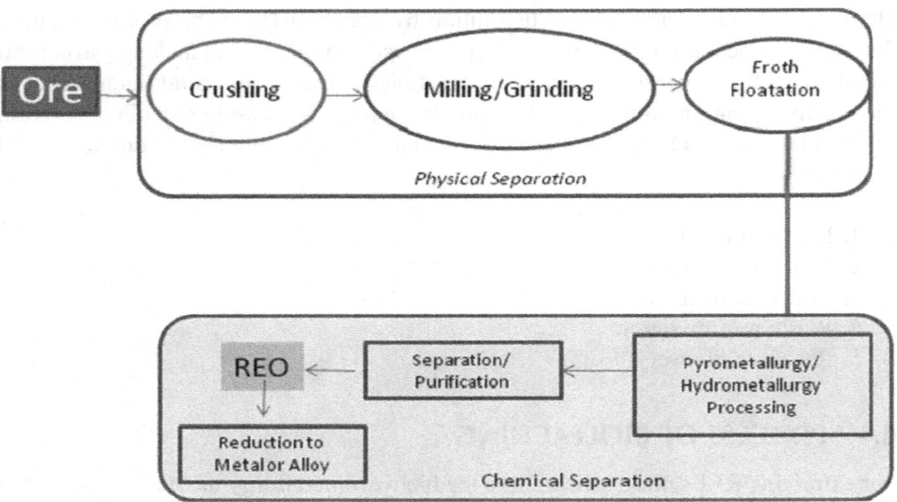

FIGURE 6.1 Diagrammatic representation of the process of bioleaching from mineral ore.

either from the oxidation of Fe^{2+} to Fe^{3+} or from the oxidation of elemental sulfur or reduced sulfur compounds to sulfates.

(ii) Indirect Bioleaching

This process takes place without the direct involvement of microorganisms, but they indirectly support the leaching by producing agents responsible for the oxidation of minerals. It can be described by the oxidation of pyrite. *Thiobacillus thiooxidans* and *Thiobacillus ferrooxidans* are usually seen as being associated with leaching dumps. In pilot plant reactors, leaching can be done continuously in a cascade series with recycling of cells and leachates (Figure 6.1).

6.4.1 Bioleaching from Mineral Ores

The bioleaching approach now proves to be an economical and eco-friendly process compared to alternative processes such as hydrometallurgy and pyrometallurgy. The bioleaching process can be defined as the extraction of metals from low-grade ores by employing microorganisms. Productions of large amounts of low-grade ores during the separation of higher-grade ores are generally discarded in waste heaps. Metals from such ores cannot be economically processed with chemical methods. There are a lot of low-grade ores, notably copper ores, that can be economically treated via bioleaching. These technological challenges are most likely due to the lack of biological methods for detecting REEs leaking from xenotime. Monazite is a phosphate mineral that is the primary source of cerium for industry. For monazite dissolution, alkali treatment is utilized, and the phosphate is recovered as a marketable by-product, trisodium phosphate, by employing caustic soda at high temperatures and pressure.

Among REE-bearing industrial residues, phosphogypsum is worthy of mention. REEs are frequently associated with phosphate deposits, and phosphogypsum wastes are produced in large quantities (100–280 Mt per year) during the wet phosphoric acid process from fertilizer production, with an estimated 21 Mt of REEs locked into the total of phosphogypsum wastes accumulated to date (Habashi, 2013). For the leaching of synthetic phosphogypsum doped with six rare earth elements, a biolixiviant generated by the bacteria *Gluconobacter oxydans* on glucose containing organic acids (yttrium, cerium, neodymium, samarium, europium, and ytterbium) was utilized. The primary organic acid present in the bioreagent was gluconic acid, and the pH of the lixiviant created by the bacteria was 2.1 (220 mM). The biolixiviant's REEs leaching yield from phosphogypsum was compared to REEs leaching results from sulfuric acid, phosphoric acid, and commercial gluconic acid. *Gluconobacter oxydans* lixiviant was more efficient in dissolving REEs than gluconic acid and phosphoric acid but less efficient than sulfuric acid (Antonick et al., 2019).

6.4.2 REE Mobilization from Solids via Biological Processes

Alternatives to physico-chemically based techniques of REE recycling include biohydrometallurgical technology. These methods, known as "bioleaching," are well-known in the mining sector and are particularly well-suited to low elemental concentrations in the materials of interest, where traditional metal recovery techniques are not cost-effective (Brandl & Faramarzi, 2006; Zhuang et al., 2015). Since the inclusion of Fe^{2+} was thought to promote bacterial tolerance to high levels of metals, leaching efficiencies were investigated in the presence and absence of ferric iron. Experiments revealed that the presence of such components inhibited bacterial development, resulting in less efficient leaching. This unfavorable impact of high powder dosage was also shown while treating CRT powder, indicating either the toxicity of CRT powder.

The leaching process was mostly caused by gluconic acid produced by bacteria (Glombitza, 2013). Cell-free tests using gluconic acid as the lixiviant yielded just 4% leaching efficiency, demonstrating that microbial metabolic activity is significant in bioleaching. A distinction between light and heavy REE was made during microbial leaching. The mobilization of light REE was preferred over heavy REE. In the absence of the microorganism, the interaction of zircon with gluconic acid revealed no distinction (Becker et al., 1986). *Acetobacter methanolicus* treatment of phosphorus furnace slag (containing 7 g of REE per kg) resulted in REE rates of up to 70% (Glombitza & Reichel, 2014; Iske et al., 1987).

A more extensive investigation on REE extraction from zircon found mobilization efficiencies of over 80% with extraction rates of 1.1 mg per hour using *Acidithiobacillus ferrooxidans* and 67% with extraction rates of 1.4 mg/hour using *Acetobacter methanolicus*. However, autotrophic *A. ferrooxidans* and heterotrophic *A. methanolicus* had major mechanistic differences in terms of recovery efficiencies. The majority of the mobilized REE was absorbed by the biomass of *A. ferrooxidans*, with just a little amount remaining in the culture fluid. The sorption of praseodymium onto *Acidithiobacillus* biomass was proven in minutes using praseodymium as a model chemical. In the culture fluids of *A. methanolicus*, however, higher levels of REE were observed, indicating that biosorption was less significant. Anaerobic bacteria have also been used to extract REE from solid materials. Sulfate-reducing

Desulfovibrio desulfuricans mobilized yttrium from phosphogypsum (a by-product of fertilizer manufacture) in a fixed-bed reactor with efficiencies of about 80% (Dudeney & Sbai, 1993). In addition, parameters impacting leaching efficiency were studied, including incubation length (1–7 days), sulfur addition (0.1%–0.5%), and mineral pre-treatment (sterilized or nonsterilized).

Mobilization increased gradually as reaction time increased, reaching a peak after 6 days. Extending the incubation period did not increase efficiency. Sulfur addition was shown to boost bacterial acid production, which had a beneficial impact on leaching efficiency. Pre-sterilization of the mineral, on the other hand, reduced leaching efficiency when compared to nonsterilized material, most likely owing to the sterilization process removing endogenous microorganisms. Total REE mobilization was 67.6% after optimization. Fungi can also help microbial REE mobilization in solid matrices. *Penicillium tricolour* (Qu & Lian, 2013) was used to remediate red mud, a waste product of bauxite processing in aluminium mining operations. Three different bioleaching processes were used: one-step bioleaching (fungal growth in the presence of sterilized red mud), two-step bioleaching (pre-cultivation of microorganisms and biomass production followed by sterilization of red mud), and cell-free spent medium (cultivation of the microorganisms followed by filtration to obtain cell-free medium, which was then added to sterilized red mud). The two-step bioleaching procedure has been demonstrated to be more appropriate for leaching red mud at high pulp densities. When looking at individual REE, the leaching efficiency of heavy REE was obviously higher than that of light REE.

Actinomycetes (including *Streptomyces fungicides*, *Streptomyces aureofaciens*, and *Streptomyces chibaensis*) have been employed to mobilize REE from sandy and silty soil samples. REE leaching efficiency of up to 37% was reported in suspensions of 10 g soil/L (cultivated for 48 hours at 30°C), depending on the strain used. For REE mobilization, mixed microbial cultures have lately been used instead of pure cultures. Acidophilic and chemolithotrophic microbial communities leached coal-derived ash-slag waste from a thermal generating plant's heaps. The underlying leaching process of metal mobilization might be either contact or noncontact bioleaching (also known as direct and indirect bioleaching). Noncontact leaching defines the separation of biomass from the material to be treated, whereas contact leaching discusses the direct physical interaction between microbes and a solid. Several species of *Aspergillus* and *Penicillium* were used to assess the efficiency of REE leaching from carbonaceous shale powder (Amin et al., 2014). Overall, *Aspergillus* sp. outperformed *Penicillium* sp. in terms of bioleaching, with *A. flavus* and *A. niger* being the most effective. Aside from a variety of solid REE-containing waste products (as discussed above), native REE-bearing minerals like monazite have also been investigated as a bioleaching substrate (Brisson et al., 2016). Three fungal species (*Aspergillus niger*, *Aspergillus terreus*, and *Paecilomyces* sp.) were grown on diverse carbon sources and on varied growth mediums. As a result, the culture fluid collected a variety of organic acids such as acetic, citric, gluconic, itaconic, lactic, oxalic, and succinic acid. The presence and concentration of these fungi were determined by the fungal strain utilized. There is evidence that acidophilic sulfur oxidizers may mobilize REE from solids such as ash-slag-waste

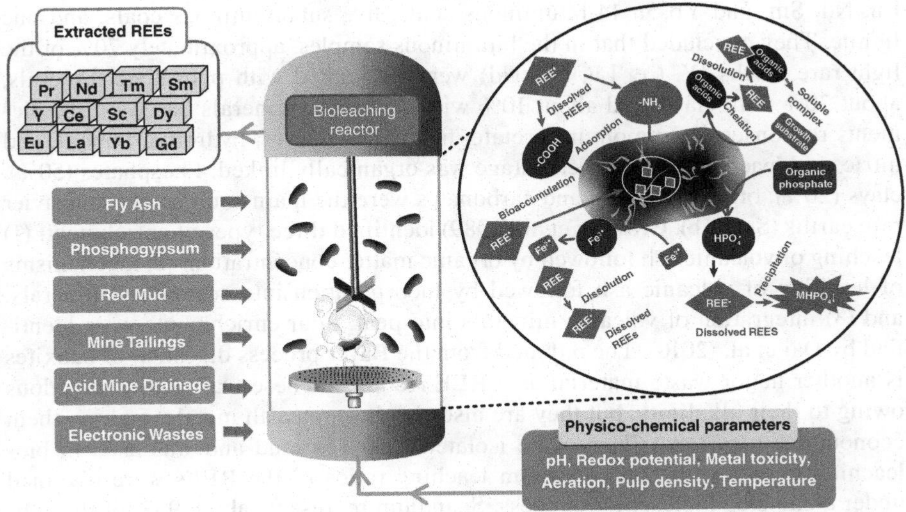

FIGURE 6.2 Bioleaching of solid waste using microorganisms.

to varying degrees. Scandium, lanthanum, cerium, and praseodymium mobilized 15%–20% of their original matrix, whereas neodymium, yttrium, samarium, gadolinium, dysprosium, erbium, and europium mobilized 25%–30% of their original matrix (Tsaplina et al., 2015). It has to be determined if this is related to the solid material's chemical properties or the metabolic preferences of the microbes used. Surprisingly, in the circumstances used, no REE sorption by microbial biomass was detected (Figure 6.2).

6.4.3 Biorecovery of Rare Earth Elements from Coal and By-Products

The genesis of rare earth elements in coal has been the subject of several studies (Birk & White, 1991; Kortenski & Bakardjiev, 1993; Pollock et al., 2000; Schatzel & Stewart, 2003). Eskenazy (1987, 1999) was able to discover organic connections in lignites that were not as visible in bituminous coals investigated elsewhere. Acidic fluids might desorb loosely bound REE on clays, with heavy REE (HREE) being preferentially desorbed, and an increase in HREE in solution would lead to an enrichment in HREE bound to organics. The high organic-bound HREE might have come from high HREE in the waters supplying the marsh as an additional or alternative source (Eskenazy, 1987). Aide and Aide (2012) confirmed that HREE-organic complexes are more stable than light rare earth elements-organic complexes, regardless of peat or coal connections. The stability of the REE-organic compounds decreases as pH decreases (Pédrot et al., 2010; Davranche et al., 2011). Finkelman (1981) calculated that only 10% of the total REE in the lignite was organically related, with the remaining 90% being connected with REE-bearing minerals. Finkelman and Palmer (Unpublished data, USGS) employed selective leaching to identify the modes of occurrence of 37 elements, including Y, Ce, La,

Lu, Nd, Sm, and Yb, in 14 bituminous coals, five subbituminous coals, and one lignite. They concluded that in the bituminous samples, approximately 70% of the light rare earths (Y, Ce, La, and Nd) were associated with phosphate minerals, about 20% with clays, and about 10% with carbonate minerals based on the elements responses to ammonium acetate, hydrochloric acid, hydrofluoric acid, and nitric acid leaches. A lower percentage was organically linked. Phosphates (50%), clays (20%), organics (30%), and carbonates were the main sources of the heavier rare earths (Sm, Yb). Crowley et al. (1989) identified three types of enrichment: (1) leaching of volcanic ash followed by organic matter concentration, (2) mechanisms of leaching of volcanic ash followed by incorporation into secondary minerals, and (3) integration of volcanic minerals into peat. Four enrichments were identified by Dai et al. (2010). The red mud from the Bayer process digestion of bauxites is another major waste material as a REE resource. These wastes are hazardous owing to their alkalinity, but they are also rich in rare earth metals, making them economically treatable. They were isolated from red mud and was used in bioleaching experiments. The optimum leaching ratios of the REEs were obtained under a one-step bioleaching process. Scandium represents about 95% of the economic value of the REEs present in red mud containing between 130 and 390 ppm (Akcil et al., 2018). *Penicillium tricolor*, a filamentous acid-producing fungus, has a pulp density of 2% due to the production of citric and oxalic acids. Nevertheless, the highest extraction yields were achieved under a two-step process at 10% (w/v) pulp density (Qu et al., 2019). The chemoheterotrophic bacterium *Acetobacter* sp. leached 53% of Lu, 61% of Y, and 52% of Sc under a one-step process at 2% pulp density. Furthermore, because the generation of organic acids excreted by the bacteria increased with the waste concentration, this bacterium proved suited for red mud bioleaching at high pulp densities (Qu et al., 2019). Electronic wastes are abandoned gadgets that have outlived their economic usefulness and can no longer be used by end users. In 2021, total worldwide e-waste generation is anticipated to reach 52.2 Mt. E-waste contains other metals in significant amounts that are still worth recovering. Copper extraction uses ferric iron produced by iron-oxidizing bacteria. Gold has been recovered from e-waste using cyanogenic bacteria or fungi. However, these procedures must be scaled and optimized. Recently, a two-step reactor has been developed to separate the production of biogenic ferric iron from the valuable metal leaching reaction, achieving a 96% recovery of Cu (Hubau et al., 2020). Magnets with a significant number of REEs (20%–30%) may be found in many e-wastes. Nd, Dy, and Pr contents in NdFeB magnets are 259.5, 42.1, and 3.4 ppm, respectively (Dev et al., 2020). When *Acidithiobacillus ferrooxidans* and *Leptospirillum ferrooxidans* were grown in the presence of magnets as well as in abiotic controls, they demonstrated excellent leaching efficiencies. Consequently, leaching mainly took place through chemical processes due to the presence of H_2SO_4. Furthermore, biodismantling is a new application of bioleaching in the recycling process of electronic waste to enhance the concentration of critical and precious materials imbedded in the electronic components. Some rare earth elements may become commercially viable by achieving a grade similar to commercial ores, assuming sufficient sorting of the components following separation (Figure 6.3).

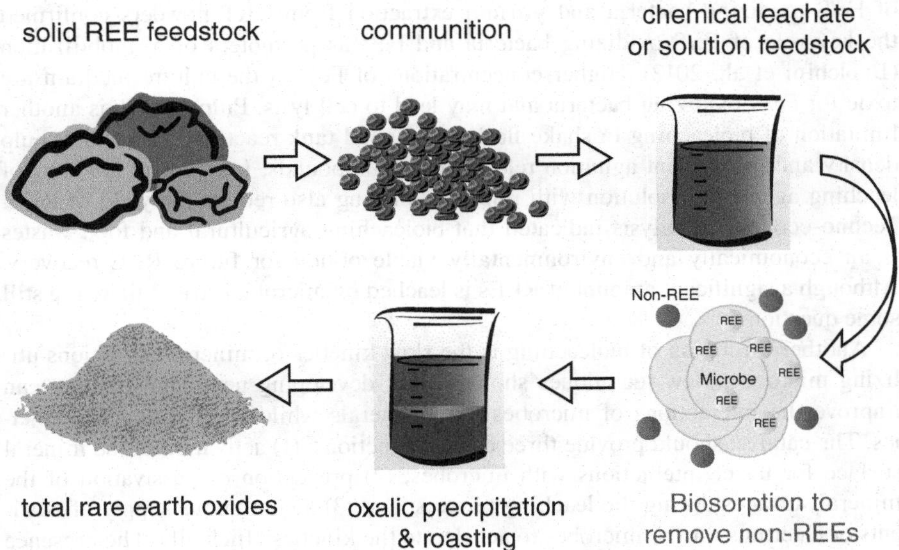

FIGURE 6.3 Biosorption process from coal.

6.5 LIMITATIONS OF BIOLEACHING

In the bioleaching of ionic compounds from REEs like oxides containing WEEE (waste from electrical and electronic equipment), ion-adsorption clays are easier to leach than phosphate and carbonate-bound materials such as maonazite and bastensite. Maintenance or choice of reaction temperature in bioleaching studies depends on the temperature tolerance of the used microorganism(s). The dependency of the microbial growth rate on temperature can be described with the Ratkowsky equation (Ratkowsky et al., 1983), which, in contrast to the Arrhenius equation of reduced reaction rates due to enzyme degradation at higher temperatures (Franzmann et al., 2005), was successfully tested on Fe^{2+} and sulfur oxidation by acidophilic microorganisms. The increased temperature is always detrimental to the growing capacity of employed microorganisms. Identification of the proper pH for the culture of microbial agents used for REE leaching is an important limitation. For example, *Aspergillus niger* at a pH value of ≤2 produces citric acid, while ≥4 produces gluconic and oxalic acids. REE oxalates correspond with lower leaching of REE (Brisson et al., 2016) because pHs act as an important factor for microbial activity as well as microbial leaching. Bioleaching is also effective even at low temperatures.

The design of a proper cultivation medium is a prerequisite for an efficient and successful bioleaching process. Application of acidophilic Fe/S-oxidizing bacteria for bioleaching of REE from nonsulfidic sources is possible. However, the addition of sulfur or pyrite as well as acidification of the cultivation medium are required for optimal microbial growth and metal solubilization (Fathollahzadeh et al., 2018). Assessment of the effect of Fe^{2+}addition on the biological activity of a mixed culture of Fe/S oxidizing bacteria and yttrium extraction from CRT powders confirmed the key role of Fe/S-oxidizing bacteria and Fe^{2+} as promoters of Y mobilization (Beolchini et al., 2012). Higher concentrations of Fe^{3+} in the culture medium are toxic for Fe/S-oxidizing bacteria and may lead to cell lysis. Pulp density is another limitation of bioleaching in shake flask and stirred tank reactors, where high pulp density and insufficient agitation release toxic compounds. Insufficient amounts of leaching agents in a solution with improper mixing also reduce the yield of REE. Techno-economic analysis indicated that bioleaching agricultural and food wastes is an economically and environmentally viable option for future REE recovery. Although a significant amount of REEs is leached by microbial strain, there are still some questions.

Another limitation of bioleaching is the slow kinetics of mineral extractions utilizing microbes. New techniques should target developing new catalysts that can improve the interactions of microbes with minerals while accelerating the kinetics. The catalyst should provide three critical functions: (1) activation of the mineral surface for faster interactions with microbes; (2) prevention of passivation of the mineral surfaces during the leaching process; and (3) a continuous supply of nutrients or electrons for the microbes to accelerate the kinetics efficiently. The presence of hydrocarbons in bioreactors also slows down the production chain by inhibiting bacterial activity. Rapidly developing techniques of bioleaching are restricted only to laboratory scale, with initiative on pilot-scale development.

6.6 HEALTH AND ENVIRONMENTAL HAZARDS OF REE

6.6.1 Effects of REE on Human Health

Despite the global expansion of the RE industries during the last decade, detailed toxicological investigations of REE-induced health problems have been relatively scarce. Exposure of people to RE toxicity can occur in various environments, such as iatrogenic, occupational, and environmental routes. The ingestion of considerable quantities of hazardous materials into the human body has increased as a result of technological developments and modern living conditions, resulting in health concerns. One of the most important concerns in the world today is the contamination of the environment by various harmful inorganic, organic, and organometallic species. Since Gd is used as a contrast agent in magnetic resonance imaging (MRI), anomalously high Gd concentrations in the femoral head bones of patients have been identified due to Gd exposure (Dushyantha et al., 2020). Lung diseases and pneumoconiosis are thus reported among workers in the rare earth industry due to the inhalation of dust containing lanthanides, especially cerium oxide (Sabbioni et al., 1982; McDonald et al., 1995). Bioaccumulation of REEs and pathological changes are reported among local residents who live near RE mining areas. In addition, it is reported that prolonged exposure to Ce is associated with lung pathologies and heart diseases in India during monazite mining and processing (Chakhmouradian and Wall, 2012). A high level of radiation has been reported in monazite mining and storage areas in India, and it has been eventually linked to different human health hazards (Padmanabhan, 2002). The genetic damage emanating from exposure to REEs is irreversible and deleterious consequences for genome stability. The outcome of mutagenic events is fundamental to understanding the extent of the toxicant's pathogenesis on human health, and the microbiological makeup of extreme environments could provide a new angle in understanding the emergence of pathogenesis (Okoh et al., 2018).

Li et al. (2013) indicated that, fortunately, the risk assessment of the consumption of vegetables containing RE should not lead to an excess of the daily intake considered harmful to human health. The highest RE concentration measured in the liver of fattening bulls amounted to 22–482 μg/kg DM for La, 37–719 μg/kg DM for Ce, and 4–73 μg/kg DM for Pr, and the lowest La, Ce, and Pr concentrations were 3–5 μg/kg DM, 5–7 μg/kg DM, and 0.5–0.7 μg/kg DM, respectively. The determination of RE concentrations reporting nontoxic and toxic effects in living organisms varies according to the total concentration of these elements in the environment and among different species and the bodyweight of individuals, which in turn depends on age with short- and long-term exposure. Technological advancements, along with current living conditions, have increased the absorption of harmful materials into the human body, resulting in health issues. The REE group represents key environmental elements that need to be investigated further in order to understand their implications for human health. The harmful effects of well-known hazardous trace elements such as As, Pb, Cd, Hg, and U have been studied by environmental scientists. There have also been multiple cases of REE occupational exposure resulting in bioaccumulation and respiratory tract problems (Sabbioni et al., 1982; McDonald et al., 1995; Yoon et al., 2005; Rim, 2017). Our knowledge of the negative impacts of REE on human health, their anthropogenic quantities and destiny in biogeochemical or anthropogenic cycles, and their individual and combined toxicological effects is still lacking. To reduce future human health hazards, further research is needed to determine anthropogenic origins, transfer pathways, bioaccumulation, and environmental behavior. As a result, the future harmful effects of REE in aquatic systems will be determined by the adoption of new public policies and the development of more effective treatment methods.

6.6.2 Effect of REE on Surroundings

Environmental consequences such as radioactive potential, acidification, eutrophication, solid waste creation, water usage, gross primary energy footprint, toxicity, and any other regional or global impact should all be considered. The majority of REEs, on the other hand, are likely to be utilized in energy conservation, efficiency, and renewable energy technologies. Many environmental issues surround the production and use of REEs. REEs having similar chemical structures are difficult to separate compounded by extremely low yield. Thus, apart from the electricity, acids, water, and resources expended in production, there may be huge amounts of waste that can be toxic and cause potential damage to the ecosystem.

Some REE minerals include high levels of radioactive elements like uranium and thorium, which can pollute the air, water, soil, and groundwater. Inadequate environmental rules and controls in the mining and processing industries are to blame for these issues. The radioactivity of some ores is one of the most serious issues. Despite some information about the environmental risks of REE when they are discharged into the environment coupled with radionuclides, the widespread usage of REE in numerous modern technologies continues to rise. The majority of the adverse effects of REE exposure on people and their possible health impacts come from studies of mine employees and those who interact with REE or its products on a daily basis, where exposure is often far higher than what the general population would encounter. According to certain research, the chemicals employed in the ore processing, extraction, and refining procedures have caused health problems for employees and surrounding populations, as well as water contamination and agricultural degradation (Rim, 2017). The health effects of REE ore processing (from both radioactive and nonradioactive pollution) have recently been mentioned as a serious worry in portions of China. The recent social and environmental controversy surrounding the establishment of the Lynas Advanced Materials Plant in Kuantan, Malaysia, elicited international outrage and accusations of environmental and social injustice (Ali, 2014).

6.6.3 Effects of E-Waste and Fertilizer

Furthermore, pollution from the dumping of massive volumes of e-waste, which releases substantial amounts of REE into the subsoils and groundwater, is rapidly growing (Haxel et al., 2002). The electronics sector produces up to 41 million tonnes of e-waste per year. Massive volumes of e-waste are being dumped, allowing considerable amounts of these elements, as well as numerous other harmful elements, to enter the subsoil and groundwater. REE has been shown to be more mobile in solutions containing F, Cl, HCO_3, CO_3, HPO_4, and PO_4. Furthermore, REE is utilized as fertilizer in agriculture to boost crop growth and output, resulting in a further increase in REE concentrations in the soil (Tyler, 2004). Mineral fertilizers (phosphate fertilizers) and soil conditioners, in general, comprise macronutrients (Ca, Mg, N, P, and S), micronutrients (such as Fe and Si), and rare earth elements (REE). REE can also come from local geological sources in soils (Liu, 1988). Furthermore, phosphate-based fertilizers are releasing substantial amounts of REE into agricultural soils. These components, which are also making their way into other environmental routes, particularly those involving ground and surface waters, are likely to contribute to pollution and human health. In China, REE-containing microfertilizers are applied directly to plants on a massive scale to improve production and quality (Diatloff et al., 1995). Following the widespread use of REE in agriculture, environmentalists are increasingly concerned that these elements may enter the food chain via plant absorption, posing a risk to human health.

6.6.4 Effect of REE on Plants

The toxicity of REE affects seed germination and plant growth with increased concentration. They enter as free ions with root absorption and are transported to the aerial part through the xylem. The main objective of phytoextraction was the depollution of contaminated soils, allowing polluted sites to be rehabilitated. Phytomining is a new domain of cultivating hyperaccumulated plants with the aim of establishing an environmentally friendly chain process. Selection of plants with suitable agronomic practices and optimization of extracting yields of economic value are the first steps in the chain, followed by the hydrometallurgical process to transform biomass into marketable REEs, which solely depends upon accumulation capacity. The first discovery of the RE accumulation in plants was reported by Robinson in 1943 in the leaves of *Carya cathayensis,* which reached 2,296 mg/kg DW of total RE (Robinson, 1943). Some publications have referred to hyperaccumulation of RE based on a criterion of 0.1 wt.% [i.e., 1,000 mg/kg of dry matter (DM)] for the sum of all 17 RE of the AP (Wang et al., 1997; Shan et al., 2003). Approximately 24 RE hyperaccumulator and potential hyperaccumulator species have been reviewed by Liu (1988) and Purwadi et al. (2020) through field investigations. These species occur across ten families of ferns, viz., Adiantaceae, Athyriaceae, Aspleniaceae, Blechnaceae, Dryopteridaceae, Gleicheniaceae, Juglandaceae, Lindsaeaceae, Phytolaccaceae, and Thelypteridaceae. The accumulation of REE in plant parts varies differently, like when the roots of *Zea mays* accumulate La concentrations 20–150 times higher than the leaves (Diatloff et al., 1999).

6.6.5 Effect of REE on Water

REE may only be available in modest amounts in natural settings via groundwater and the atmosphere, but its rising usage has increased the quantity of REE available and produced various novel bioaccumulation pathways (in plants, animals, and human beings). The background amount of REE concentration in surface and subsurface fluids varies greatly and is mostly determined by local geology. Al-Rimawi et al. (2013) found exceptionally high amounts of REE and numerous other metals in ground water samples from the south West Bank/Palestine in recent research. The authors are concerned since the majority of them have no maximum permissible levels and there is little information on their toxicity to human health. On the other hand, they identified the maximum allowable values of certain REE in drinking water. These limitations, according to the authors, were calculated using ecotoxicology and environmental chemistry data. In nations like China, the United States, India, Malaysia, and Brazil, intensive REE mining and manufacturing operations have had major environmental and health consequences. Cutting, drilling, blasting, shipping, stockpiling, and processing are all examples of mining processes that can discharge dust containing REE, other dangerous metals, and chemicals into the adjacent water bodies, affecting local soil, animals, and flora as well as humans.

6.7 IMPORTANCE OF RECYCLING REE

As the global demand for REE has escalated with the progression of time since 1960 with the launch of color television, REE sources are limited and have a risk of depletion in nearby communities, so judiciously producing primary REE and recycling and reusing secondary REE may be the substitute for sustainable exploitation. Due to the supply risk of REEs and the monopoly of the REEs market, REE recycling is currently considered an effective method to alleviate market fluctuations (Dushyantha et al., 2020). Electronic waste comes from scrapped devices that are at the end of their economic life and consumers cannot utilize them anymore. The total generation of e-waste in 2021 is expected to reach 52.2 Mt worldwide (Castro et al., 2021). Due to the limited availability of primary resources, an increase in REE recycling is urgently needed to expand REE supply and reduce waste generation and environmental pollution caused by mining activities. Thus, bioleaching represents a potential technology to recover precious metals from unconventional sources, including mine tailings, electronic scrap, and spent catalysts, as well as for waste treatment purposes. E-waste is a fast-growing waste stream throughout the world, growing at a rate of 3%–5% per year and accounting for about 5% of municipal waste (Tan et al., 2015; Baldé et al., 2015). Recycling discarded cell phones and printed circuit boards through bioleaching is an alternative option for secondary REE. Korean journal databases show the recovery of minerals like Cu, Co, Al, Ni, and Zn from electronic waste by the use of bacteria and fungi, but further research is required to strengthen the bioleaching of e-waste in a cost-effective manner as a good source of secondary REEs. Additionally, some post-consumer wastes can be recycled due to their significant quantities of REE, such as magnets (38%), lamp phosphors (32%), and metal alloys (13%). Modern fluorescent lamps typically retain more than 20% (*w/w*) REE (Ce, Eu, La, Tb, and Y) (Castro et al., 2021). Moreover, biodismantling is a new application of bioleaching in the recycling of e-waste to enhance the concentration of critical and precious materials contained in the electronic components. When achieving sufficient classification of the components after separation, some rare earth elements may become economically viable because their grade would be similar to commercial ores. A concentration of 9000 μg/g of dysprosium was detected in one of these separated fractions (Monneron-Enaud et al., 2020).

6.8 CONCLUSION

Bioleaching, an environmentally benign process, will be effective for the extraction of REEs on an industrial scale by selectively attacking the exposed ores in the near future, overcoming the scarcity of ore deposits. We addressed the past, present, and future scenarios of bioleaching of REEs with a focus on past history, prospective over chemical leaching, methodology, limitations, recycling, and environmental and health hazards. The alliance between government organizations, private sector, academia, and research institutions working together following the government framework or norms of biomining REEs may establish a secure future for developing countries with self-sufficiency. The direction of future revolutionary research is toward the extraction of specific bioenzymes with acute processes and equipment. An important note for the environment is that bioleaching minimizes the gases from green houses in the atmosphere.

REFERENCES

Aide, M.T. and Aide, C. "Rare Earth Elements: Their Importance in Understanding Soil Genesis". *International Scholarly Research Network ISRN Soil Science* 2012 (2012): 783876. doi:10.5402/2012/783876.

Akcil, A., Abdulvaliyev, N., Abhilash, R. and Meshram, P. "Overview on Extraction and Separation of Rare Earth Elements from Red Mud: Focus on Scandium." *Mineral Processing and Extractive Metallurgy Review* 39, no. 3 (2018): 145–151.

Ali, S.H. "Social and Environmental Impact of the Rare Earth Industries." *Resources* 3, no. 1 (2014): 123–134. doi:10.1016/j.scitotenv.2011.11.017

Alori, E. T., Glick, B.R. and Babalola, O.O. "Microbial Phosphorus Solubilization and its Potential for Use in Sustainable Agriculture." *Frontiers in Microbiology* 8 (2017): 971.

Al-Rimawi, F., Kanan, K. and Qutob, M. "Analysis of Different Rare Metals, Rare Earth Elements, and Other Common Metals in Groundwater of South West Bank/Palestine by ICP/MS-Data and Health Aspects". *Journal of Environmental Protection* 4, no. 10 (2013): 1157–1164. DOI: 10.4236/jep.2013.410132

Amin, M.M., El-Aassy, I.E., El-Feky, M.G., Sallam A.M., El-Sayed, E.M., Nada, A.A. and Harpy, N.M. "Fungal Leaching of Rare Earth Elements from Lower Carboniferous Shales, Southwestern Sinai, Egypt". *Romanian Journal of Biophysics* 24, no. 1 (2014): 25–41.

Antonick, P.J., Hu, Z., Fujita, Y., Reed, D.W. Das, G., Wu, L., Shivaramaiah, R., Kim, P., Eslamimanesh, A., Lencka, M.M., Jiao, Y., Anderko, A., Navrotsky, A., Riman, R.E. "Bio- and Mineral Acid Leaching of Rare Earth Elements from Synthetic Phosphogypsum." *The Journal of Chemical Thermodynamics* 132 (2019): 491–496.

Australian Industry Commission. "New and Advanced Materials." (1995).

Baldé, C. P., Kuehr, R., Blumenthal, K., Gill, S. F., Huisman, J., Kern, M., Micheli, P. and Magpantay, E. *E-Waste Statistics: Guidelines on Classifications, Reporting and Indicators*. United Nations University, Bonn, Germany (2015).

Barnett, M.J., Palumbo-Roe, B. and Gregory, S.P. "Comparison of Heterotrophic Bioleaching and Ammonium Sulfate Ion Exchange Leaching of Rare Earth Elements from a Madagascan Ion-Adsorption Clay." *Minerals* 8, no. 6 (2018): 236. doi:10.3390/min8060236

Becker, S., Bullmann, M., Dietze, H.J. and Iske, U. "Mass Spectrographic Determination of Selected Chemical Elements by Microbial Leaching of Zircon". *Fresenius' Zeitschrift für Analytische Chemie* 324 (1986): 37–42.

Beolchini, F., Fonti, V., Dell'Anno, A., Rocchetti, L. and Vegliò, F. "Assessment of Biotechnological Stratagies for the Valorization of Metal Bearing Wastes". *Waste Management* 32, no. 5 (2012): 949–956. doi:10.1016/j.wasman.2011.10.014

Birk, D. and White, J.C. "Rare Earth Elements in Bituminous Coals and Underclays of the Sydney Basin, Nova Scotia: Element Sites, Distribution, Mineralogy". *International Journal of Coal Geology* 19 (1991): 219–251.

Brandl, H. and Faramarzi, M.A. "Microbe-Metal-Interactions for the Biotechnological Treatment of Metal-Containing Solid Waste". *China Particuology* 4, no. 2 (2006):93–97.

Brisson, V. L., Zhuang, W.-Q. and Alvarez-Cohen, L. "Bioleaching of Rare Earth Elements from Monazite Sand". *Biotechnology and Bioengineering* 113, no. 2 (2016):339–348. doi:10.1002/bit.25823

Castro, L., Blázquez, M.L., González, F and Muñoz, J.A. "Rare Earth Elements Biorecovery from Mineral Ores and Industrial Wastes". *Heavy Metals- Their Environmental Impacts and Mitigation*, 2021. doi:10.5772/intechopen.94594

Chakhmouradian, A. R. and Wall, F. "Rare Earth Elements: Minerals, Mines, Magnets (and more)." *Elements* 8, no. 5 (2012): 333–340. doi:10.2113/gselements.8.5.33

Crowley, S.S., Stanton, R.W., Ryer, T.A. "The Effects of Volcanic Ash on the Maceral and Chemical Composition of the C Coal Bed, Emery Coal Field, Utah". *Organic Geochemistry* 14, no. 3 (1989): 315–331. doi:10.1016/0146-6380(89)90059-4

Dai, S., Zhou, Y., Zhang, M., Wang, X., Wang, J., Song, X., Jiang, Y., Luo, Y., Song, Z., Yang, Z. and Ren, D. "A New Type of Nb (Ta)-Zr(Hf)-REE-Ga Polymetallic Deposit in the Late Permian Coal-Bearing Strata, Eastern Yunnan, Southwestern China: Possible Economic Significance and Genetic Implications. *International Journal of Coal Geology* 83, no. 1 (2010): 55–63. doi:10.1016/j.coal.2010.04.002

Davranche, M., Grybos, M., Gruau, G., Pédrot, M., Dia, A. and Marsac, R. "Rare Earth Element Patterns: A Tool for Identifying Trace Metal Sources During Wetland Soil Reduction". *Chemical Geology* 284, no. 1–2 (2011): 127–137. doi:10.1016%2Fj.chemgeo.2011.02.014

De Baar, H.J.W., Michael, P. B., Brewer, P.G. and Bruland, K.W. "Rare Earth Elements in the Pacific and Atlantic Oceans." *Geochimica et Cosmochimica Acta* 49, no. 9 (1985): 1943–1959.

De Lima, I. B. and Filho, W. L. (eds.). *Rare Earths Industry: Technological, Economic, and Environmental Implications*. Elsevier (2015).

Delvasto, P., Valverde, A., Ballester, A.J.A., Munoz, F., González, M.L., Igual, B.J.M. and García-Balboa, C. "Diversity and Activity of Phosphate Bioleaching Bacteria from a High-Phosphorus Iron Ore." *Hydrometallurgy* 92, no. 3–4 (2008): 124–129.

Dev, S., Sachan, A., Dehghani, F., Ghosh, T., Briggs, B. R. and Aggarwal, S. "Mechanisms of Biological Recovery of Rare-Earth Elements from Industrial and Electronic Wastes: A review." *Chemical Engineering Journal* 397 (2020): 124596.

Devi, K., Syamala, O.S. and Singh, T.C. "Hazardous Waste Management in India-A Review." *International Journal of Creative Research Thoughts* 6, no. 1 (2018): 1547–1555.

Diatloff, E., Smith, F. W. and Asher, C. J. "Rare-earth Elements and Plantgrowth. 3. Responses of Corn and Mungbean to Low Concentrations of Cerium in Dilute, Continuously flowing Nutrient Solutions." *Journal of Plant Nutrition* 18 (1995): 1991–2003.

Diatloff, E., Asher, C. J. and Smith, F. W. "The Effects of Rare Earth Elements on the Growth and Nutrition of Plants," In: *Materials Science Forum*. Trans Tech Publications Ltd., Switzerland, pp. 354–360 (1999). doi:10.4028/www.scientific.net/msf.315-317.354

Dudeney, A.W.L. and Sbai, M.L. "Bioleaching of Rare-Earth-Bearing Phosphogypsum". In: Torma, A. E., Wey, J. E. and Lakshmanan, V. L. (eds.) *Biohydrometallurgical Technologies. The Minerals, Metals, & Materials Society*. Jackson Hole, Warrendale, PA, pp. 39–47 (1993).

Dushyantha, N., Batapola, N., Ilankoon, I. M. S. K., Rohitha, S., Premasiri, R., Abeysinghe, B., Ratnayake, N. and Dissanayake, K. "The Story of Rare Earth Elements (REEs): Occurrences, Global Distribution, Genesis, Geology, Mineralogy and Global Production." *Ore Geology Reviews* 122 (2020): 103521.

Ehrlich, H.L. "Past, Present and Future of Biohydrometallurgy." *Hydrometallurgy* 59, no. 2–3 (2001): 127–134.

Eskenazy, G.M. "Rare Earth Elements and Yttrium in Lithotypes of Bulgarian Coals". *Organic Geochemistry* 11, no. 2 (1987): 83–89.

Eskenazy, G.M. "Aspects of the Geochemistry of Rare Earth Elements in Coal: An Experimental Approach". *International Journal of Coal Geology* 38, no. 3–4 (1999): 285–295.

Fathollahzadeh, H., Becker, T., Eksteen, J.J., Kaksonen, A.H. and Watkin, E.L.J. "Microbial Contact Enhances Bioleaching of Rare Earth Elements". *Bioresource Technology Reports* 3 (2018):102–108. doi:10.1016/j.biteb.2018.07.004

Finkelman, R.B. "The Origin, Occurrence and Distribution of the Inorganic Constituents in Low-Rank Coals". In: Schobert, H. H. (ed.) *Proceedings of the Basic Coal Science Workshop, Houston, TX, USA, 8–9 December 1981*. Grand Forks Energy Technology Center, Grand Forks, ND, pp. 70–90 (1981).

Franzmann, P.D., Haddad, C.M., Hawkes, R.B., Robertson, W.J. and Plumb, J.J. "Effects of Temperature on the Rates of Iron and Sulfur Oxidation by Selected Bioleaching Bacteria and Archaea: Application of the Ratkowsky Equation. *Minerals Engineering* 18, no. 13–14 (2005):1304–1314. doi:10.1016/j.mineng.2005.04.006

Glombitza F., and Reichel S. Metal-containing Residues from Industry and in the Environment: Geobiotechnological Urban Mining. *Geobiotechnology* 141 (2013): 49–107. doi: 10.1007/10_2013_254

Goyne, K. W., Brantley, S.L. and Chorover, J. "Rare Earth Element Release from Phosphate Minerals in the Presence of Organic Acids." *Chemical Geology* 278, no. 1–2 (2010): 1–14.

Gupta, C.K. and Krishnamurthy, N. "Extractive Metallurgy of Rare Earths." *International Materials Reviews* 37, no. 1 (1992): 197–248.

Habashi, F. "Extractive Metallurgy of Rare Earths." *Canadian Metallurgical Quarterly* 52, no. 3 (2013): 224–233.

Hassanien, W.A.G., Desouky, O.A.N. and Hussien, S.S.E. "Bioleaching of Some Rare Earth Elements from Egyptian Monazite using *Aspergillus ficuum* and *Pseudomonas aeruginosa*." *Walailak Journal of Science and Technology (WJST)* 11, no. 9 (2014): 809–823.

Haxel, G. *Rare Earth Elements: Critical Resources for High Technology*. Vol. 87, no. 2. (2002). US Department of the Interior, US Geological Survey.

Haxel, G., Hedrick, J. and Orris, G. *Rare Earth Elements: Critical Resources for High Technology*. US Department of the Interior, US Geological Survey (2002).

Hubau, A., Minier, M., Chagnes, A., Joulian, C., Silvente, C. and Guezennec, A.G. "Recovery of Metals in a Double-Stage Continuous Bioreactor for Acidic Bioleaching of Printed Circuit Boards (PCBs)." *Separation and Purification Technology* 238 (2020): 116481.

Ilankoon, I.M.S.K., Tang, Y., Ghorbani, Y., Northey, S.M.Y., Deng, X. and McBride, D. "The Current State and Future Directions of Percolation Leaching in the Chinese Mining Industry: Challenges and opportunities." *Minerals Engineering* 125 (2018): 206–222.

Iske, U., Bullmann, M. and Glombitza, F. "Organoheterotrophic Leaching of Resistant Materials." *Acta Biotechnologica* 7, no. 5 (1987): 401–407.

Kortenski, J. and Bakardjiev, S. "Rare Earth and Radioactive Elements in Some Coals from the Sofia, Svoge and Pernik Basins, Bulgaria." *International Journal of Coal Geology* 22, no. 3–4 (1993): 237–246.

Kronholm, B., Anderson, C.G. and Taylor, P.R. "A Primer on Hydrometallurgical Rare Earth Separations." *JOM* 65, no. 10 (2013): 1321–1326.

Li, X., Chen, Z., Chen, Z and Zhang, Y. "A Human Health Risk Assessment of Rare Earth Elements in Soil and Vegetables from a Mining Area in Fujian Province, Southeast China." *Chemosphere* 93, no. 6 (2013): 1240–1246. doi:10.1016/j.chemosphere.2013.06.085

Liu, Z. "The Effects of Rare Earth Elements on Growth of Crops V." In *Proceedings of the International Symposium. New Results in the Research of Hardly Known Trace Elements and Their Role in Food Chain.* University of Horticulture and Food Industry, Budapest, p. 23 (1988).

Mancheri, N.A., Sprecher, B., Bailey, G., Ge, J. and Tukker, A. "Effect of Chinese Policies on Rare Earth Supply Chain Resilience." *Resources, Conservation and Recycling* 142 (2019): 101–112.

McDonald, J.W., Ghio, A.J., Sheehan, C. E., Bernhardt, P.F. and Roggli, V.L. "Rare Earth (Cerium Oxide) Pneumoconiosis: Analytical Scanning Electron Microscopy and Literature Review." *Modern Pathology: An Official Journal of the United States and Canadian Academy of Pathology, Inc* 8, no. 8 (1995): 859–865.

Monneron-Enaud, B., Wiche, O. and Schlömann, M. "Biodismantling, A Novel Application of Bioleaching in Recycling of Electronic Wastes." *Recycling* 5, no. 3 (2020): 22. doi: 10.3390/recycling5030022

Muravyov, M.I., Bulaev, A.G., Melamud, V.S. and Kondrat'eva, T.F. "Leaching of Rare Earth Elements from Coal Ashes using Acidophilic Chemolithotrophic Microbial Communities." *Microbiology* 84, no. 2 (2015): 194–201.

Okoh, M.P., Olobayetan, I.W. and Machunga-Mambula, S. S. "Bioleaching, a Technology for Metal Extraction and Remediation: Mitigating Health Consequences for Metal Exposure." *International Journal of Development and Sustainability* 7, no.7 (2018): 2103–2118.

Padmanabhan, V.T. "Radioactive Minerals and Private Sector Mining." *Economic and Political Weekly* (2002): 4365–4367.

Park, S. and Liang, Y. "Bioleaching of Trace Elements and Rare Earth Elements from Coal Fly Ash." *International Journal of Coal Science & Technology* 6, no. 1 (2019): 74–83. doi:10.1007/s40789-019-0238-5

Pédrot, M., Dia, A. and Davranche, M. "Dynamic Structure of Humic Substances: Rare Earth Elements as a Fingerprint." *Journal of Colloid and Interface Science* 345, no. 2 (2010): 206–213.

Peelman, S., Sun, Z.H.I., Sietsma, J. and Yang, Y. "Leaching of Rare Earth Elements: Review of Past and Present Technologies." *Rare Earths Industry* (2016): 319–334.

Pollock, S.M., Goodarzi, F. and Riediger, C.L. "Mineralogical and Elemental Variation of Coal from Alberta, Canada: An Example from the No. 2 Seam, Genesee Mine." *International Journal of Coal Geology* 43, no. 1–4 (2000): 259–286.

Prabhu, N., Borkar, S. and Garg, S. "Phosphate solubilization by microorganisms: Overview, mechanisms, applications and advances (Chapter 11)." (2019).

Purwadi, I., Gei, V., Erskine, P. D., Echevarria, G., Mesjasz-Przybyłowicz, J., Przybyłowicz, W. J. and van der Ent, A. (2020). "Tools for the discovery of hyperaccumulator plant species in the field and in the herbarium." In: van der Ent, A., Baker, A. J. M., Echevarria, G., Simonnot, M-O. and Morel, J. L. (eds.), Agromining: Farming for Metals: Extracting Unconventional Resources from Plants. Springer International Publishing, Cham, pp. 183–195. doi: 10.1007/978-3-030-58904-2_9

Qu, Y. and Lian, B. "Bioleaching of Rare Earth and Radioactive Elements from Red Mud Using *Penicillium tricolor* RM-10." *Bioresource Technology* 136 (2013): 16–23.

Qu, Y., Li, H., Wang, X., Tian, W., Shi, B., Yao, M., et al. "Bioleaching of major, rare earth, and radioactive elements from red mud by using indigenous chemoheterotrophic bacterium Acetobacter sp." *Minerals* 9, no. 3 (2019): 67.

Ramprasad, C., Gwenzi, W., Chaukura, N., Azelee, N.W., Rajapaksha, A. U., Naushad, M. and Rangabhashiyam, S. "Strategies and Options for the Sustainable Recovery of Rare Earth Elements from Electrical and Electronic Waste". *Chemical Engineering Journal* 442, no. 1 (2022): 135992. doi:10.1016/j.cej.2022.135992

Rasoulnia, P., Barthen, R. and Lakaniemi, A.M. "A Critical Review of Bioleaching of Rare Earth Elements: The Mechanisms and Effect of Process Parameters." *Critical Reviews in Environmental Science and Technology* 51, no. 4 (2021): 378–427.

Ratkowsky, D.A., Lowry, R. K., McMeekin, T. A., Stokes, A.N. and Chandler, R.E. "Model for Bacterial Culture Growth Rate throughout the Entire Biokinetic Temperature Range." *Journal of Bacteriology* 154, no. 3 (1983): 1222–1226. doi:10.1128/JB.154.3.1222-1226.1983

Rim, K.T. A Book review: "Rare Earth Elements in Human and Environmental Health; At the Crossroads between Toxicity and Safety" *Journal of Applied Biological Chemistry* 60, no. 3 (2017): 207–211.

Robinson, W.O. "The Occurrence of Rare Earths in Plants and Soils." *Soil Science* 56, no. 1 (1943): 1–6. doi:10.1097/00010694-194307000-00001

Rosenblum, S. and Fleischer, M. "The Distribution of Rare-Earth Elements in Minerals of the Monazite Family." No. 2140. US Government Printing Office (1995).

Roskill. Rare Earths: "Market Outlook to 2020". (2015)

Sabbioni, E., Pietra, R., Gaglione, P., Vocaturo, G., Colombo, F., Zanoni, M. and Rodi, F. "Long-Term Occupational Risk of Rare-Earth Pneumoconiosis A Case Report as Investigated by Neutron Activation Analysis." *Science of the Total Environment* 26, no. 1 (1982): 19–32.

Schatzel, S. J. and Stewart, B.W. "Rare Earth Element Sources and Modification in the Lower Kittanning Coal Bed, Pennsylvania: Implications for the Origin of Coal Mineral Matter and Rare Earth Element Exposure in Underground Mines." *International Journal of Coal Geology* 54, no. 3–4 (2003): 223–251.

Science History Institute. *History and Future of Rare Earth Elements. What Are the Rare Earth Elements, and Where Do They Come From?* Science History Institute (2022). https://www.sciencehistory.org/learn/science-matters/case-of-rare-earth-elements-history-future.

Shan, X.Q., Wang, H., Zhang, S., Zhou, H., Zheng, Y., Yu, H. and Wen, B. "Accumulation and Uptake of Light Rare Earth Elements in a Hyperaccumulator *Dicropteris dichotoma*." *Plant Science* 165, no. 6 (2003): 1343–1353.

Shin, D., Jiwoong, K., Byung-su, K., Jinki, J. and Jae-chun, L. "Use of Phosphate Solubilizing Bacteria to Leach Rare Earth Elements from Monazite-Bearing Ore." *Minerals* 5, no. 2 (2015): 189–202.

Tan, Z., Zhuang, Q. and Anthony K.W. "Modeling Methane Emissions from Arctic Lakes: Model Development and Site-Level Study." *Journal of Advances in Modeling Earth Systems* 7, no. 2 (2015): 459–483.

Theodore, G. *The Elements: A visual exploration of Every Known Atom in the Universe*. Black Dog & Leventhal Publishers, New York, p. 240 (2009). ISBN 978-1-57912-814-2

Tsaplina, I.A., Panyushkina, A.E., Grigor'eva, N. V., Bulaev, A. G. and Kondrat'eva, T. F. "Growth of Acidophilic Chemolithotrophic Microbial Communities and Sulfur Oxidation in the Presence of Coal Ashes." *Microbiology* 84, no. 2 (2015): 177–189.

Tyler, G. "Ionic charge, Radius, and Potential Control Root/Soil Concentration Ratios of Fifty Cationic Elements in the Organic Horizon of a Beech (*Fagus sylvatica*) Forest Podzol." *Science of the Total Environment* 329, no. 1–3 (2004): 231–239.

United States Geological Survey (USGS), 2018. Mineral Commodit Summaries Gov. Print. Off. Washington, DC. (2018) https://www.usgs.gov/centers/nmic/rare-earthsstatistics-and-information (accessed 09.23.2019)

Voncken, J.H.L. *The Rare Earth Elements: An Introduction*. Springer International Publishing, Cham, Switzerland (2016).

Walawalkar, M., Nichol, C.K. and Azimi, G. "Process Investigation of the Acid Leaching of Rare Earth Elements from Phosphogypsum Using HCl, HNO_3, and H_2SO_4." *Hydrometallurgy* 166 (2016): 195–204.

Wang, Y.Q., Sun, J.X., Chen, H.M. and Guo, F.Q. "Determination of the Contents and Distribution Characteristics of REE in Natural Plants by NAA". *Journal of Radioanalytical and. Nuclear Chemistry* 219, no. 1 (1997): 99–103. doi:10.1007/BF02040273

Weng, Z.H., Jowitt, S.M., Mudd, G.M. and Haque, N. "Assessing Rare Earth Element Mineral Deposit Types and Links to Environmental Impacts." *Applied Earth Science* 122, no. 2 (2013): 83–96.

Yoon, H.K., Moon, H.S., Park, S.H., Song, J.S., Lim, Y. and Kohyama, N. "Dendriform Pulmonary Ossification in Patient with Rare Earth Pneumoconiosis." *Thorax* 60, no. 8 (2005): 701–703

Zhou, B., Li, Z. and Chen, C. "Global Potential of Rare Earth Resources and Rare Earth Demand from Clean Technologies." *Minerals* 7, no. 11 (2017): 203.

Zhuang, W.Q., Fitts, J.P., Ajo-Franklin, C.M., Alvarez-Cohen, S.L. and Hennebel, T. "Recovery of Critical Metals Using Biometallurgy." *Current Opinion in Biotechnology* 33 (2015): 327–335.

7 An Understanding of the Underlying Mechanisms Involved in the Bacterial Remediation of Arsenic, Cadmium, and Lead

Bhagyudoy Gogoi, Pranjal Bharali,
Viphrezolie Sorhie, and Lemzila Rudithongru
Nagaland University

7.1 INTRODUCTION

Heavy metals (HMs), once assumed to be naturally occurring substances, are now the cause of various anthropogenic activities that get distributed into the various environmental compartments. The HMs are regarded as 'metals with densities more than 5 g/cm^3 which form extremely dangerous chemicals upon reaction' (Cook, 1977). Numerous natural and anthropogenic factors contribute to the distribution of HMs into the environment, which includes weathering of rocks (Bradl, 2005), volcanic activities (Szumińska et al., 2018; Press and Sievers, 1994), mineral exploration (Saha and Paul, 2016), and allied industrial activities (Bradl, 2005) or agricultural practices (Cai et al., 2012) involving chemicals and fertilizers (Yu and Tsunoda, 2004). In general, pollution occurring in the environment can be due to both natural and anthropogenic origin, such as natural rock weathering (Li et al., 2022), bacterial pollutant (Umana et al., 2022; Mahmoud et al., 2022), volcanic eruptions (Szumińska et al., 2018), and anthropogenic sources such as mining (Sun et al., 2022) and several others as described in Table 7.1. However, the term 'pollution' is attributed mostly to human beings, as most anthropogenic sources deliberate negative impacts on the ecosystem and biodiversity, human well-being, disruption of the environment, as well as the reduction of facilities and infrastructure (Masindi and Muedi, 2018). In recent years, HM pollution has become a global environmental challenge affecting human health (Rajendran et al., 2022). Thus, active research pertaining to HM pollution is crucial to reducing HM pollution at a global level (Bayraktar et al., 2022).

The most common HMs are arsenic (As), cadmium (Cd), lead (Pb), zinc (Zn), nickel (Ni), mercury (Hg), chromium (Cr), and copper (Cu) (Masindi and Muedi, 2018). HM clean-up needs extra caution in order to safeguard the quality of soil, air, and water; human and animal well-being, and all sectors in general. HM clean-up

DOI: 10.1201/9781003451457-7

TABLE 7.1
Different Sources of Environmental Contamination by Arsenic, Cadmium, and Lead

Sl. No.	Sources of Contamination	Type of Heavy Metals	Reference
1	Mining	Pb, Cd, and As	Sun et al. (2022); Bouida et al. (2022); Wen et al. (2022)
2	Ore smelting	Pb	Ludolphy et al. (2022)
3	Coal industry	Pb	Yan et al. (2022)
4	Storage battery industry	Pb	Kumar et al. (2022)
5	Automobile industry	Pb	Parsoya and Perwej (2022)
6	Metal-plating industry	Pb	Khan et al. (2022)
7	Leather-tanning industry	Pb	Bouida et al. (2022)
8	Fertilizer industry	Pb, Cd, As	Rosariastuti and Astuti (2022); Suhani et al. (2021); Khosravi-Darani et al. (2022)
9	Pesticide industry	Pb, As	Rani et al. (2022); Tarfeen et al. (2022)
10	Pigments additive	Pb	Masri et al. (2022)
11	Gasoline	Pb	Rubio et al. (2022)
12	Chlor-alkali	Cd	Wu et al. (2022)
13	Paint	Cd	Goyal et al. (2021)
14	Copper alloys	Cd	Gujre et al. (2021)
15	Alkaline batteries	Cd	Suhani et al. (2021)
16	Zinc refining	Cd	Zhou et al. (2021)
17	Natural geochemical processes	As	Raju (2022)
18	Wood preservatives	As	Hassan et al. (2022)

technologies have been developed; however, they are prohibitively expensive, time-consuming, and contaminate the environment (Masindi and Muedi 2018). Bacterial bioremediation systems have recently been discovered to be exceptionally successful in dealing with heavy pollution in most aquatic and terrestrial habitats. Numerous studies have indicated the potential utility of bacterial systems in the remediation of As, Cd, and Pb polluted environments. Bacteria such as *Lactobacillus acidophilus* (Singh and Sarma, 2010), *Acidithiobacillus ferrooxidans* (Yan et al., 2010), *Staphylococcus xylosus* (Aryal et al., 2010; 2011), *Rhodococcus* sp. (Prasad et al., 2011), *Bacillus cereus* (Mohd Bahari et al., 2013; Titah et al., 2018) and *Arthrobacter globiformis* (Titah et al., 2018), are known to be very effective as biosorbents of arsenic (Sahmoune, 2016). Likewise, *Enterobacter* sp. (Lu et al., 2006; Feria-Cáceres et al., 2022), *Pseudomonas aeruginosa* (Wang and Can, 2009), *P. putida* (Shamim and Rehman, 2015; Deng et al., 2022), *P.* sp. (Wang and Can, 2009), *Ochrobactrum anthropi* (Girawale et al., 2022), *Sphingomonas paucimobilis* (Tangaromsuk et al., 2002), *Aeromonas* sp. (Ibrahim et al., 2020), and *S. xylosus* (Lata et al., 2019) are reported to be very proficient biosorbents in the remediation of cadmium-polluted environments. Similarly, *B. firmus* (Shiomi, 2015), *Corynebacterium glutamicum* (Tiquia-Arashiro, 2018), *Enterobacter* sp. (Okpara-Elom, 2022), *P. putida* (Gürgan et al., 2022), and *Streptomyces rimosus* (Timková et al., 2018) are found to be good biosorbents in lead-polluted environments. Moreover, bacteria such as *Rhizobium metallidurans* can remediate cadmium-polluted environments by secreting exopolysaccharides (EPS) (Kowalkowski et al., 2019). Furthermore, bacteria such as *Clostridium thermoaceticum* can cause the bioprecipitation of toxic cadmium ions (Dalei, 2014). Moreover, bacteria such as *Delftia* sp. BAs29 are reported to be effective in arsenic-polluted groundwater remediation by 'arsenite oxidation' (Biswas et al., 2019).

However, the mechanism of action of bacterial remediation for all types of HMs has yet to be thoroughly explored, necessitating more study. The present chapter summarizes the inherent mechanisms adapted by the bacterial communities in response to the toxicity imposed by the HMs viz. As, Cd, and Pb metal ions.

7.2 ENVIRONMENTAL POLLUTION BY HMs (As, Cd and Pb)

HM pollution occurs both directly and indirectly through effluent discharges from industries, refineries, and waste treatment facilities, as well as pollutants that enter the aquatic system through soil or groundwater and the atmosphere through rainfall (Vijayaraghavan and Yun, 2008).

Arsenic is found in minerals such as As_2O_3 and may be found in a variety of oxidation states (–3, 0, +3, +5) in the atmosphere, natural waters, rocks, soil, and organisms (Ullah et al. 2022). Furthermore, in aerobic conditions, As (V) predominates in the form of arsenate (AsO_4^{3-}) in a variety of protonation states, including hydrogen arsenate ($HAsO_4^{2-}$), arsenic acid (H_3AsO_4), arsenate (AsO_4^{3-}), and dihydrogen arsenate ($H_2AsO_4^-$) (Wuana and Okieimen, 2011). Some of the issues associated with arsenic exposure in plants include a reduction in yield and biomass, along with serious impacts such as cell leakage and electrolyte imbalance (Ghosal et al., 2022). Likewise, in animals, arsenic has been known to cause serious diseases such as hyperglycemia (Singh et al., 2020), diabetes (Liu et al., 2014), and elevated blood pressure (Wang et al., 2021). At a global scale, at least 105 developing and developed countries, including more than 200 million people, are at risk of contracting As poisoning from consuming contaminated drinking water (Bundschuh et al., 2022). In oxic aquatic environments, the pentavalent oxyanion arsenate ($H_2AsO_4^-$ or $HAsO_4^{2-}$) predominates, whereas trivalent arsenite [$As(OH)_3$] or infrequently $H_2AsO_3^-$ predominates in reducing anoxic environments (Kruger et al., 2013). Arsenic can be released into the environment by natural geochemical processes (Raju, 2022), human activities such as wood preservatives (Hassan et al., 2022), mining activities (Wen et al., 2022), and some other sources as described in Table 7.1 (Kruger et al., 2013).

Cadmium typically forms complicated oxides with ores of lead, copper, and zinc, sulfides, and carbonates since it is seldom found in pure form in nature (Monachese et al., 2012). Cadmium oxide (CdO) is less soluble in water than cadmium sulfate ($CdSO_4$) and cadmium chloride ($CdCl_2$) (Darwish et al., 2016). Cadmium is distributed into the environment through a variety of natural processes, including the earth's crust and mantles, as well as rock weathering and volcanic activity (Kesler and Simon, 2015). Moreover, Cd is widely used in a variety of sectors, such as chlor-alkali (Wu et al., 2022), paints (Goyal et al., 2021), copper alloys (Gujre et al., 2021), and several others, as described in Table 7.1. Cadmium-based quantum dots have recently been widely employed as biomedical sensors (Altintas et al., 2018), during drug delivery systems (Abdelhamid, 2022), and also in imaging (Mo et al., 2017). Cadmium is also a common HM, which can lead to a variety of disorders such as reduced bone density (Suhani et al., 2021), gastric cancer, and lung cancer (Jyothi, 2020), and is thus considered toxic to human health (Amjad et al., 2022).

Lead is found naturally and has a bluish-grey color. It is often found as a mineral and is frequently associated with other elements such as sulfur (PbS and $PbSO_4$) or oxygen ($PbCO_3$) (Wuana and Okieimen, 2011).

Lead phosphates [$Pb_3(PO_4)_2$], lead carbonates ($PbCO_3$), and lead hydroxides [$Pb(OH)_2$] are the most common insoluble lead compounds (Wuana and Okieimen, 2011). Among the insoluble lead compounds, $PbCO_3$ usually forms at a pH greater than 6 (Wuana and Okieimen, 2011). Divalent lead compounds are ionic, such as lead sulfate ($Pb^{2+}SO_4^{2-}$), whereas tetravalent lead compounds are covalent, such as tetraethyl lead [$Pb(C_2H_5)_4$] (Wuana and Okieimen, 2011). Lead is the most dangerous pollutant, and its widespread usage has resulted in environmental and health concerns in many regions of the world. Mining (Sun et al., 2022), smelting of ores (Ludolphy et al., 2022), coal combustion (Yan et al., 2022), effluents from storage battery industries (Kumar et al., 2022), and several others as described in Table 7.1 are all common anthropogenic sources of Pb contamination in the environment (Kushwaha et al., 2018). Over the past three centuries, the amount of lead in the environment has increased more than 1,000-fold as a result of extensive human activity (Ruikar and Pawar, 2022). Lead is a persistent environmental pollutant that accumulates in food chains over time, causing biomagnification at various tropical levels, and is thus referred to as a cumulative poison (Naik and Dubey, 2013).

7.3 BACTERIAL REMEDIATION OF ARSENIC

Bacterial arsenic remediation is mostly due to its natural capacity to mobilize and convert arsenic via various processes such as oxidation, reduction, and methylation pathways and its resistance to arsenic (Irshad et al., 2021). The arsenic remediation mechanism is presented in Figure 7.1.

7.3.1 Detoxification of Arsenite by Bacterial Oxidation

Bacteria can oxidize arsenite naturally (Kruger et al., 2013). Arsenic may be quickly oxidized from its poisonous form, As (III), to a less toxic form, As (V), which can be efficiently used for arsenic water treatment (Biswas et al., 2019; Okibe and Fukano, 2019). The catalytic activity of a periplasmic enzyme called 'As (III) oxidase,' also known as 'AioAB' (Lett et al., 2012), which is also a member of the DMSO (Dimethyl sulfoxide) reductase family, is primarily responsible for As (III) oxidation (Mobar and Bhatnagar, 2022). Furthermore, another member of the DMSO reductase

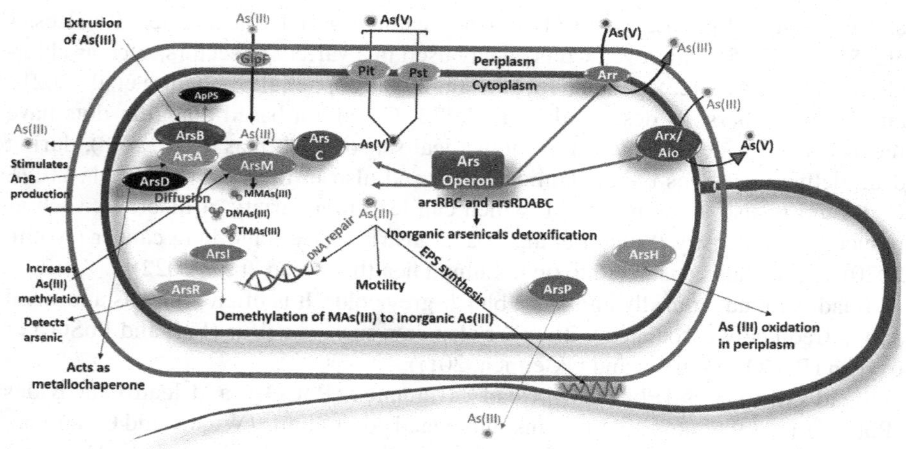

FIGURE 7.1 Mechanism of arsenic detoxification pathways in bacteria. (Adapted from Kruger et al., 2013.)

family that is widely present in nature, i.e., arsenite oxidase 'ArxAB,' is capable of generating anaerobic oxidation of arsenite with nitrate (Kruger et al., 2013; Zargar et al., 2012).

7.3.2 Detoxification of Arsenate by Bacterial Reduction

There are two types of arsenate reduction systems. One system is Ars C's cytoplasmic arsenate reduction, which results in Ars B's arsenite extrusion. The periplasmic respiratory arsenate reductase ArrA is another system (Kruger et al., 2013). Arsenate resistance in bacteria, such as *Shewanella* sp. strain ANA-3, is attributed to two distinct pathways: detoxification and arsenate respiratory reduction (Lizárraga et al., 2022). The respiratory arsenate reductase (ArrA and ArrB) enzymes have a molybdopterin core in ArrA and a Fe-S center in ArrB (Irshad et al., 2021). *Geospirillum arsenophilus* strain MIT-13 was one of the first dissimilatory arsenate respiring prokaryotes (DARPs) discovered (Páez-Espino et al., 2009).

7.3.3 Bacterial Detoxification of Trivalent Organo-Arsenicals

7.3.3.1 Bacterial Methylation of Arsenite to Trivalent Organo-Arsenicals

The presence of ArsM has been linked with the methylation of arsenite by S-adenosylmethionine (SAM) methyltransferase (AsMTs), expressed by the *ArsM* gene (Viacava et al., 2020). Bacteria methylate As (III) to produce methyarsenite [MAs (III)], dimethylarsenite [DMAs (III)], and gaseous trimethylarsenites [TMAs (III)] (Yang and Rosen, 2016). The *ArsM* gene, which encodes 283 amino acid residues and is found in both prokaryotes and eukaryotes, was first cloned from the bacterium *Rhodopseudomonas palustris* (Li et al., 2016a; Zhu et al., 2017). This gene might be a helpful molecular marker in the study of arsenic-methylating bacteria (Jia et al., 2013). Recently, reports based on the metagenome-assembled genomes of microorganisms expressing *ArsM* indicated that *Paraclostridium* sp. strain EML actively methylated As anaerobically (Viacava et al., 2022).

7.3.3.2 Bacterial Oxidation of Trivalent Methylarsenicals and Volatilization of TMAs (III)

Trimethylarsenite, or TMAs (III), is typically nontoxic due to its ability to be volatilized from cells (Mondal et al., 2021). Methyl arsenite, DMAs (III), and TMAs (III) can, however, be oxidized into less-toxic pentavalent molecules like methylarsonic acid [MAs(V)], dimethyl arsenic acid [DMAs (V)], and trimethyl arsenic oxide [TMAsO(V)] (Yan et al., 2019). According to research, the *ArsH* gene can accelerate the conversion of methylation arsenic intermediates such as MAs (III) into less harmful pentavalent MAs (V) (Chen et al., 2015a). It has also been claimed that the *ArsH* gene functions as a reductase, facilitating the reduction of oxygen (O_2) and azo dyes to hydrogen peroxide (H_2O_2) (Xue et al., 2014). The *ArsH* gene, which was identified from the Ars operon in *Yersinia enterocolitica*, was thought to offer arsenic resistance in *Y. enterocolitica* and *Sinorhizobium meliloti*, but the evidence was insufficient to support the notion (Chen and Rosen, 2020). However, a recent investigation of *P. putida* strain VTw33 revealed that the *ArsH* gene was involved in arsenite oxidation in the periplasm (Chang et al., 2018).

7.3.3.3 Bacterial Extrusion of Trivalent Methylarsenicals

Methylarsenite, or MAs (III), may be removed from bacterial cells via the methylarsenite transporter ArsP, which is also an MAs (III) efflux permease initially observed in *Campylobacter jejuni* (Yan et al., 2019). This ArsP protects the bacterial cell from MAs (III) and other trivalent organoarsenic chemicals (Chen et al., 2015b). *ArsP*, encoded in an Ars operon from *C. jejuni*, was found to be expressed in *Escherichia coli* for MAs (III) resistance and is recognized to be the first discovered efflux mechanism for trivalent organoarsenicals (Chen et al., 2015b).

7.3.4 Pentavalent Organoarsenical Detoxification

7.3.4.1 Bacterial Reduction of Methylarsonic Acid and Demethylation or Efflux of Methyl Arsenite

Herbicides and antimicrobial growth promoters in chicken farms have significantly enhanced the dispersion of organoarsenicals in the environment (Zhu et al., 2014). Pentavalent organoarsenicals, such as monosodium methylarsonic acid, are absorbed by bacteria via phosphate permeases or aquaglyceroporins and are eliminated from the bacterial cell via a two-step detoxification mechanism (Yoshinaga and Rosen, 2014). The first phase includes the conversion of organoarsenicals to MAs (III) by an undiscovered enzyme, followed by detoxification via demethylation to inorganic As (III) by the *ArsI* gene, as shown in Figure 7.2 (Nadar et al., 2014; Yoshinaga and Rosen, 2014). Inorganic As (III) can occasionally be extruded from the bacterial membrane via the *ArsP* gene (Yan et al., 2019).

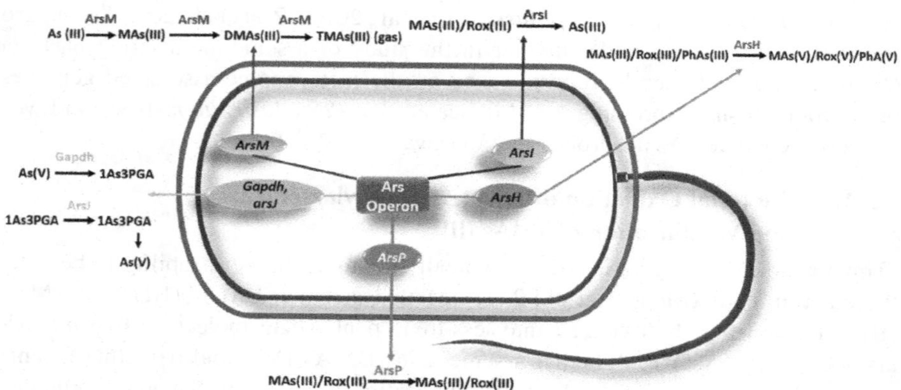

FIGURE 7.2 Arsenic methylation pathways and organic arsenic detoxification pathways. (Adapted from Yan et al., 2019.)

7.3.4.2 Integrated Bacterial Extrusion of Pentavalent Organic Arsenicals

Bacterial resistance to inorganic As (V) may also be obtained by a combination of As (V) participating in glycolysis with pentavalent organoarsenical elimination (Chen et al., 2016). This occurs in a two-step detoxification process that is regulated by the *Gapdh* and *arsJ* genes as found in the Ars operon of *P. aeruginosa* DK2 (Chen et al., 2016), as shown in Figure 7.2. Glyceraldehyde-3-phosphate dehydrogenase is transcribed by the *Gapdh* gene, which catalyzes the biotransformation of inorganic phosphate, d-glyceraldehyde-3-phosphate (G3P), and NAD^+ to 1,3-bisphosphoglycerate (Yan et al., 2019). By replacing phosphate, inorganic As (V) can impede glycolysis and generate a very unstable molecule, 1-arseno-3-phosphoglycerate (1As3PGA), which can be eliminated from the bacterial cell with the assistance of an efflux permease, Ars J, as well as fast dissociation into As (V) and 3-phosphoglycerate (3PGA) (Chen et al., 2016).

7.3.5 Arsenic Resistance System (ARS)

The ars system of inorganic arsenic detoxification is mostly ascribed to Ars operons found in chromosomes and plasmids of both Gram-positive and Gram-negative bacteria (Yan et al., 2019). The most prevalent Ars operons are the 'arsRBC' and 'arsRDABC' gene operons, which promote bacterial detoxification of inorganic arsenicals as well as other ars-related genes (Yan et al., 2019).

7.3.5.1 Bacterial Resistance System in Arsenite Detoxification

Arsenite ions enter the bacterial cell mostly through the glycerol transport system, which is assisted by the glyceroporin channel GlpF, which is encoded by the *glpF* gene (Mohsin et al., 2021). The arsenite efflux permease 'ArsB,' expressed by the *arsB* gene of the Ars operons, facilitates the removal of trivalent arsenic from the bacterial cell (Yan et al., 2019). However, the presence of ArsA, an As (III)-stimulated ATPase produced by the *arsA* gene, increases the production of ArsB

(Yan et al., 2019). ArsB functions as a membrane anchor protein for the catalytic subunit ArsA and as an efflux channel for the elimination of As (III) (Darma et al., 2022). Furthermore, interaction of As (III) with ArsA's three cysteine residues results in allosteric activation (Jiang et al., 2019).

ArsR is an essential As(III)-responsive trans-acting repressor protein that may detect arsenic in the environment and negatively control the transcription of other genes in the Ars operon (Haque et al., 2022; Ali et al., 2022). ArsD is yet another protein of the Ars operon that was thought to be a weak As(III)-responsive transcriptional repressor with the ability to bind at the same position on the Ars promoter region of ArsR with very little affinity (Yan et al., 2019). Research indicates that ArsD operates as an As(III) metallochaperone by transferring As(III) to the metal-binding site of ArsA of the Ars efflux pump (Darma et al., 2022).

7.3.5.2 Bacterial Resistance System in Arsenate Detoxification

Arsenate is accumulated by bacterial cells via two phosphate transporters, Pst (phosphate-specific transport) and Pit (inorganic phosphate transport), which are transcribed by the *pst* and *pit* genes (Garbinski et al., 2019). Phosphate-specific transport (Pst) is responsible for phosphate transport with high affinity but limited efficacy, whereas Pit is a nonspecific phosphate transporter with better efficiency (Garbinski et al., 2019). Arsenate detoxification is mediated by either cytoplasmic reduction of pentavalent arsenate to trivalent arsenite by the ArsC protein or extrusion of trivalent arsenite via the ArsB or ArsAB pumps. Furthermore, the ArsC protein, encoded by the arsC gene in the Ars operon, triggers the transformation of inorganic arsenate to arsenite (Yan et al., 2019). Aquaglyceroporin channel AqpS was first identified in the Ars operon of *Sinorhizobium meliloti*, which gives arsenate resistance by interacting with arsenate reducer ArsC and substituting ArsB's role of eliminating arsenite from cells (Chen and Rosen, 2021; Yan et al., 2019).

7.4 MECHANISMS OF CADMIUM REMEDIATION

Processes including bacterial precipitation, protein and EPS binding, related Cd(II) absorption and efflux systems, and oxidative stress reduction and damage repair systems are all part of the Cd(II) ions remediation mechanism, as illustrated in Figure 7.3.

7.4.1 Bioprecipitation of Divalent Cadmium Ions

Bacteria can interact with divalent cadmium ions to produce hydrogen sulfide (H_2S), as in sulfate-reducing bacteria, resulting in internal or extracellular cadmium sulfide (CdS) precipitates (Gallardo-Benavente et al., 2019). Furthermore, urease-producing bacteria can use the enzyme urease to convert urea to carbonate in order to precipitate divalent cadmium ions (Zhao et al., 2019). Furthermore, certain bacteria increase the quantities of soil-soluble phosphate (PO_4^{3-}) and phosphoric acid (H_3PO_4), which can react with divalent cadmium to generate insoluble cadmium phosphate [$Cd_3(PO_4)_2$] (Gadd, 2010). Thus, the bioprecipitation process of bacteria includes the generation of carbonates, phosphates, and H_2S while interacting with divalent cadmium ions to create cadmium salts.

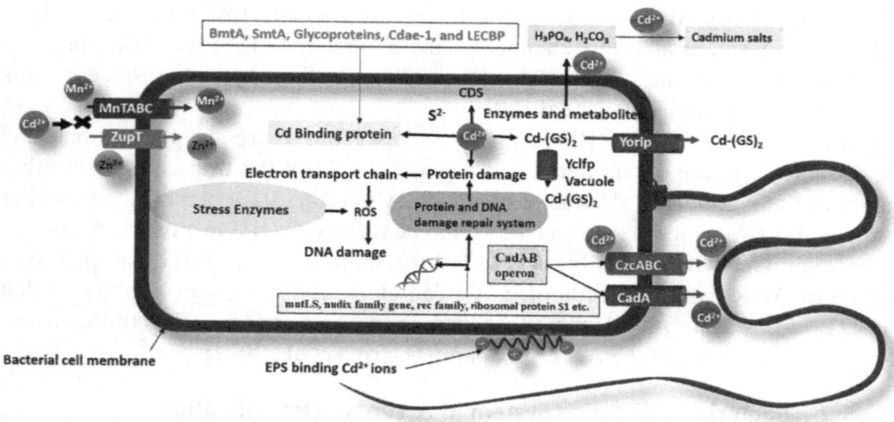

FIGURE 7.3 Mechanism of cadmium detoxification in bacteria. (Adapted from Xia et al., 2021.)

7.4.2 Divalent Cadmium Ion Binding by Proteins and EPS

To counteract the effects of divalent cadmium ions, bacteria can create metal-binding proteins such as metallothionein (BmtA) (Ziller and Fraissinet-Tachet, 2018). It has been reported that SmtA (a typical BmtA protein) expression is suppressed in the presence of SmtB but activated in the presence of divalent cadmium ions (Busenlehner et al., 2003). Glycoproteins, Cdae-1 (a cadmium binding protein), and LECBP (a non-anticadmium binding protein) are some of the most recently discovered Cd (II) binding proteins (Xia et al., 2021). Furthermore, the bacterial cell wall can bind Cd (II) ions by secreting EPSs, which can effectively trap Cd (II) ions. This is mostly due to changes in pH near EPSs, complex formation with amine and hydroxyl groups of EPSs, proton exchange, and negative charge adsorption (Comte et al. 2008). According to Mathivanan et al. (2021), EPS synthesis in the *B. cereus* KMS3-1 strain increased in the presence of cadmium metal stress. Furthermore, the hazardous Cd (II) ions were biotransformed into a less toxic and less soluble form, such as CdS, assisting in the detoxification procedure (Mathivanan et al., 2021). Another bacterium, *Rhizobium metallidurans* SIB-Pb-R3, produced more EPS under Cd (II) stress, with alterations in EPS morphology (Kowalkowski et al., 2019).

7.4.3 Bacterial System of Divalent Cadmium Ion Uptake and Efflux Pathways

Reduced Cd (II) uptake and enhanced Cd (II) efflux via manganese [Mn (II)] and zinc [Zn (II)] transporters, which have comparable chemical characteristics to Cd (II) ions, are another effective resistance mechanism used by bacteria to minimize Cd (II) toxicity (Ammendola et al., 2014). Divalent cadmium efflux proteins, which are involved in Cd (II) resistance in bacteria, can belong to a variety of membrane protein families, including the ATP binding cassette superfamily, the resistance-nodulation-division

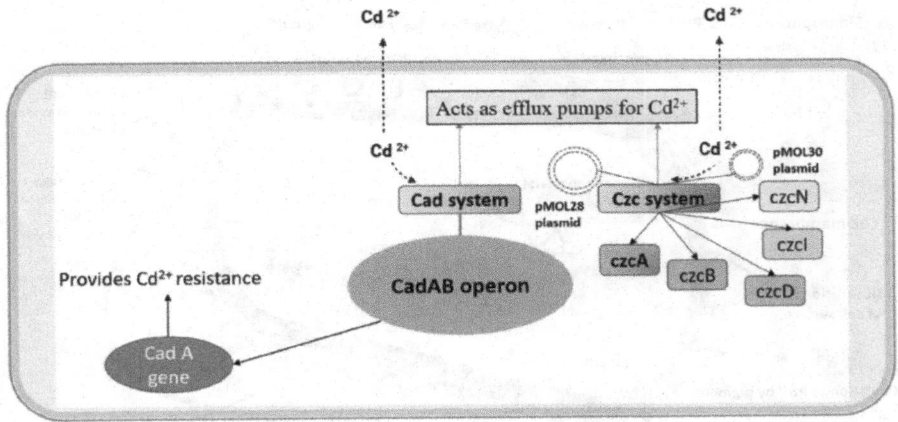

FIGURE 7.4 Cad system and Czc system of Cd efflux pathways and resistance in a bacterium. (Adapted from Yan et al., 2019.)

superfamily, and the multidrug and toxin extrusion family (Aygül, 2015; Abbas et al., 2018). Cadmium resistance is produced in *Alcaligenes eutrophus* and *Ralstonia* sp. CH34 through plasmid-mediated efflux pumps such as the Czc and Cad systems. The CadAB operon in the Cad system contains the *CadA* gene, which provides Cd (II) resistance, as shown in Figure 7.4 (Abbas et al., 2018). Similarly, in *A. eutrophus*, the Czc system, which is transcribed by plasmids pMOL28 and pMOL30, aids in the outflow of Cd (II), as well as zinc and cobalt. The Czc system is made up of numerous genes, including *czcA*, *czcB*, *czcD*, *czcI*, and *czcN*, as shown in Figure 7.4 (Abbas et al., 2018).

7.4.4 Divalent Cadmium Ions: Oxidative Stress Reduction and Damage Repair Systems

Bacterial systems can produce stress enzymes such as glutathione reductase, superoxide dismutase, glutathione, and catalase to reduce oxidative stress caused by Cd (II) ions (Xia et al., 2021). To alleviate the oxidative stress caused by Cd (II), microorganisms such as *Rhizobium* sp. can convert aldehyde to aldehyde-respective alcohol (Matos et al., 2019). Furthermore, for maintaining cellular integrity under Cd (II) stress, a number of damage repair systems are involved, including DNA repair genes (Joe et al., 2011) (mutLS, nudix family gene, rec family) as well as proteins responsible for specific synthesis and repair such as heat shock proteins, ribosomal protein S1, and energy metabolism proteins (Xia et al., 2021).

7.5 MECHANISM OF LEAD RESISTANCE IN BACTERIA

Bacterial lead resistance develops predominantly as a result of inherent mechanisms such as biosorption of lead on the bacterial cell surface or on extracellular polymeric substances, lead metal ion bioprecipitation, lead bioaccumulation, and lead binding by

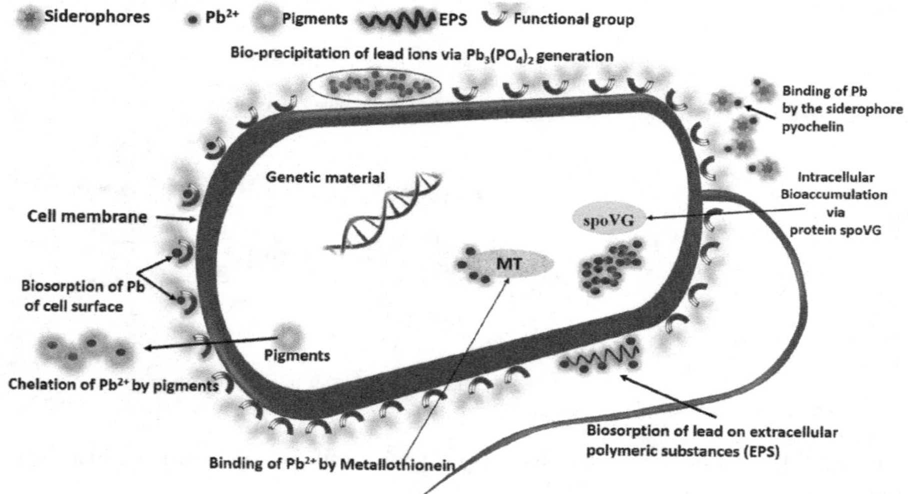

FIGURE 7.5 Mechanism of lead detoxification in bacteria. (Adapted from Tiquia-Arashiro, 2018.)

siderophores (Naik and Dubey, 2013; Tiquia-Arashiro, 2018). Furthermore, changes in cell structure, pigmentation, biotransformation of organo-lead compounds, and extrusion mechanisms for lead metal ion transport can all be attributed to lead resistance in bacteria (Kushwaha et al., 2018). The schematic illustration in Figure 7.5 depicts the processes of lead resistance in bacterial cells.

7.5.1 Biosorption of Lead on Bacterial Cell Surface

Bacteria can do biosorption efficiently due to their tiny size and quick production time. According to Ayangbenro et al. (2019), bacterial lead (Pb) biosorption is reliant on the adsorption of Pb on the bacterial surface. The bacterial Pb biosorption process may be divided into two phases: the first includes fast Pb binding to vacant binding sites, followed by a saturated binding phase (Jin et al., 2016). Bacterial cell walls are naturally negatively charged, allowing them to hold metal cations and control the flow of metal ions flow the membrane (Sultan and Haq, 2022). EPS and bacterial cell walls are rich in functional groups such as amino, carbonyl, amide, hydroxyl, and phosphate, which give Pb binding sites (Jin et al., 2016). However, the type of functional group present in the bacterial cell can play a major role in the bacterial cell's capacity to bind with cations such as Pb metal ions, as most Gram-positive bacteria have peptidoglycan carboxyl groups as potent metal cation binding sites, whereas Gram-negative bacteria may have a phosphate group for efficient metal binding (Gadd, 2009).

According to studies, pH and preliminary Pb concentration govern Pb^{2+} adsorption on bacterial cell surfaces, with biosorption of Pb peaking at an ideal pH when the bacterial surface's negative charges are at their highest (Çolak et al., 2011).

At low pH, however, Pb biosorption is reduced because Pb^{2+} ions compete with H^+ ions and positive charges are shown on the cell surface (Çolak et al., 2011). In a study, a lead-resistant *B. subtilis* X3, which was isolated from a lead mine plant, showed maximum adsorption of Pb^{2+} at pH 4 (Qiao et al., 2019). In another study, at pH 6, *B. subtilis* enhanced the bioleaching capacity of Pb^{2+} in alkaline soil, thereby enhancing Pb^{2+} mobility in the soil (Cai et al., 2018). The biosorption process, on the other hand, might vary depending on the bacteria or HMs involved. Pb biosorption was observed by Rahman et al. (2019) in both live and dead biomass of *Staphylococcus hominis* strain AMB2, although the biosorption was greater in living biomass.

7.5.2 Biosorption of Lead on Exopolysaccharides

EPS are found on the outer surface of bacterial cells to shield them from HM stress, and they are also involved in the bioremediation process (Sardar et al., 2018). These are capsule-like structures made up of a combination of proteins, carbohydrates, acids, lipids, and organic and inorganic substances (Sardar et al., 2018). Most investigations have found that tyrosine and tryptophan-like proteins play a key function in protecting against Pb metal ions (Teng et al., 2019). Bacterial EPS has been shown to help clean up Pb-contaminated environments in a variety of ways. Extracellular sequestration of Pb is also connected to EPS, with the proteins in EPS assisting bacteria in acquiring resistance to Pb metal ions (Teng et al., 2019). Furthermore, investigations show that soluble EPS has the maximum Pb binding (Teng et al., 2019).

7.5.3 Bioprecipitation of Lead Metal Ions

Bacteria can aid in the reduction of hazardous metal bioavailability and toxicity by generating precipitates of insoluble complexes (Naik and Dubey, 2013). Divalent lead has been reported to form numerous insoluble precipitates with chlorides, phosphates, and hydroxyl ions, as well as carbonates, either intracellularly or extracellularly (Levinson et al., 1996). In a study by Mwandira et al. (2017), bioremediation of a Pb-contaminated mine area was done using *Pararhodobacter* sp. via the microbially induced calcium carbonate precipitation (MICP) process. Research findings on ureolytic bacteria such as *Pararhodobacter* sp. indicate complete removal of Pb^{2+} from soil when present in a concentration of 0.5 mM (Mwandira et al., 2017). Other similar bacteria, such as *Sporosarcina pasteurii* (Mugwar and Harbottle, 2016), *Terrabacter tumescens* (Li et al., 2016b), and *Rhodobacter spharoides* (Li et al., 2016c), have shown Pb removal efficiency greater than 90% (Tiquia-Arashiro, 2018).

Furthermore, bacteria like *Providentia alcalifaciens* strain 2EA, which is resistant to lead, may bioprecipitate Pb^{2+} by generating lead phosphate (Naik et al., 2013). Similarly, bacteria such as *Klebsiella* sp., *B. iodinium* GP13, and *B. pumilus* S3 have been reported to form precipitates of lead(II) sulfide (Pan et al., 2017). Thus, bioprecipitation of lead metal ions can be successfully done with the help of bacteria in a much more economic and greener way.

7.5.4 Intracellular Bioaccumulation of Lead

The bioaccumulation of lead metal ions inside the bacterial cell occurs through the binding of metallothionein with the lead metal ions (Joutey et al., 2015). The synthesis of bacterial metallothionein is, however, associated with increased metal ion exposure (Tiquia-Arashiro, 2018). The metallothioneins are basically cysteine-rich proteins that can immobilize the HMs and minimize microbial stress (Tiquia-Arashiro, 2018). Silverberg et al. (1976) first reported the intracellular accumulation of Pb metal ions when the bacterial cells were exposed to trimethyl lead acetate. According to Sharma et al. (2006), when exposed to lead, the bacterial strain *P. fluorescens* produced 18 distinct proteins, but only one of them was related to a spoVG protein, which is crucial for Pb metal ion absorption.

7.5.5 Siderophore-Binding with Lead

Siderophores are regarded as microbial chelators that act as mediators for iron transport in cells along with other metals (Saha et al., 2016). Moreover, lead-resistant bacterial strains get enhanced siderophore production on exposure to lead nitrate (Tiquia-Arashiro, 2018). Siderophores like 'pyochelin' are reported to be involved in the process of Pb^{2+} binding in the bacterial strain *P. aeruginosa* PA01 (Braud et al., 2010). This inherent capacity of bacterial siderophores decreases Pb^{2+} ion mobility through the formation of stable metal–ligand complexes and thus helps in the detoxification of Pb-polluted environments.

7.5.6 Modifications in Cell Morphology

When bacteria are frequently exposed to hazardous HMs and organic compounds, their cell structure changes dramatically (Naik and Santosh, 2013). Bacterial morphology change is one of the techniques they use to deal with environmental challenges. When a Pb-resistant bacterial strain, *P. aeruginosa* 4EA, was exposed to a 0.8 mM lead nitrate solution, Naik and Dubey (2011) noticed a significant change in bacterial cell shape and size. Changes in cell shape in lead-exposed *P. aeruginosa* strain 4EA appear to indicate the bacterial cells' response to toxic lead (Naik and Dubey, 2011). Similarly, another Pb-resistant bacterial strain, *Enterobacter cloacae* P2B, when exposed to 1.6 mM lead nitrate, showed significant changes in cell morphology, such as cell size and shrinkage (Naik et al., 2012).

7.5.7 Pigmentations in Pb-Resistant Bacteria

Bacterial pigmentation is most commonly seen in areas where there are large concentrations of organic pollutants or inorganic pollutants such as HMs (Naik and Dubey, 2013). The yellowish-green pigmentation observed in *P. chlororaphis* is due to the pigment 'pyoverdine' is responsible for the breakdown of triphenyltin (Kushwaha et al., 2018). Moreover, siderophores pigments such as pyoverdine and pyochelin, produced from *P. aeruginosa,* are responsible for the solubilization of Fe^{3+} as well its transfer into bacterial cells with the aid of partial receptors (Naik and Dubey, 2013).

HM ions such as Pb^{2+}, Cd^{2+}, and Zn^{2+} can also form stable complexes with these microbial siderophores (Naik and Dubey, 2013). Moreover, it has been observed that the Pb-resistant strains *P. vesicularis* and *Streptomyces* sp. produce red and red-brown coloring on exposure to lead nitrate (Naik and Dubey, 2013).

7.5.8 Biotransformation of Organo-Lead Compounds

Organo-lead compounds such as tetraethyl lead and tetramethyl lead are widely used as antiknocking agents in gasoline (Adarakatti and Ashoka, 2022). Moreover, these compounds are extremely mutagenic and teratogenic (Priyadarshanee et al., 2022). Tetraethyl lead and TML compounds are very volatile and photolyzed (Naik and Dubey, 2013). Microbial consortiums or certain endogenic bacteria are capable of decomposing organo-lead compounds via the biotransformation pathway; the degradation of the organo-lead compounds usually occurs from tri-alkyl to di-alkyl molecules and finally to inorganic lead (Naik and Dubey, 2013).

7.5.9 Efflux Mechanisms within Bacteria

Many lead-resistant bacteria have an efflux mechanism for lead detoxification that is guided by cell membrane proteins and includes the active transport of toxic Pb^{2+} ions from the cell and into the surrounding environments (Kushwaha et al., 2018). Efflux proteins are broadly classified into three families: (1) Capsule Biogenesis Assembly transporters, (2) Cation Diffusion Facilitators, and (3) P-type Adenosine Triphosphatases (ATPases) (Rehan and Alsohim, 2019). The Capsule Biogenesis Assembly transporters extrude toxic cations extracellularly, from the cytoplasm and periplasm via the plasma membrane. Transporters are only found in divalent metals (Sevak et al., 2021). When HM concentrations in the cell cytoplasm are low, cation diffusion facilitators play a significant function (Rehan and Alsohim, 2019). Efflux transporters' major function is to maintain the equilibrium of essential metal ions in the cell while also protecting the organism from toxic HMs like Pb (Hynninen et al., 2009; Hynninen, 2010). Because of the invariant proline (P) that is followed and/or preceded by cysteine (C), P-type ATPases are also known as CPx-type ATPases (Sevak et al., 2021).

7.6 CONCLUSION

Arsenic, cadmium, and lead have been identified as the principal contributors of HM pollution in both terrestrial and aquatic environments, and their levels have been steadily escalating with the growth of contemporary industrial and agricultural activities. This, in turn, has a major detrimental influence on the various ecosystem components, which need utmost attention. Among the existing remediation approaches, microbial remediation, particularly bacterial-based systems, has played an important role in the decontamination of HM pollutants. Such organisms have a variety of detoxification pathways against toxic metal ions, including sequestration, adsorption to the bacterial cell surface via functional groups, accumulation (intracellularly or extracellularly), oxidation, and reduction into less toxic or unstable compounds. Furthermore, such bacteria may generate EPS, siderophores, metal complexes, or engage in bioprecipitation to capture or eliminate detrimental toxic metal ions. In certain instances, lead-resistant bacterial species exhibit pigmentation or morphological alterations. In general, most of the metallophilic bacteria have been reported to be effective bisorbents against harmful HM ions favored by various physical and chemical elements. The present chapter highlights contemporary bioremediation research findings and acts as a link between bacterial detoxification mechanisms and their adaptation in a contaminated environment with arsenic, cadmium, and lead. Although numerous resistance/detoxification mechanisms have been discovered, the precise interactions between HM ions and microbes remain unknown and vary among bacterial species. Hence, additional study is needed to eliminate the research gaps so that the technology may be implemented more successfully in large-scale field applications.

ACKNOWLEDGEMENTS

We thank the Department of Environmental Science at Nagaland University and the UGC Non-NET and NFST fellowships (Ministry of Tribal Affairs) for their financial assistance in carrying out our study. We also thank Miss Sangeeta Gogoi Gohain for her guidance and support during the study.

REFERENCES

Abbas, S. Z., M. Rafatullah, K. Hossain, N. Ismail, H. A. Tajarudin, and H. P. S. Abdul Khalil. "A review on mechanism and future perspectives of cadmium-resistant bacteria." *International Journal of Environmental Science and Technology* 15, no. 1 (2018): 243–262.

Abdelhamid, Hani Nasser. "Quantum dots hybrid systems for drug delivery." In *Hybrid Nanomaterials for Drug Delivery*, Prashant Kesharwani and N.K. Jain Sawston (Eds.), pp. 323–338. Woodhead Publishing, Sawston, Cambridge, 2022.

Adarakatti, Prashanth Shivappa, and S. Ashoka. "Nanocomposite materials interface for heavy metal ions detection." In *Nanocomposite Materials for Sensor*, Manorama Singh, Vijai K Rai, and Ankita Rai (Eds.), p. 163. Bentham Science Publishers, Sharjah, 2022.

Ali, Sajad, Anshika Tyagi, Muntazir Mushtaq, Henda Al-Mahmoudi, and Hanhong Bae. "Harnessing plant microbiome for mitigating arsenic toxicity in sustainable agriculture." *Environmental Pollution* 300 (2022): 118940.

Altintas, Zeynep, Frank Davis, and Frieder W. Scheller. "Applications of quantum dots in biosensors and diagnostics." *Biosensors and Nanotechnology: Applications in Health Care Diagnostics* (2018): 183–199.

Amjad, Muhammad, Muhammad Mohsin Iqbal, Ghulam Abbas, Abu Bakar Umer Farooq, Muhammad Asif Naeem, Muhammad Imran, Behzad Murtaza, Muhammad Nadeem, and Sven-Erik Jacobsen. "Assessment of cadmium and lead tolerance potential of quinoa (*Chenopodium quinoa* Willd) and its implications for phytoremediation and human health." *Environmental Geochemistry and Health* 44, no. 5 (2022): 1487–1500.

Ammendola, Serena, Mauro Cerasi, and Andrea Battistoni. "Deregulation of transition metals homeostasis is a key feature of cadmium toxicity in *Salmonella*." *Biometals* 27, no. 4 (2014): 703–714.

Aryal, M., M. Ziagova, and M. Liakopoulou-Kyriakides. "Study on arsenic biosorption using Fe (III)-treated biomass of *Staphylococcus xylosus*." *Chemical Engineering Journal* 162, no. 1 (2010): 178–185.

Aryal, M., M. Ziagova, and M. Liakopoulou-Kyriakides. "Comparison of Cr (VI) and As (V) removal in single and binary mixtures with Fe (III)-treated *Staphylococcus xylosus* biomass: thermodynamic studies." *Chemical Engineering Journal* 169, no. 1–3 (2011): 100–106.

Ayangbenro, Ayansina Segun, Olubukola Oluranti Babalola, and Oluwole Samuel Aremu. "Bioflocculant production and heavy metal sorption by metal resistant bacterial isolates from gold mining soil." *Chemosphere* 231 (2019): 113–120.

Aygül, Abdurrahman. "The importance of efflux systems in antibiotic resistance and efflux pump inhibitors in the management of resistance." *Mikrobiyoloji bulteni* 49, no. 2 (2015): 278–291.

Bayraktar, Emine Pirinc, Oznur Isinkaralar, and Kaan Isinkaralar. "Usability of several species for monitoring and reducing the heavy metal pollution threatening the public health in urban environment of Ankara." *World Journal of Advanced Research and Reviews* 14, no. 3 (2022): 276–283.

Biswas, Rimi, Vivekanand Vivekanand, Anima Saha, Ashok Ghosh, and Angana Sarkar. "Arsenite oxidation by a facultative chemolithotrophic *Delftia* spp. BAs29 for its potential application in groundwater arsenic bioremediation." *International Biodeterioration & Biodegradation* 136 (2019): 55–62.

Bouida, Leila, Mohd Rafatullah, Abdelfateh Kerrouche, Mohammad Qutob, Abeer M. Alosaimi, Hajer S. Alorfi, and Mahmoud A. Hussein. "A review on cadmium and lead contamination: Sources, fate, mechanism, health effects and remediation methods." *Water* 14, no. 21 (2022): 3432.

Bradl, H. B. "Sources and origins of heavy metals." In *Interface Science and Technology*, vol. 6, pp. 1–27. Elsevier, 2005.

Braud, Armelle, Valérie Geoffroy, Françoise Hoegy, Gaëtan L. A. Mislin, and Isabelle J. Schalk. "Presence of the siderophores pyoverdine and pyochelin in the extracellular medium reduces toxic metal accumulation in *Pseudomonas aeruginosa* and increases bacterial metal tolerance." *Environmental Microbiology Reports* 2, no. 3 (2010): 419–425.

Bundschuh, Jochen, Nabeel Khan Niazi, Mohammad Ayaz Alam, Michael Berg, Indika Herath, Barbara Tomaszewska, Jyoti Prakash Maity, and Yong Sik Ok. "Global arsenic dilemma and sustainability." *Journal of Hazardous Materials* 436 (2022): 129197.

Busenlehner, Laura S., Mario A. Pennella, and David P. Giedroc. "The SmtB/ArsR family of metalloregulatory transcriptional repressors: structural insights into prokaryotic metal resistance." *FEMS Microbiology Reviews* 27, no. 2-3 (2003): 131–143.

Cai, Limei, Zhencheng Xu, Mingzhong Ren, Qingwei Guo, Xibang Hu, Guocheng Hu, Hongfu Wan, and Pingan Peng. "Source identification of eight hazardous heavy metals in agricultural soils of Huizhou, Guangdong Province, China." *Ecotoxicology and Environmental Safety* 78 (2012): 2–8.

Cai, Yue, Xiaoping Li, Dongying Liu, Changlin Xu, Yuwei Ai, Xuemeng Sun, Meng Zhang et al. "A novel Pb-resistant *Bacillus subtilis* bacterium isolate for co-biosorption of hazardous Sb (III) and Pb (II): thermodynamics and application strategy." *International Journal of Environmental Research and Public Health* 15, no. 4 (2018): 702.

Chang, Jin-Soo, In-Ho Yoon, and Kyoung-Woong Kim. "Arsenic biotransformation potential of microbial arsH responses in the biogeochemical cycling of arsenic-contaminated groundwater." *Chemosphere* 191 (2018): 729–737.

Chen, Jian, Hiranmoy Bhattacharjee, and Barry P. Rosen. "ArsH is an organoarsenical oxidase that confers resistance to trivalent forms of the herbicide monosodium methylarsenate and the poultry growth promoter roxarsone." *Molecular Microbiology* 96, no. 5 (2015a): 1042–1052.

Chen, Jian, Mahendra Madegowda, Hiranmoy Bhattacharjee, and Barry P. Rosen. "ArsP: a methylarsenite efflux permease." *Molecular Microbiology* 98, no. 4 (2015b): 625–635.

Chen, Jian, Venkadesh Sarkarai Nadar, and Barry P. Rosen. "Aquaglyceroporin AqpS from *Sinorhizobium meliloti* conducts both trivalent and pentavalent methylarsenicals." *Chemosphere* 270 (2021): 129379.

Chen, Jian, and Barry P. Rosen. "The arsenic methylation cycle: how microbial communities adapted methylarsenicals for use as weapons in the continuing war for dominance." *Frontiers in Environmental Science* 8 (2020): 43.

Chen, Jian, Masafumi Yoshinaga, Luis D. Garbinski, and Barry P. Rosen. "Synergistic interaction of glyceraldehydes-3-phosphate dehydrogenase and ArsJ, a novel organoarsenical efflux permease, confers arsenate resistance." *Molecular Microbiology* 100, no. 6 (2016): 945–953.

Çolak, Ferdağ, Necip Atar, Demet Yazıcıoğlu, and Asim Olgun. "Biosorption of lead from aqueous solutions by *Bacillus* strains possessing heavy-metal resistance." *Chemical Engineering Journal* 173, no. 2 (2011): 422–428.

Comte, S., Gilles Guibaud, and Michel Baudu. "Biosorption properties of extracellular polymeric substances (EPS) towards Cd, Cu and Pb for different pH values." *Journal of Hazardous Materials* 151, no. 1 (2008): 185–193.

Cook, J. "Environmental pollution by heavy metals." *International Journal of Environmental Studies* 10, no. 4 (1977): 253–266.

Dalei, Kalpana. "Synthesis and characterization of cadmium sulphide nanoparticles and its utilization in removal of cadmium from aqueous solution." PhD diss., 2014.

Darma, Aminu, Jianjun Yang, Peiman Zandi, Jin Liu, Katarzyna Możdżeń, Xing Xia, Ali Sani, Yihao Wang, and Ewald Schnug. "Significance of *shewanella* species for the phytoavailability and toxicity of arsenic-a review." *Biology* 11, no. 3 (2022): 472.

Darwish, Ayman M., Wael H. Eisa, Ali A. Shabaka, and Mohamed H. Talaat. "Investigation of factors affecting the synthesis of nano-cadmium sulfide by pulsed laser ablation in liquid environment." *Spectrochimica Acta Part A: Molecular and Biomolecular Spectroscopy* 153 (2016): 315–320. .

Deng, Min, Kai Li, Yu-Jian Yan, Fei Huang, and Dan Peng. "Enhanced cadmium removal by growing *Bacillus cereus* RC-1 immobilized on different magnetic biochars through simultaneous adsorption and bioaccumulation." *Environmental Science and Pollution Research* 29, no. 13 (2022): 18495–18507.

Feria-Cáceres, Pedro F., Lucas Penagos-Velez, and Claudia X. Moreno-Herrera. "Tolerance and cadmium (Cd) immobilization by native bacteria isolated in cocoa soils with increased metal content." *Microbiology Research* 13, no. 3 (2022): 556–573.

Gadd, Geoffrey Michael. "Biosorption: critical review of scientific rationale, environmental importance and significance for pollution treatment." *Journal of Chemical Technology & Biotechnology: International Research in Process, Environmental & Clean Technology* 84, no. 1 (2009): 13–28.

Gadd, Geoffrey Michael. "Metals, minerals and microbes: geomicrobiology and bioremediation." *Microbiology* 156, no. 3 (2010): 609–643.

Gallardo-Benavente, Carla, Ornella Carrión, Jonathan D. Todd, Joana C. Pieretti, Amedea B. Seabra, Nelson Durán, Olga Rubilar, José M. Pérez-Donoso, and Andrés Quiroz. "Biosynthesis of CdS quantum dots mediated by volatile sulfur compounds released by Antarctic *Pseudomonas fragi*." *Frontiers in Microbiology* (2019): 1866.

Garbinski, Luis D., Barry P. Rosen, and Jian Chen. "Pathways of arsenic uptake and efflux." *Environment International* 126 (2019): 585–597.

Ghosal, Kavita, Moumita Chatterjee, Sharmistha Ganguly, Subhamita Sen Niyogi, and Dwaipayan Sinha. "Arsenic induced responses in plants: impacts on different plant groups, from cyanobacteria to higher plants." In *Arsenic in Plants: Uptake, Consequences and Remediation Techniques*, Prabhat Kumar Srivastava, Rachana Singh, Parul Parihar, and Sheo Mohan Prasad (Eds.), pp. 64–98. John Wiley & Sons Ltd., Hoboken, NJ, 2022.

Girawale, Savita D., Surya N. Meena, Vinod S. Nandre, Suresh B. Waghmode, and Kisan M. Kodam. "Biosynthesis of vanillic acid by *Ochrobactrum anthropi* and its applications." *Bioorganic & Medicinal Chemistry* 72 (2022): 117000.

Goyal, Taru, Prasenjit Mitra, Preeti Singh, Shailja Sharma, and Praveen Sharma. "Assessement of blood lead and cadmium levels in occupationally exposed workers of Jodhpur, Rajasthan." *Indian Journal of Clinical Biochemistry* 36, no. 1 (2021): 100–107.

Gujre, Nihal, Latha Rangan, and Sudip Mitra. "Occurrence, geochemical fraction, ecological and health risk assessment of cadmium, copper and nickel in soils contaminated with municipal solid wastes." *Chemosphere* 271 (2021): 129573.

Gürgan, Muazzez, Eylül İrem İrez, and Sevinç Adiloğlu. "Understanding bioremediation of metals and metalloids by genomic approaches." In *Omics Insights in Environmental Bioremediation*, Vineet Kumar, Indu Shekhar Thakur (Eds.), pp. 375–392. Springer, Singapore, 2022.

Haque, Farhana, Ishrat Jabeen, Chaman Ara Keya, and Sabbir R. Shuvo. "Whole-genome sequencing and comparative analysis of heavy metals tolerant *Bacillus anthracis* FHq strain isolated from tannery effluents in Bangladesh." *AIMS Microbiology* 8, no. 2 (2022): 227.

Hassan, Syeda Sara, Syed Shane Zehra, Zubair Ahmed, Mohammad Younis Younis Talpur, and Sallahuddin Panhwar. "Removal of arsenic contaminants from water using iron oxide nanoparticles by inductively coupled plasma mass spectrometery." (2022).

Hynninen, Anu. "Zinc, cadmium and lead resistance mechanisms in bacteria and their contribution to biosensing." (2010). https://helda.helsinki.fi/bitstream/handle/10138/20645/zinccadm.pdf?sequence=1. Accessed 29 April 2022.

Hynninen, Anu, Thierry Touzé, Leena Pitkänen, Dominique Mengin-Lecreulx, and Marko Virta. "An efflux transporter PbrA and a phosphatase PbrB cooperate in a lead-resistance mechanism in bacteria." *Molecular Microbiology* 74, no. 2 (2009): 384–394.

Ibrahim, U. B., S. Yahaya, I. Yusuf, and A. H. Kawo. "Cadmium (Cd) and lead (Pb) uptake potential and surface properties of *Aeromonas* spp. isolated from soil of local mining site." *Microbiology Research Journal International* 30, no. 3 (2020): 36–47.

Irshad, Sana, Zuoming Xie, Sajid Mehmood, Asad Nawaz, Allah Ditta, and Qaisar Mahmood. "Insights into conventional and recent technologies for arsenic bioremediation: a systematic review." *Environmental Science and Pollution Research* 28, no. 15 (2021): 18870–18892.

Jia, Yan, Hai Huang, Min Zhong, Feng-Hua Wang, Li-Mei Zhang, and Yong-Guan Zhu. "Microbial arsenic methylation in soil and rice rhizosphere." *Environmental Science & Technology* 47, no. 7 (2013): 3141–3148.

Jiang, Nan, Hongyan Li, and Hongzhe Sun. "Recognition of proteins by metal chelation-based fluorescent probes in cells." *Frontiers in Chemistry* 7 (2019): 560.

Jin, Yu, Xin Wang, Tingting Zang, Yang Hu, Xiaojing Hu, Guangming Ren, Xiuhong Xu, and Juanjuan Qu. "Biosorption of lead (II) by *Arthrobacter* sp. 25: process optimization and mechanism." *Journal of Microbiology and Biotechnology* 26, no. 8 (2016): 1428–1438.

Joe, Min-Ho, Sun-Wook Jung, Seong-Hun Im, Sang-Yong Lim, Hyun-Pa Song, Oh-Suk Kwon, and Dong-Ho Kim. "Genome-wide response of *Deinococcus radiodurans* on cadmium toxicity." *Journal of Microbiology and Biotechnology* 21, no. 4 (2011): 438–447.

Joutey, Nezha Tahri, Hanane Sayel, Wifak Bahafid, and Naïma El Ghachtouli. "Mechanisms of hexavalent chromium resistance and removal by microorganisms." *Reviews of Environmental Contamination and Toxicology* 233 (2015): 45–69.

Jyothi, Narjala Rama. "Heavy metal sources and their effects on human health." In *Heavy Metals-Their Environmental Impacts and Mitigation*, Mazen Nazal and Hongbo Zhao (Eds.), IntechOpen, London, 2020.

Kesler, Stephen E., Adam C. Simon, and Adam F. Simon. *Mineral Resources, Economics and The Environment*. Cambridge University Press, Cambridge, 2015.

Khan, Abdul Ahad, Jawad Gul, Salman Raza Naqvi, Imtiaz Ali, Wasif Farooq, Rabia Liaqat, Hamad AlMohamadi, Libor Štěpanec, and Dagmar Juchelková. "Recent progress in microalgae-derived biochar for the treatment of textile industry wastewater." *Chemosphere* (2022): 135565.

Khosravi-Darani, Kianoush, Yasir Rehman, Ioannis A. Katsoyiannis, Evgenios Kokkinos, and Anastasios I. Zouboulis. "Arsenic exposure via contaminated water and food sources." *Water* 14, no. 12 (2022): 1884.

Kowalkowski, Tomasz, Aneta Krakowska, Michał Złoch, Katarzyna Hrynkiewicz, and Bogusław Buszewski. "Cadmium-affected synthesis of exopolysaccharides by rhizosphere bacteria." *Journal of Applied Microbiology* 127, no. 3 (2019): 713–723.

Kruger, Martin C., Philippe N. Bertin, Hermann J. Heipieper, and Florence Arsène-Ploetze. "Bacterial metabolism of environmental arsenic-mechanisms and biotechnological applications." *Applied Microbiology and Biotechnology* 97, no. 9 (2013): 3827–3841.

Kumar, Sazal, Md Aminur Rahman, Md Rashidul Islam, Md Abul Hashem, and Mohammad Mahmudur Rahman. "Lead and other elements-based pollution in soil, crops and water near a lead-acid battery recycling factory in Bangladesh." *Chemosphere* 290 (2022): 133288.

Kushwaha, Anamika, Nidhi Hans, Sanjay Kumar, and Radha Rani. "A critical review on speciation, mobilization and toxicity of lead in soil-microbe-plant system and bioremediation strategies." *Ecotoxicology and Environmental Safety* 147 (2018): 1035–1045.

Lata, Sneh, Hemant Preet Kaur, and Tulika Mishra. "Cadmium bioremediation: a review." *International Journal of Pharmaceutical Sciences and Research* 10, no. 9 (2019): 4120–4128.

Lett, Marie-Claire, Daniel Muller, Didier Lièvremont, Simon Silver, and Joanne Santini. "Unified nomenclature for genes involved in prokaryotic aerobic arsenite oxidation." *Journal of Bacteriology* 194, no. 2 (2012): 207–208.

Levinson, Hillel S., Inga Mahler, Patricia Blackwelder, and Terri Hood. "Lead resistance and sensitivity in *Staphylococcus aureus*." *FEMS Microbiology Letters* 145, no. 3 (1996): 421–425.

Li, Jiaojiao, Shashank S. Pawitwar, and Barry P. Rosen. "The organoarsenical biocycle and the primordial antibiotic methylarsenite." *Metallomics* 8, no. 10 (2016a): 1047–1055.

Li, Liang, Bo Zhang, Binhui Jiang, Yan Zhao, Guangsheng Qian, and Xiaomin Hu. "Potentially toxic elements in weathered waste-rocks of Fushun western opencast mine: distribution, source identification, and contamination assessment." *Environmental Geochemistry and Health* 44, no. 6 (2022): 1813–1826.

Li, Meng, Xiaohui Cheng, Hongxian Guo, and Zuan Yang. "Biomineralization of carbonate by Terrabacter tumescens for heavy metal removal and biogrouting applications." *Journal of Environmental Engineering* 142, no. 9 (2016b): C4015005.

Li, Xiaomin, Weihua Peng, Yingying Jia, Lin Lu, and Wenhong Fan. "Bioremediation of lead contaminated soil with *Rhodobacter sphaeroides*." *Chemosphere* 156 (2016c): 228–235.

Liu, Su, Xuechao Guo, Bing Wu, Haiyan Yu, Xuxiang Zhang, and Mei Li. "Arsenic induces diabetic effects through beta-cell dysfunction and increased gluconeogenesis in mice." *Scientific Reports* 4, no. 1 (2014): 1–10.

Lizárraga, Wendy C., Carlo G. Mormontoy, Hedersson Calla, Maria Castañeda, Mario Taira, Ruth Garcia, Claudia Marín, Michel Abanto, and Pablo Ramirez. "Complete genome sequence of *Shewanella algae* strain 2NE11, a decolorizing bacterium isolated from industrial effluent in Peru." *Biotechnology Reports* 33 (2022): e00704.

Lu, Wei-Bin, Jun-Ji Shi, Ching-Hsiung Wang, and Jo-Shu Chang. "Biosorption of lead, copper and cadmium by an indigenous isolate *Enterobacter* sp. J1 possessing high heavy-metal resistance." *Journal of Hazardous Materials* 134, no. 1–3 (2006): 80–86.

Ludolphy, Catharina, Uwe Kierdorf, and Horst Kierdorf. "Antlers of European roe deer (*Capreolus capreolus*) as monitoring units to assess lead pollution in a floodplain contaminated by historical metal ore mining, processing, and smelting in the Harz Mountains, Germany." *Environmental Pollution* 302 (2022): 119080.

Mahmoud, Mohamed E., Nesma A. Fekry, and Sally M. S. Mohamed. "Effective removal of Pb (II)/4-nitroaniline/*E. faecalis* and *E. coli* pollutants from water by a novel unique graphene quantum dots@ gemifloxacin@ double-layered Fe/Al nanocomposite." *Journal of Water Process Engineering* 46 (2022): 102562.

Masindi, Vhahangwele, and Khathutshelo L. Muedi. "Environmental contamination by heavy metals." In *Heavy Metals*, Hosam M. Saleh and Refaat Aglan (Eds.). IntechOpen, London, 2018.

Masri, Shahir, Alana M. W. LeBrón, Michael D. Logue, Patricia Flores, Abel Ruiz, Abigail Reyes, Juan Manuel Rubio, and Jun Wu. "Use of radioisotope ratios of lead for the identification of historical sources of soil lead contamination in Santa Ana, California." *Toxics* 10, no. 6 (2022): 304.

Mathivanan, Krishnamurthy, Jayaraman Uthaya Chandirika, Thangavel Mathimani, Rajendran Rajaram, Gurusamy Annadurai, and Huaqun Yin. "Production and functionality of exopolysaccharides in bacteria exposed to a toxic metal environment." *Ecotoxicology and Environmental Safety* 208 (2021): 111567.

Matos, Diana, Carina Sá, Paulo Cardoso, Adília Pires, Sílvia M. Rocha, and Etelvina Figueira. "The role of volatiles in *Rhizobium* tolerance to cadmium: effects of aldehydes and alcohols on growth and biochemical endpoints." *Ecotoxicology and Environmental Safety* 186 (2019): 109759.

Mo, Dan, Liang Hu, Guangming Zeng, Guiqiu Chen, Jia Wan, Zhigang Yu, Zhenzhen Huang, Kai He, Chen Zhang, and Min Cheng. "Cadmium-containing quantum dots: properties, applications, and toxicity." *Applied Microbiology and Biotechnology* 101, no. 7 (2017): 2713–2733.

Mobar, Sanjoli, and Pradeep Bhatnagar. "Arsenic bioremediation potential of arsenite oxidizing bacteria isolated from geogenic and anthropogenically contaminated soil." *Pollution* 8, no. 4 (2022): 1137–1149.

Mohd Bahari, Zaratulnur, Wahid Ali Hamood Altowayti, Zaharah Ibrahim, Jafariah Jaafar, and Shafinaz Shahir. "Biosorption of As (III) by non-living biomass of an arsenic-hypertolerant *Bacillus cereus* strain SZ2 isolated from a gold mining environment: equilibrium and kinetic study." *Applied Biochemistry and Biotechnology* 171, no. 8 (2013): 2247–2261.

Mohsin, Hareem, Maria Shafique, and Yasir Rehman. "Genes and biochemical pathways involved in microbial transformation of arsenic." In *Arsenic Toxicity: Challenges and Solutions*, Nitish Kumar (Ed.), pp. 391–413. Springer, Singapore, 2021.

Monachese, Marc, Jeremy P. Burton, and Gregor Reid. "Bioremediation and tolerance of humans to heavy metals through microbial processes: a potential role for probiotics?." *Applied and Environmental Microbiology* 78, no. 18 (2012): 6397–6404.

Mondal, Sayanta, Krishnendu Pramanik, Sudip Kumar Ghosh, Priyanka Pal, Tanushree Mondal, Tithi Soren, and Tushar Kanti Maiti. "Unraveling the role of plant growth-promoting rhizobacteria in the alleviation of arsenic phytotoxicity: a review." *Microbiological Research* 250 (2021): 126809.

Mugwar, Ahmed J., and Michael J. Harbottle. "Toxicity effects on metal sequestration by microbially-induced carbonate precipitation." *Journal of Hazardous Materials* 314 (2016): 237–248.

Mwandira, Wilson, Kazunori Nakashima, and Satoru Kawasaki. "Bioremediation of lead-contaminated mine waste by *Pararhodobacter* sp. based on the microbially induced calcium carbonate precipitation technique and its effects on strength of coarse and fine grained sand." *Ecological Engineering* 109 (2017): 57–64.

Nadar, S. Venkadesh, Masafumi Yoshinaga, Palani Kandavelu, Banumathi Sankaran, and Barry P. Rosen. "Crystallization and preliminary X-ray crystallographic studies of the ArsI C-As lyase from *Thermomonospora curvata*." *Acta Crystallographica Section F: Structural Biology Communications* 70, no. 6 (2014): 761–764.

Naik, M. M., D. Khanolkar, and S. K. Dubey. "Lead-resistant *Providencia alcalifaciens* strain 2 EA bioprecipitates Pb^{+2} as lead phosphate." *Letters in Applied Microbiology* 56, no. 2 (2013): 99–104.

Naik, Milind Mohan, and Santosh Kumar Dubey. "Lead-enhanced siderophore production and alteration in cell morphology in a Pb-resistant *Pseudomonas aeruginosa* strain 4EA." *Current Microbiology* 62, no. 2 (2011): 409–414.

Naik, Milind Mohan, and Santosh Kumar Dubey. "Lead resistant bacteria: lead resistance mechanisms, their applications in lead bioremediation and biomonitoring." *Ecotoxicology and Environmental Safety* 98 (2013): 1–7.

Naik, Milind Mohan, Anju Pandey, and Santosh Kumar Dubey. "Biological characterization of lead-enhanced exopolysaccharide produced by a lead resistant *Enterobacter cloacae* strain P2B." *Biodegradation* 23, no. 5 (2012): 775–783.

Okibe, Naoko, and Fukano Yuken. "Bioremediation of highly toxic arsenic via carbon-fiber-assisted indirect As (III) oxidation by moderately-thermophilic, acidophilic Fe-oxidizing bacteria." *Biotechnology letters* 41, no. 12 (2019): 1403–1413.

Okpara-Elom, Ifeoma Anthonia, Charles Chike Onochie, Michael Okpara Elom, Emmanuel Ezaka, and Ogbonnaya Elom. "Bioremediation of heavy metal polluted soil using plant growth promoting bacteria: an assessment of response." *Bioremediation Journal* (2022): 1–20.

Páez-Espino, David, Javier Tamames, Víctor de Lorenzo, and David Cánovas. "Microbial responses to environmental arsenic." *Biometals* 22, no. 1 (2009): 117–130.

Pan, Xiaohong, Zhi Chen, Lan Li, Wenhua Rao, Zhangyan Xu, and Xiong Guan. "Microbial strategy for potential lead remediation: a review study." *World Journal of Microbiology and Biotechnology* 33, no. 2 (2017): 1–7.

Parsoya, Shubham, and Asif Perwej. "Current needs of making changes in transportation and energy policies to mitigate the bad and harmful impacts of environmental pollution; an indian perspective." *An Indian Perspective* 10 (2022).

Prasad, Kumar Suranjit, Prashant Srivastava, Vaidyanathan Subramanian, and Jaishree Paul. "Biosorption of As (III) ion on *Rhodococcus* sp. WB-12: biomass characterization and kinetic studies." *Separation Science and Technology* 46, no. 16 (2011): 2517–2525.

Press, F., and R. Sievers. *Fundamentals of Geology*. Freeman and Company, New York, 1994

Priyadarshanee, Monika, Uma Mahto, and Surajit Das. "Mechanism of toxicity and adverse health effects of environmental pollutants." In *Microbial Biodegradation and Bioremediation*, Surajit Das and Hirak Ranjan Dash (Eds.), pp. 33–53. Elsevier, Amsterdam, 2022.

Qiao, Weichuan, Yunhao Zhang, Hao Xia, Yang Luo, Si Liu, Shiyu Wang, and Weihan Wang. "Bioimmobilization of lead by *Bacillus subtilis* X3 biomass isolated from lead mine soil under promotion of multiple adsorption mechanisms." *Royal Society Open Science* 6, no. 2 (2019): 181701.

Rahman, Zeeshanur, Lebin Thomas, and Ved Pal Singh. "Biosorption of heavy metals by a lead (Pb) resistant bacterium, *Staphylococcus hominis* strain AMB-2." *Journal of Basic Microbiology* 59, no. 5 (2019): 477–486.

Rajendran, Saravanan, T. A. K. Priya, Kuan Shiong Khoo, Tuan K. A. Hoang, Hui-Suan Ng, Heli Siti Halimatul Munawaroh, Ceren Karaman, Yasin Orooji, and Pau Loke Show. "A critical review on various remediation approaches for heavy metal contaminants removal from contaminated soils." *Chemosphere* 287 (2022): 132369.

Raju, N. Janardhana. "Arsenic in the geo-environment: a review of sources, geochemical processes, toxicity and removal technologies." *Environmental Research* 203 (2022): 111782.

Rani, Raksha, Preeti Sharma, Rajesh Kumar, and Younis Ahmad Hajam. "Effects of heavy metals and pesticides on fish." In *Bacterial Fish Diseases*, Gowhar Hamid Dar, Rouf Ahmad Bhat, Humaira Qadri, Khalid M. Al-Ghamdy and Khalid Rehman (Eds.), pp. 59–86. Academic Press, Hakeem, Cambridge, MA, 2022.

Rehan, Medhat, and Abdullah S. Alsohim. "Bioremediation of heavy metals." In *Environmental Chemistry and Recent Pollution Control Approaches*, Hugo Saldarriaga-Noreña, Mario Alfonso Murillo-Tovar, Robina Farooq, Rajendra Dongre, Sara Riaz (Eds.), p. 145. IntechOpen, London, 2019.

Rosariastuti, Retno, and Hapsari Ayu Astuti. "Study of potential plant of Biduri (*Calotropis gigantea*) with combination of bacteria, organic matter, and inorganic fertilizer for bioremediation of lead (Pb) contaminated soil." In *IOP Conference Series: Earth and Environmental Science*, vol. 1016, no. 1, p. 012016. IOP Publishing, 2022.

Rubio, Juan Manuel, Shahir Masri, Ivy R. Torres, Yi Sun, Keila Villegas, Patricia Flores, Michael D. Logue, Abigail Reyes, Alana MW LeBrón, and Jun Wu. "Use of historical mapping to understand sources of soil-lead contamination: case study of Santa Ana, CA." *Environmental Research* 212 (2022): 113478.

Ruikar, Aditya, and Hitesh S. Pawar. "Diversity and Interaction of Microbes in Biodegradation." In *Microbial Community Studies in Industrial Wastewater Treatment*, Aditya Ruikar and Hitesh S. Pawar (Eds.), pp. 185–213. CRC Press, London, 2022.

Saha, Maumita, Subhasis Sarkar, Biplab Sarkar, Bipin Kumar Sharma, Surajit Bhattacharjee, and Prosun Tribedi. "Microbial siderophores and their potential applications: a review." *Environmental Science and Pollution Research* 23, no. 5 (2016): 3984–3999.

Saha, Priti, and Biswajit Paul. "Assessment of heavy metal pollution in water resources and their impacts: a review." *Journal of Basic and Applied Engineering Research* 3 (2016): 671–675.

Sahmoune, Mohamed Nasser. "The role of biosorbents in the removal of arsenic from water." *Chemical Engineering & Technology* 39, no. 9 (2016): 1617–1628.

Sardar, Usha R., Erravelli Bhargavi, Indrama Devi, Biswanath Bhunia, and Onkar Nath Tiwari. "Advances in exopolysaccharides based bioremediation of heavy metals in soil and water: a critical review." *Carbohydrate Polymers* 199 (2018): 353–364.

Sevak, Pooja Inder, Bhupendra Kishanprasad Pushkar, and Pooja Nana Kapadne. "Lead pollution and bacterial bioremediation: a review." *Environmental Chemistry Letters* 19, no. 6 (2021): 4463–4488.

Shamim, Saba, and Abdul Rehman. "Antioxidative enzyme profiling and biosorption ability of *Cupriavidus metallidurans* CH34 and *Pseudomonas putida* mt2 under cadmium stress." *Journal of Basic Microbiology* 55, no. 3 (2015): 374–381.

Sharma, Surbhi, C. S. Sundaram, Pratibha M. Luthra, Yogendra Singh, Ravi Sirdeshmukh, and W. N. Gade. "Role of proteins in resistance mechanism of *Pseudomonas fluorescens* against heavy metal induced stress with proteomics approach." *Journal of Biotechnology* 126, no. 3 (2006): 374–382.

Shiomi, Naofumi. "An assessment of the causes of lead pollution and the efficiency of bioremediation by plants and microorganisms." In *Advances in Bioremediation of Wastewater and Polluted Soil*, Naofumi Shiomi (Ed.), pp. 247–274. IntechOpen, London, 2015.

Silverberg, B. A., P. T. Wong, and Y. K. Chau. "Ultrastructural examination of Aeromonas cultured in the presence of organic lead." *Applied and Environmental Microbiology* 32, no. 5 (1976): 723–725.

Singh, Asha Lata, and P. N. Sarma. "Removal of arsenic (III) from waste water using *Lactobacillus acidophilus*." *Bioremediation Journal* 14, no. 2 (2010): 92–97.

Singh, Manish Kumar, Shailendra Dwivedi, Suraj Singh Yadav, Rajesh Singh Yadav, and Sanjay Khattri. "Anti-diabetic effect of *Emblica-officinalis* (Amla) against arsenic induced metabolic disorder in mice." *Indian Journal of Clinical Biochemistry* 35, no. 2 (2020): 179–187.

Suhani, Ibha, Sinha Sahab, Vaibhav Srivastava, and Rajeev Pratap Singh. "Impact of cadmium pollution on food safety and human health." *Current Opinion in Toxicology* 27 (2021): 1–7.

Sultan, Insha, and Qazi Mohd Rizwanul Haq. "Bacterial mechanisms for remediation." In *Bioremediation of Toxic Metal (loid) s*, Anju Malik, Mohd. Kashif Kidwai, and Vinod Kumar Garg (Eds.), pp. 115–136. CRC Press, Boca Raton, FL, 2022.

Sun, Rongguo, Yue Gao, and Yang Yang. "Leaching of heavy metals from lead-zinc mine tailings and the subsequent migration and transformation characteristics in paddy soil." *Chemosphere* 291 (2022): 132792.

Szumińska, Danuta, Sebastian Czapiewski, Małgorzata Szopińska, and Żaneta Polkowska. "Analysis of air mass back trajectories with present and historical volcanic activity and anthropogenic compounds to infer pollution sources in the South Shetland Islands (Antarctica)." *Bulletin of Geography. Physical Geography Series* 15, no. 1 (2018): 111–137.

Tangaromsuk, J., P. Pokethitiyook, M. Kruatrachue, and E. S. Upatham. "Cadmium biosorption by *Sphingomonas paucimobilis* biomass." *Bioresource Technology* 85, no. 1 (2002): 103–105.

Tarfeen, Najeebul, Khair Ul Nisa, Burhan Hamid, Zaffar Bashir, Ali Mohd Yatoo, Mohd Ashraf Dar, Fayaz Ahmad Mohiddin, Zakir Amin, Rabi'atul Adawiyah Ahmad, and R. Z. Sayyed. "Microbial remediation: a promising tool for reclamation of contaminated sites with special emphasis on heavy metal and pesticide pollution: a review." *Processes* 10, no. 7 (2022): 1358.

Teng, Zedong, Wen Shao, Keyao Zhang, Yaoqiang Huo, Jing Zhu, and Min Li. "Pb biosorption by *Leclercia adecarboxylata*: protective and immobilized mechanisms of extracellular polymeric substances." *Chemical Engineering Journal* 375 (2019): 122113.

Timková, Ivana, Jana Sedláková-Kaduková, and Peter Pristaš. "Biosorption and bioaccumulation abilities of actinomycetes/streptomycetes isolated from metal contaminated sites." *Separations* 5, no. 4 (2018): 54.

Tiquia-Arashiro, Sonia M. "Lead absorption mechanisms in bacteria as strategies for lead bioremediation." *Applied Microbiology and Biotechnology* 102, no. 13 (2018): 5437–5444.

Titah, Harmin Sulistiyaning, Siti Rozaimah Sheikh Abdullah, Mushrifah Idris, Nurina Anuar, Hassan Basri, Muhammad Mukhlisin, Bieby Voijant Tangahu, Ipung Fitri Purwanti, and Setyo Budi Kurniawan. "Arsenic resistance and biosorption by isolated rhizobacteria from the roots of *Ludwigia octovalvis*." *International Journal of Microbiology* 2018 (2018).

Ullah, S., D. F. Shams, S. A. Ur Rehman, S. A. Khattak, M. Noman, G. Rukh, H. Bibi et al. "Application of visible light activated thiolated cobalt doped ZnO nanoparticles towards arsenic removal from aqueous systems." *Digest Journal of Nanomaterials & Biostructures (DJNB)* 17, no. 2 (2022).

Umana, Ime Michael, Peter Amba Neji, and John Akwagioge Agwupuye. "Assessment of underground water quality in Okobo local government area of Akwa Ibom State, Nigeria." *Applied Water Science* 12, no. 5 (2022): 1–12.

Viacava, Karen, Karin Lederballe Meibom, David Ortega, Shannon Dyer, Arnaud Gelb, Leia Falquet, Nigel P. Minton, Adrien Mestrot, and Rizlan Bernier-Latmani. "Variability in arsenic methylation efficiency across aerobic and anaerobic microorganisms." *Environmental Science & Technology* 54, no. 22 (2020): 14343–14351.

Viacava, Karen, Jiangtao Qiao, Andrew Janowczyk, Suresh Poudel, Nicolas Jacquemin, Karin Lederballe Meibom, Him K. Shrestha, Matthew C. Reid, Robert L. Hettich, and Rizlan Bernier-Latmani. "Meta-omics-aided isolation of an elusive anaerobic arsenic-methylating soil bacterium." *The ISME Journal* (2022): 1–10.

Vijayaraghavan, K., and Yeoung-Sang Yun. "Bacterial biosorbents and biosorption." *Biotechnology Advances* 26, no. 3 (2008): 266–291.

Wang, Jianlong, and Can Chen. "Biosorbents for heavy metals removal and their future." *Biotechnology Advances* 27, no. 2 (2009): 195–226.

Wang, Xin, Yi Wu, Xiaojie Sun, Qing Guo, Wei Xia, Yongning Wu, Jingguang Li, Shunqing Xu, and Yuanyuan Li. "Arsenic exposure and metabolism in relation to blood pressure changes in pregnant women." *Ecotoxicology and Environmental Safety* 222 (2021): 112527.

Wen, Qiqian, Xiao Yang, Xiulan Yan, and Linsheng Yang. "Evaluation of arsenic mineralogy and geochemistry in gold mine-impacted matrices: speciation, transformation, and potential associated risks." *Journal of Environmental Management* 308 (2022): 114619.

Wu, Zhiyuan, Dan Zhang, Tianxiang Xia, and Xiaoyang Jia. "Characteristics, sources and risk assessments of heavy metal pollution in soils of typical chlor-alkali residue storage sites in northeastern China." *PLoS One* 17, no. 9 (2022): e0273434.

Wuana, Raymond A., and Felix E. Okieimen. "Heavy metals in contaminated soils: a review of sources, chemistry, risks and best available strategies for remediation." *International Scholarly Research Notices* 2011 (2011): 402647.

Xia, Xian, Shijuan Wu, Zijie Zhou, and Gejiao Wang. "Microbial Cd (II) and Cr (VI) resistance mechanisms and application in bioremediation." *Journal of Hazardous Materials* 401 (2021): 123685.

Xue, Xi-Mei, Yu Yan, Hui-Juan Xu, Ning Wang, Xiao Zhang, and Jun Ye. "ArsH from *Synechocystis* sp. PCC 6803 reduces chromate and ferric iron." *FEMS Microbiology Letters* 356, no. 1 (2014): 105–112.

Yan, Ge, Xingxiang Chen, Shiming Du, Zixin Deng, Lianrong Wang, and Shi Chen. "Genetic mechanisms of arsenic detoxification and metabolism in bacteria." *Current Genetics* 65, no. 2 (2019): 329–338.

Yan, Lei, Huanhuan Yin, Shuang Zhang, Feifan Leng, Wenbin Nan, and Hongyu Li. "Biosorption of inorganic and organic arsenic from aqueous solution by *Acidithiobacillus ferrooxidans* BY-3." *Journal of Hazardous Materials* 178, no. 1–3 (2010): 209–217.

Yan, Tingting, Weijun Zhao, Xinyang Yu, Hongxi Li, Zhikang Gao, Min Ding, and Junsheng Yue. "Evaluating heavy metal pollution and potential risk of soil around a coal mining region of Tai'an City, China." *Alexandria Engineering Journal* 61, no. 3 (2022): 2156–2165.

Yang, Hung-Chi, and Barry P. Rosen. "New mechanisms of bacterial arsenic resistance." *Biomedical Journal* 39, no. 1 (2016): 5–13.

Yoshinaga, Masafumi, and Barry P. Rosen. "AC· As lyase for degradation of environmental organoarsenical herbicides and animal husbandry growth promoters." *Proceedings of the National Academy of Sciences* 111, no. 21 (2014): 7701–7706.

Yu, Ming-Ho, and Humio Tsunoda. *Environmental Toxicology: Biological and Health Effects of Pollutants*. CRC Press, Boca Raton, FL, 2004.

Zargar, Kamrun, Alison Conrad, David L. Bernick, Todd M. Lowe, Viktor Stolc, Shelley Hoeft, Ronald S. Oremland, John Stolz, and Chad W. Saltikov. "ArxA, a new clade of arsenite oxidase within the DMSO reductase family of molybdenum oxidoreductases." *Environmental Microbiology* 14, no. 7 (2012): 1635–1645.

Zhao, Xingqing, Min Wang, Hui Wang, Ding Tang, Jian Huang, and Yu Sun. "Study on the remediation of Cd pollution by the biomineralization of urease-producing bacteria." *International Journal of Environmental Research and Public Health* 16, no. 2 (2019): 268.

Zhou, Tong, Zhaoyang Wang, Peter Christie, and Longhua Wu. "Cadmium and lead pollution characteristics of soils, vegetables and human hair around an open-cast lead-zinc mine." *Bulletin of Environmental Contamination and Toxicology* 107, no. 6 (2021): 1176–1183.

Zhu, Yong-Guan, Masafumi Yoshinaga, Fang-Jie Zhao, and Barry P. Rosen. "Earth abides arsenic biotransformations." *Annual Review of Earth and Planetary Sciences* 42 (2014): 443–467.

Zhu, Yong-Guan, Xi-Mei Xue, Andreas Kappler, Barry P. Rosen, and Andrew A. Meharg. "Linking genes to microbial biogeochemical cycling: lessons from arsenic." *Environmental Science & Technology* 51, no. 13 (2017): 7326–7339.

Ziller, Antoine, and Laurence Fraissinet-Tachet. "Metallothionein diversity and distribution in the tree of life: a multifunctional protein." *Metallomics* 10, no. 11 (2018): 1549–1559.

8 Mechanism of Long-Term Arsenic Remediation by Microbial Intervention

Debanjana Sengupta and Arup Kumar Mitra
St. Xavier's College (Autonomous)

Arunima Bhattacharya
St. Xavier's College (Autonomous)
Omit Université de Bordeaux

8.1 INTRODUCTION

8.1.1 Toxicity of Arsenic

The 33rd element of the periodic table of group 15(A), arsenic (As) is a metalloid with both the characteristic features of metal and nonmetal (Mandal and Suzuki, 2002). The name has been derived from the Persian word "Zarnikh," which means "Yellow Orpiment" (Vahidnia et al., 2007; Sengupta, 2021). It is present in the environment in the two major forms As(V) (arsenate) and As(III) (arsenite). Among these two states, arsenate is the predominant oxidation state, which is thermodynamically stable in the environment (Cullen and Reimer, 1989), and being an analogue of phosphate, it takes part in phosphate uptake and metabolism. Arsenite is more ambulant and soluble in nature and can interact with thiol (-SH) groups of amino acids, degrading the protein compounds (Tamaki and Frankenberger, 1992; Hughes, 2002; Sengupta, 2021). Additionally, organic species of arsenic, such as DMA (dimethylarsinic acid), MMA (monomethylarsonic acid), and TMAO (trimethyl arsine oxide), are also found in soils, although at a much lesser percentage (<5% of total As) (Huang et al., 2011; Molina et al., 2019; Sengupta, 2021). Arsenic is a toxic metalloid as per the United States Environmental Protection Agency and the Agency for Toxic Substances and Disease Registry and has continuously topped the Agency for Toxic Substances and Disease Registry hazardous substances' list. Human exposure to arsenic mainly occurs from geological sources, mining, coal burning, copper smelting, and volcanic eruptions. Apart from these, arsenic also originates from anthropogenic sources like herbicides, pesticides, animal feeds, wood-preservatives, paints, dyes, etc. (Molina et al., 2019; Sengupta, 2021). In the ecosystem, arsenic meets the food chain in the arsenic-contaminated drinking water derived from the arsenic-rich soil.

DOI: 10.1201/9781003451457-8

8.1.2 Adverse Effects of Arsenic Contamination

Arsenic is detrimental to the health of living organisms when exposed via dietary means, water, etc. Considering humans, especially inorganic As (iAs) forms are the most toxic and, as per the European Food Safety Authority (2014) and the World Health Organization, are major concerns in human health, besides being confirmed carcinogens; in fact, in 2016, the World Health Organization (WHO) considered 10 µg/L as the safe limit for concentration of arsenic in drinking water. The WHO demarcates between acute and long-term effects of exposure to arsenic—acute arsenic poisoning causes immediate symptoms of nausea, pain in the abdomen, and diarrhea, followed by numbness and a tingling sensation in the extremities, cramps in muscles, and ultimately, in extreme cases, death (WHO, 2018); long-term exposure to high levels of iAs via food and drinking water usually causes the first manifestation of symptoms in the skin, including skin lesions, changes in pigmentation, hyperkeratosis, etc., after a minimum exposure of approximately 5 years, which are probable indicators of skin cancer. Besides, tumors in the kidney, liver, bladder, and lung have also been reported, with the most common cause of mortality related to arsenic being cancer of the lungs (WHO, 2018). Further, As is permeable through the blood–brain barrier, central and peripheral neuropathies typically affecting the peripheral nervous systems, oxidative stress, cytoskeletal disorganization, neuronal apoptosis, etc. are also observed (Vahidnia et al., 2007; Molina et al., 2019; Sengupta, 2021). Additionally, cardiovascular diseases, diabetes, myriad hepatic diseases, renal toxicity, reproductive failure and infertility, gonad dysfunction and hormonal imbalances, teratogenic effects on developing fetuses in pregnant females leading to increased infant mortality, etc. are also effects of chronic exposure to inorganic arsenic (Wu et al., 2011; Dávila-Esqueda et al., 2012; Molina et al., 2019; Sengupta, 2021). Chronic arsenic exposure may also affect immunity against the respiratory influenza A (H_1N_1) virus (Kozul et al., 2009; Sengupta, 2021). Arsenic may pose a threat to public health not only by being present in hazardous concentrations in groundwater (e.g., above the USEPA, 2001 permissible limit of 10 ppb in drinking water), but also in crops irrigated with that contaminated water and/or those growing in soils with As contamination and/or food prepared with such water or crops. These effects are particularly observed in countries like Bangladesh, India, China, Chile, Mexico, Argentina, the United States, etc. (Ahmad et al., 2018).

However, exposure is not just detrimental to humans–plants and animals are also adversely affected. In plants, arsenic toxicity leads to inhibition of seed germination, the stunted height of plants, necrosis, and reduction in growth of both root and shoot, decreased yield of fruit and grains, withering and yellowing of leaves, etc. Especially in rice (*Oryza sativa* L.) irrigated with water contaminated by arsenic, arsenic accumulation is common due to the high mobility of the metalloid in flooded conditions (Barrachina et al., 1995; Abedin et al., 2002; Shrivastava et al., 2015). Autopsies of animals suspected to have been arsenic-poisoned have shown erythema and edema of gastric and intestinal mucosa, edema of the lungs, degeneration of capillaries in the kidney and gastrointestinal tract, etc., as well as signs of behavioral changes like depression, anorexia, dehydration, increased frequency of urination, etc. Besides, abnormalities in dermal consistency (dried and dead skin), irritation in the eyes, and nasal flow discharge have also been observed on cutaneous exposure to and inhalation of arsenic, respectively (National Research Council, 1977; Eisler, 1988; Shrivastava et al., 2015).

8.1.3 Need for Bioremediation

Owing to the aforementioned detrimental effects, remediation of arsenic-polluted soil and groundwater is essential; physical and chemical methods such as oxidation, precipitation, adsorption, and ion exchange are fast and effective, but not environment-friendly, for which biological remediation became necessary – bioremediation can occur via plants, i.e., phytoremediation, by phytoextraction, phytostabilization, rhizofiltration, phytovolatilization, etc., or via microbes, biostimulation and/or bioaugmentation (Glick, 2003; Ghosh et al., 2014; Shrivastava et al., 2015). This chapter discusses such bioremediation mechanisms in arsenic-tolerant bacteria, including the chromosomally encoded arsenical resistance (*ars*) operon, with special emphasis on long term arsenic mitigation through volatilization by specific strains of *Exiguobacterium indicum* and *Bacillus cereus*, based on a study conducted at the Dhapa dumping ground in Kolkata, West Bengal, India.

8.2 MECHANISM OF ARSENIC UPTAKE BY BACTERIA AND ARSENIC-TOLERANT BACTERIA

8.2.1 Arsenic Uptake Mechanisms by Living Cells

Arsenate and arsenite are adsorbed in cells by different uptake systems. Most organisms use phosphate channels for arsenate uptake since arsenate is a phosphate analogue; e.g., *E. coli* uses two phosphate transporters in arsenate uptake, the main, low affinity Pit, and high affinity Pst phosphate transporters (Rosenberg et al., 1977; Sengupta, 2021); in the case of *Saccharomyces cerevisiae*, arsenate is adsorbed through the Pho84 and Pho87 phosphate transporters (Bun-Ya et al., 1991; Bun-ya et al., 1996), while mammalian cells use the phosphate channel NaPiIIb in As(V) uptake (Villa-Bellosta and Sorribas, 2010; Sengupta, 2021).

Arsenite is a strong oxyacid with a pKa value of 9.2 in water (Ramírez-Solís et al., 2004), and owing to its structural similarity with glycerol (Porquet and Filella, 2007), it is adsorbed by the aqua glyceroporin GlpF in *E. coli* (Sanders et al., 1997; Meng et al., 2004), which is an important member of the intrinsic protein (MIP) superfamily that participates in water and small solute (like glycerol and urea) transport, independent of energy requirements. *S. cerevisiae* (Wysocki et al., 2001) uses Fps1p similar to GlpF in As(III) uptake and hexose transporters like HXT1, HXT3, HXT4, HXT5, HXT7, and HXT9 in $As(OH)_3$ uptake (Liu et al., 2004); humans use the four aqua glyceroporins AQP9, AQP7, AQP3, and AQP10 for arsenite uptake. Further, the mammalian glucose transporter GLUT1 (which is similar to yeast HXT hexose permeases), along with GLUT2 in hepatocytes and GLUT5 in small intestines, perform As(III) and arsenite uptake (Liu et al., 2006; Drobná et al., 2010; Calatayud et al., 2012; Sengupta, 2021). Another two organic anionic transmembrane proteins of humans, OATPB (Calatayud et al., 2012) and OAPTC (Liu et al., 2006), are also involved in endogenous and exogenous transport of organic compounds, along with arsenite accumulation (Hagenbuch and Meier, 2004; Sengupta, 2021).

8.2.2 Arsenic-Tolerant Bacteria

According to Sengupta (2021), arsenic-tolerant microbes use different arsenic detoxification mechanisms such as cytoplasmic arsenite export, chelation inside the microbial cells, etc. Given that the first such living organism was found in an anaerobic environment highly concentrated with heavy metals and metalloids, including arsenic (Zhu et al., 2014), an inherent arsenic tolerance mechanism has been hypothesized. Therefore, the bacterial arsenic tolerance mechanism has been generated based on the arsenite transporters with the help of *arsB* (arsenic permease). Additionally, several arsenate-reducing and arsenite-oxidizing microorganisms have been found in the environment, which accumulate arsenic as a terminal respiratory electron acceptor (Oremland and Stolz, 2005; Sengupta, 2021). For example, *Geobacter sulfurreducens* is majorly found to accumulate arsenic through enzymatic metabolism (Islam et al., 2005; Sengupta, 2021), with an ability to grow in 500 μM arsenate, with its arsenic tolerance ability depending on intracellular arsenate reduction and arsenite efflux. Similarly, several arsenates (up to 100 mM) and arsenites (20 mM) can be tolerated by bacterial isolates of *Aeromonas, Exiguobacterium, Acinetobacter, Bacillus*, and *Pseudomonas* (Anderson and Cook, 2004; Sengupta, 2021); other arsenic tolerant bacterial isolates include *Acidothiobacillus, Alcaligenes, Arsenicicoccus bolidensis, Arthrobacter agilis, Deinococcus, Desulfitobacterium, Kocuria erythromyxa, Microbacterium hydrocarbonoxydans, Oceanimonas doudoroffii, Staphylococcus succinus, Variovorax paradoxus,* etc. (Oremland et al., 2004; Suresh et al., 2004; Bachate et al., 2009; Sengupta, 2021). Shakoori et al. (2010) isolated three arsenic-tolerant bacteria, *Klebsiella oxytoca, Citrobacter freundii,* and *Bacillus anthracis*, which were able to reduce arsenate up to 290 mg/L, 290 mg/L, and 240 mg/L, respectively (Sengupta, 2021); Paul et al. (2015) isolated and identified arsenic-tolerant *Pseudomonas, Flavobacterium, Brevundimonas, Polaromonas, Rhodococcus, Methyloversatilis,* and *Methylotenera* from groundwater in West Bengal, and the bacteria played an important role in arsenate reduction (Sengupta, 2021). Further, 16S rDNA analyses by Goswami et al. (2015) identified six highly arsenic-tolerant bacterial strains: *Microbacterium oleivorans, Acinetobacter soli, Acinetobacter venetianus, Acinetobacter junii, Acinetobacter baumannii,* and *Acinetobacter calcoaceticus* (Sengupta, 2021).

In summary, various microbial arsenic detoxification mechanisms have been identified, such as cell-mediated arsenite efflux, arsenate reduction, arsenite oxidation involving electron transport systems, and the use of the *ars* operon (Oremland and Stolz, 2005), of which the *ars* operon has been discussed in detail in the next section.

8.3 THE *ARS* OPERON AND ITS COMPONENTS

Arsenic-resistant genes in microorganisms are either present in the plasmid or in the genomic DNA. According to Cai et al. (1998), the five genes containing the *ars* operon, *arsRDABC* were observed in *Escherichia coli*, and the three genes containing *arsRBC* were reported in *Staphylococcus aureus* (Rosenstein et al., 1992; Silver et al., 1993).

In the *arsRBC* operon, *arsR* encodes a regulatory repressor, *arsB* is a membrane-associated arsenic permease involved in arsenate and arsenite resistance, and *arsC* is responsible for cellular arsenate reductase. In some Gram-negative bacteria, the *ars RDABC* operon exists, where two additional genes, *arsA* and *arsD* are observed. *arsA* codes for an intracellular ATPase that forms a dimer by binding with the membrane-associated ArsB protein – the arsenite permease can either function alone or with an ATPase by the formation of an ArsAB complex; *arsD* encodes a metallochaperone, which is a trans-acting co-repressor and transports arsenite to ArsA (Lin et al., 2007; Sengupta, 2021). The process has been elucidated schematically in Figure 8.1.

8.3.1 ArsR: Arsenic Repressor Protein

ArsR (13 kD), found in the *E. coli* R773 plasmid, has 117 residues and encodes an inducer-dependent trans-acting repressor that plays an important role in *ars* operon expression (Francisco et al., 1990; Xu and Rosen, 1997; Sengupta, 2021). Three active domains are found in ArsR – a domain for metal ligation, one for DNA interaction, and another for dimerization – ArsR regulates transcription of the operon by binding with the inducer, resulting in an unbound operator (Wu and Rosen, 1993; Sengupta, 2021).

8.3.2 ArsD: Metallochaperone

ArsD (13 kD) functions as the secondary transcription regulator in the *ars* operon. ArsR and ArsD both help in the inhibition of *ars* operon expression, albeit in different ways. While ArsR is repressed in the presence of 10 μM sodium arsenite, ArsD

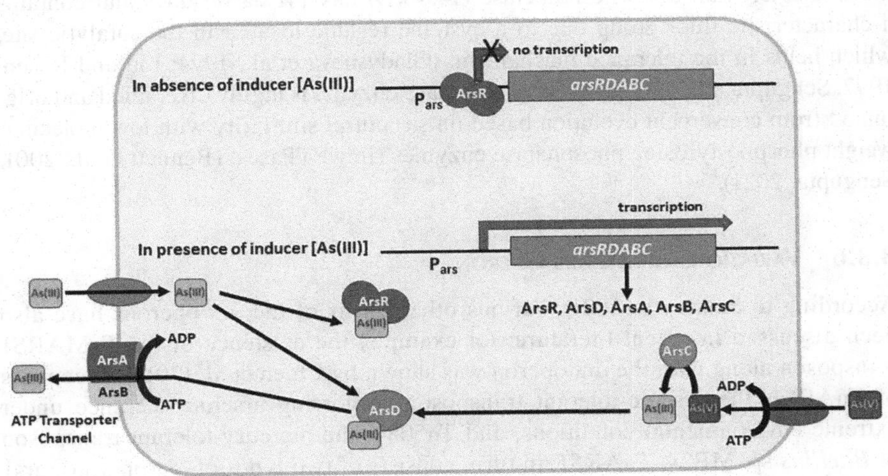

FIGURE 8.1 *ars* Operon of *E. coli* pR773 and R46 plasmids. (Adapted and modified from Chen and Rosen 2014.)

is repressed in the presence of 100 μM sodium arsenite, which indicates that ArsR is more highly associated with the inducer than ArsD. ArsD also has regulatory action on the expression of ArsB (Sato and Kobayashi, 1998; Kalia and Joshi, 2009; Sengupta, 2021).

8.3.3 ArsB: Arsenic Permease

ArsB (45 kD), the transmembrane arsenite transporter situated at the inner membrane, plays an important role in arsenic tolerance and As(III) efflux with the help of the ArsA ATPase. ArsB consists of 12 membrane domains; among these domains, five are cytoplasmic and seven are periplasmic coils (Chen et al., 1985; Dey et al., 1994; Carlin et al., 1995; Sengupta, 2021).

8.3.4 ArsA: ATPase

The ability of *E. coli* to tolerate arsenic is based on the ATP transporter channel (formed by the ArsB-ArsA complex) mediated arsenite efflux and As(III) extrusion from the cell (Dey et al., 1994). The functional part of the As(III) transporter is the 63 kD ArsA-ATPase. It has two homologous nucleotide binding domains – the N-terminal A1 domain (1–281 aa residues) and the C-terminal A2 domain (321–583 residues), which are joined by a pliable linker (residues 283–320) (Dey et al., 1994; Ramaswamy and Kaur, 1998; Sengupta, 2021).

8.3.5 ArsC: Arsenate Reductase

The main arsenic detoxification mechanism of bacteria and yeast is arsenate reduction, and the arsenate reductase enzyme involved in the conversion of arsenate to arsenite is crucial; arsenate reductase (14.8 kD) has 141 aa residues and contains a characteristic thiol group due to a cysteine residue located in the catalytic site, which helps in the tolerance mechanism. (Gladysheva et al., 1994; Liu and Rosen, 1997; Sengupta, 2021). The arsenate reductase enzyme is highly diversified and originated from convergent evolution based on structural similarity with low molecular weight phospho-tyrosine phosphatase enzymes (lmwPTPases) (Bennett et al., 2001; Sengupta, 2021).

8.3.6 Variations of the *ars* Operon

According to Sengupta (2021), various other forms of the *ars* operons have also been discussed in recent literature; for example, the existence of the TnMARS1 transposon along with the *ars* operon was shown by Chien et al. (2019). It consists of TnARS1, the arsenic-tolerant transposon conferring arsenic tolerance under extreme environmental conditions, and Tn5084, the mercury-tolerant transposon of *Bacillus* sp. MB24. TnARS1, in turn, consists of two types of *ars* operons, ars1 and *ars2*, of which *ars2* is more useful. Further, Ordoñez et al. (2015) showed that *Exiguobacterium* S17 has the Acr3 transporter that plays an important role in arsenite efflux, rendering the bacteria arsenate (150 mM) and arsenite (10 mM) tolerant. Another arsenic-tolerant bacterium, *Corynebacterium glutamicum*, was found by Mateos et al. (2006), and whole genome sequencing showed the existence of two *ars* operons, *ars1* and *ars2*, which played a significant role in arsenate and arsenite tolerance up to >400 mM and 12 mM, respectively. These *ars1* and *ars2* operons were part of the genomic DNA, with the operon consisting of the arsenic regulatory protein encoded by *arsR*, the arsenite permease transporter encoded by *arsB*, and the arsenate reductase encoded by *arsC* (Ordoñez et al., 2015). Similarly, according to Castro-Severyn et al. (2019), the extremophilic *Exiguobacterium* isolated from Salar de Huascois tolerant to both arsenate and arsenite and has an *arsA*-encoded ATPase arsenite efflux pump, which, under arsenic stress, is co-overexpressed with Cdr, Hpf, LuxS, GLpX, GlnE, and Fur coenzymes. Besides, arsenic-tolerant *Exiguobacterium* sp. PS (NCIM 5463), which utilizes the phosphate ABC transporter ATPase subunit and pyridine nucleotide disulfide oxidoreductase, two unique proteins, in fighting arsenic stress, has also been reported. It is interesting to note that the usual mechanism of *ars*operon is absent in *Exiguobacterium* sp. PS (NCIM 5463) (Sancheti et al., 2013).

8.4 LONG-TERM ARSENIC MITIGATION THROUGH VOLATILIZATION

8.4.1 Arsenic Volatilization

Volatilization of arsenic refers to the formation of organic and inorganic arsenicals enzymatically by reduction of As(V) to As(III), followed by a sequential cascade of biomethylation reactions. Most microorganisms utilize S-adenosine methionine (SAM), the complex of methionine and adenosine produced by the enzyme methionine adenosyl transferase, which acts as a universal methyl donor, resulting in the formation of S-adenosyl homocysteine (SAH) after methylation. (Mukhopadhyay et al., 2002; Liu et al., 2011; Sengupta, 2021). Arsenic methylation has been widely found in bacteria, archaea, and eukaryotic organisms – for example, Zhang et al. (2015) reported that *Pseudomonas alcaligenes* NBRC14159 uses the arsenite S-adenosylmethionine methyltransferase enzyme to instantly methylate arsenite to dimethyl arsenate and small amounts of trimethyl arsenic oxide.

In 1890, Bartolomeo Gosio, an Italian physician, observed arsenic biomethylation for the first time in the different fungi that lived on arsenic-contaminated organic substances – the fungi produced a gas that had the smell of garlic and was named after him as 'Gosio Gas'. In 1935, Challenger, along with Higginbottom, described that the 'Gosio Gas' was trimethyl arsine, $(CH3)_3As$, or TMAs(III) (Challenger and Higginbottom, 1935). He first suggested the arsenic methylation mechanism in 1947, where pentavalent arsenate was reduced to trivalent methylated arsenic forms by the formation of MMAs(V), DMAs(V), TMAs(V)O, As(III), MMAs(III), DMAs(III) and TMAs(III), using SAM as the methyl donor (Sengupta, 2021). Bacteria have an arsenite S-adenosylmethionine methyltransferase (*arsM*) gene, which is homologous to eukaryotic As_3MT (arsenite methyltransferase), and among the earliest examples, *Escherichia coli* expressing *arsM* from *Rhodopseudomonas palustris* has been shown to convert inorganic arsenic to trimethyl arsine [TMAs(III)], a volatile gas, and this methylation reaction results in arsenic removal from the cells due to the volatilization (Yuan et al., 2008; Liu et al., 2011; Sengupta, 2021). The *arsM* gene has four conserved cysteine residues, which form the active site of the methyltransferase enzyme, and its expression is regulated by *arsR* of the *ars* operon (Dheeman et al., 2014; Sengupta, 2021).

Arsine is a toxic gas as well; however, since it is distributed in the atmosphere instead of being concentrated at a particular point, its toxicity is not considered acute. Besides, when the conversion of states occurs, the greater evil, the toxicity of drinking water, can be prevented. Thus, arsenic volatilization by methylation is suggested as one of the important arsenic detoxifying mechanisms (Sengupta, 2021).

8.4.2 Case Study from the Dhapa Dumping Ground of Kolkata, West Bengal, India

Experiments on a novel strain of *Exiguobacterium* sp., *Exiguobacterium indicum* Strain DSAM 62 (isolated and identified by the author in Sengupta et al., 2020), and *Bacillus cereus* Strain DSAM 01 (Sengupta et al., 2013; 2014), revealed a possible strong mechanism of arsenic uptake and volatilization mediated by a mass array of stress-related proteome (Sengupta, 2021). The mechanism of upregulation of enzymes and genes during arsenic stress has been discussed.

8.4.2.1 Outline of the Methods of the Experiments

8.4.2.1.1 Sludge Sample Collection and Arsenic Analysis

The collections of the sludge samples were done from the subsurface area (0–15 cm in depth) of three different places in Dhapa (22.5373°N, 88.4334°E), Dumping Ground, Kolkata. The samples were then retained in sterile polythene bags at 4°C until further analysis. The ICP-OES (inductively coupled plasma–optical emission spectrometry) method was applied to determine the arsenic contents of the collected sludge samples (Sengupta, 2021). ICP-OES is a method that employs plasma energy generated by argon gas and high-frequency electric current to atomize sample solutions, excite the component elements (atoms), which return to lower energy and release emission rays (spectrum rays), and then measure the corresponding photon wavelengths. The type and content of each element are thereby determined based on the position and intensity of the photon rays (Figure 8.2).

8.4.2.1.2 Arsenic Volatilization

In the arsenic chemotrapping setup, the bacterial consortia consisting of *Exiguobacterium indicum* Strain DSAM 62 (S6C2) and *Bacillus cereus* Strain DSAM 01 (C-7) were inoculated in two sets, i.e., autoclaved and unautoclaved, in 500 g of arsenic-contaminated sludge in 1 L glass round bottom flasks, and the flask was connected with an aerator pump for the air supply through the glass inlet pipe; on the other hand, there were two pipes attached to the cork of the flask and to the glass column filled with 1% HNO_3 and 10% $AgNO_3$-treated silica beads for capturing the volatilized arsine gas (Zhang et al., 2015). Among the two pipes,

FIGURE 8.2 Sludge sample collection site, Dhapa Dumping Ground, Kolkata. (Sengupta 2021.)

one was used for the passing of the arsine gas to the silica beads, and the other was used for adding 1% HNO_3 to the silica beads for washing. After 72 hours of incubation, the eluant was collected from the column by opening the stopcock. According to the method of Kundu et al. (2002), 3 mL of sodium dodecyl sulfate (10^{-2} M), 50 μL of methylene blue (0.5×10^{-3} M) and 100 μL of sodium borohydride (10^{-1} M) were mixed with 850 μL of the eluant in a stoppered cuvette. The mixture was incubated for 3 minutes in the dark at room temperature, and then the absorbance was recorded spectrophotometrically at 660 nm (Optizen) (Sengupta, 2021) (Figure 8.3).

8.4.2.1.3 Arsenic Analysis after the Chemo Trapping Study

Quantification of the residual arsenic content of the biologically treated sludge sample was done using the ICP-OES technique (Sengupta, 2021).

8.4.2.2 Results

8.4.2.2.1 Sludge Sample Collection and Its Arsenic Content Analysis

See Table 8.1.

8.4.2.2.2 Volatilization of Arsenic

See Table 8.2.

8.4.2.2.3 Arsenic Analysis after the Chemo Trapping Experiment

See Table 8.3.

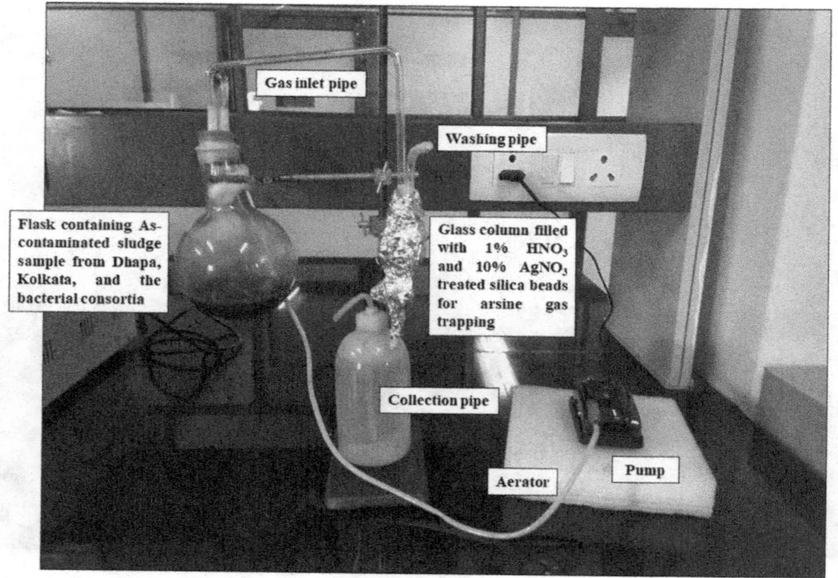

FIGURE 8.3 Arsenic chemotrapping setup. (Sengupta, 2021.)

TABLE 8.1
Concentration of Arsenic in the Collected Sludge Sample (Sengupta, 2021)

Serial Number	Name of the Sample	Arsenic Concentration (mg/L or ppm.)
1.	Sample-A	3.19 ± 0.001
2.	Sample-B	3.52 ± 0.001
3.	Sample-C	3.63 ± 0.001

All values calculated as mean ± SD of triplicates.

8.4.2.3 Conclusions

Among the three samples, the highest concentration of arsenic was observed at 3.63 mg/L in Sample C. Further, upon incubation for 72 hours, the highest amount of volatilized arsenic or arsine gas (62.67% of reduced arsenate in arsine gas and 4.32% of reduced arsenite in arsine gas) was found in the unautoclaved sludge sample with bacterial consortia *Exiguobacterium indicum* Strain DSAM 62 (S6C2) and *Bacillus cereus* Strain DSAM 01 (C-7). implying that the S6C2 and C-7 bacterial consortium could volatilize arsenic and/or enhance or trigger the volatilization process. Besides that, after 72 hours of incubation, the same sample showed the highest amount of arsenic uptake, which led to the inference that the consortium and the sludge microbes showed the maximum arsenic uptake capability when compared with the other conditions of the setup (Sengupta, 2021).

TABLE 8.2
Arsine Gas Concentration in the Arsenic Chemo Trapping Experiment (Sengupta2021)

Condition of the Setup	Initial Concentration of Arsenic of the Arsenic Contaminated Sludge (mg/L or ppm)	Final Concentration of Arsenic of the Arsenic Contaminated Sludge (mg/L or ppm)	Reduced Arsenate Concentration of Arsine Gas (mg/L or ppm)	% of Reduced Arsenate of Arsine Gas	Reduced Arsenite Concentration of Arsine Gas (mg/L or ppm)	% of Reduced Arsenite of Arsine Gas
Control	3.63	2.49 ± 0.001	0.134 ± 0.001	3.69%	0.021 ± 0.001	0.58%
Unautoclaved Sludge Sample with Bacterial Consortia (S6C2 and C-7)	3.63	0.77 ± 0.001	2.275 ± 0.001	62.67%	0.157 ± 0.001	4.32%
Autoclaved Sludge Sample with Bacterial Consortia (S6C2 and C-7)	3.63	0.948 ± 0.001	2.228 ± 0.001	61.38%	0.153 ± 0.001	4.21%

All values calculated as mean ± SD of triplicates.

TABLE 8.3
Arsenic Analysis after the Chemo Trapping Experiment: Remnant Concentration of Arsenic in the Sludge Sample after the Impact of Arsenic Chemo trapping Experiment in the Setups (Sengupta, 2021)

Serial Number	Name of the Sample	Condition of the Set Up	Concentration of Remnant Arsenic (mg/L or ppm)
1.	Sample-C	Control	2.49 ± 0.001
2.	Sample-C	Unautoclaved Sludge Sample with Bacterial Consortia (S6C2 and C-7)	0.77 ± 0.001
3.	Sample-C	Autoclaved Sludge Sample with Bacterial Consortia (S6C2 and C-7)	0.948 ± 0.001

All values calculated as mean ± SD of triplicates.

8.5 CONCLUSION AND FUTURE PERSPECTIVES

The health hazards of arsenic contamination of soil and groundwater are immense and, in most cases, debilitating, if not fatal. While testing for the presence of the toxic metalloid is rampant and awareness has increased, it is essential that remedial measures that are also eco-friendly and do not have collateral ill effects be discussed and developed for widespread use. Bio-volatilization and biomethylation of arsenic are such pathways of bioremediation that are currently being explored in further detail, both at the lab scale and at the probable larger *in situ* scale. Although the toxic effects of the end products of biomethylation remain to be deciphered and studied, the mechanism is being explored in other bacteria as well as yeast and fungi; genetic engineering of plants to produce transgenics expressing microbial methyltransferases has been performed and studied in *Arabidopsis thaliana* and rice (Verma et al., 2016; 2019), among others. However, the complex network of interactions between various components of the different types of bacterial *ars* operons, their eukaryotic homologs, and associated methylation machinery needs to be elucidated and probed into in further detail to expand the scope of innovative metabolic engineering. Besides, recent integrated approaches like phytobial bioremediation (phytoremediation assisted by soil microbes, especially bioremediating bacteria), simultaneous biosorption and bioaccumulation (SBB), subsurface flow constructed wetlands, phytosuction separation, etc. are being studied (Irshad et al., 2021). The experiments discussed in this chapter, with *Exiguobacterium indicum* Strain DSAM 62 (S6C2) and *Bacillus cereus* Strain DSAM 01 (C-7), re-confirm the ability of microbes in the remediation of arsenic, especially from dumping grounds, and indicate scope for further scientific and logistical research regarding the possibilities of inoculating microbial consortia at such sites for combating arsenic contamination of soil and related health hazards. Finally, the effects of such reactions on the arsenic biogeochemical cycles and the existence of similar or more efficient microbial communities need to be better understood and explored, both in shallow and deep arsenic-rich sediments, to get a clear view of the putative potential of microbial bioremediation and its allied fields of phytoremediation, mycoremediation, and phytobial remediation.

ACKNOWLEDGEMENTS

The authors would like to acknowledge the Department of Science & Technology, Government of West Bengal, for funding the research work and Rev. Fr. Dominic Savio SJ, Principal, St. Xavier's College (Autonomous), Kolkata, for his active encouragement.

REFERENCES

Abedin, Md Joinal, J. Cotter-Howells, and Andy A. Meharg. 2002. "Arsenic Uptake and Accumulation in Rice (*Oryza sativa* L.) Irrigated with Contaminated Water." *Plant and Soil* 240 (2): 311–19. https://www.jstor.org/stable/24121150.

Ahmad, Sk Akhtar, Manzurul Haque Khan, and Mushfiqul Haque. 2018. "Arsenic Contamination in Groundwater in Bangladesh: Implications and Challenges for Healthcare Policy." *Risk Management and Healthcare Policy* 11: 251–61. https://doi.org/10.2147/RMHP.S153188.

Anderson, Craig R., and Gregory M. Cook. 2004. "Isolation and Characterization of Arsenate-Reducing Bacteria from Arsenic-Contaminated Sites in New Zealand." *Current Microbiology* 48 (5): 341–47. https://doi.org/10.1007/s00284-003-4205-3.

Bachate, S. P., L. Cavalca, and V. Andreoni. 2009. "Arsenic-Resistant Bacteria Isolated from Agricultural Soils of Bangladesh and Characterization of Arsenate-Reducing Strains." *Journal of Applied Microbiology* 107 (1): 145–56. https://doi.org/10.1111/j.1365-2672.2009.04188.x.

Barrachina, A. Carbonell, F. Burlo Carbonell, and J. Mataix Beneyto. 1995. "Arsenic Uptake, Distribution, and Accumulation in Tomato Plants: Effect of Arsenite on Plant Growth and Yield." *Journal of Plant Nutrition* 18 (6): 1237–50. https://doi.org/10.1080/01904169509364975.

Bennett, Matthew S., Zhi Guan, Martin Laurberg, and Xiao-Dong Su. 2001. "*Bacillus subtilis* Arsenate Reductase Is Structurally and Functionally Similar to Low Molecular Weight Protein Tyrosine Phosphatases." *Proceedings of the National Academy of Sciences of the United States of America* 98 (24): 13577–82. https://doi.org/10.1073/pnas.241397198.

Bun-ya, M., K. Shikata, Shinji Nakade, Chulee Yompakdee, Satoshi Harashima, and Y. Oshima. 1996. "Two New Genes, PHO86 and PHO87, Involved in Inorganic Phosphate Uptake in *Saccharomyces cerevisiae*." *Current Genetics* 29 (4): 344–51. https://doi.org/10.1007/s002940050055.

Bun-Ya, Masanori, Mamoru Nishimura, Satoshi Harashima, and Y. Oshima. 1991. "The *PHO84* Gene of *Saccharomyces cerevisiae* Encodes an Inorganic Phosphate Transporter." *Molecular and Cellular Biology* 11 (6): 3229–38. https://doi.org/10.1128/mcb.11.6.3229.

Cai, Jie, Kirsty Salmon, and Michael S. DuBow. 1998. "A Chromosomal Ars Operon Homologue of *Pseudomonas aeruginosa* Confers Increased Resistance to Arsenic and Antimony in *Escherichia coli*." *Microbiology (Reading, England)* 144 (Pt 10) (10): 2705–29. https://doi.org/10.1099/00221287-144-10-2705.

Calatayud, Marta, Julio A. Barrios, Dinoraz Vélez, and Vicenta Devesa. 2012. "In Vitro Study of Transporters Involved in Intestinal Absorption of Inorganic Arsenic." *Chemical Research in Toxicology* 25 (2): 446–53. https://doi.org/10.1021/tx200491f.

Carlin, Arthur, Weiping Shi, Saibal Dey, and Barry P. Rosen. 1995. "The Ars Operon of *Escherichia coli* Confers Arsenical and Antimonial Resistance." *Journal of Bacteriology* 177 (4): 981–86. https://doi.org/10.1128/jb.177.4.981-986.1995.

Castro-Severyn, Juan, Coral Pardo-Esté, Yoelvis Sulbaran, Carolina Cabezas, Valentina Gariazzo, Alan Briones, Naiyulin Morales, et al. 2019. "Arsenic Response of Three Altiplanic *Exiguobacterium* Strains with Different Tolerance Levels against the Metalloid Species: A Proteomics Study." *Frontiers in Microbiology* 10: 2161. https://doi.org/10.3389/fmicb.2019.02161.

Challenger, Frederick, and Constance Higginbottom. 1935. "The Production of Trimethylarsine by *Penicilliumbrevicaule (Scopulariopsisbrevicaulis)*." *The Biochemical Journal* 29 (7): 1757–78. https://doi.org/10.1042/bj0291757.

Chen, Chih-Ming, H. L. Mobley, and Barry P. Rosen. 1985. "Separate Resistances to Arsenate and Arsenite (Antimonate) Encoded by the Arsenical Resistance Operon of R Factor R773." *Journal of Bacteriology* 161 (2): 758–63. https://doi.org/10.1128/jb.161.2.758-763.1985.

Chen, Jian, and Barry P. Rosen. 2014. "Biosensors for Inorganic and Organic Arsenicals." *Biosensors* 4 (4): 494–512. https://doi.org/10.3390/bios4040494.

Chien, Mei-Fang, Ying-Ning Ho, Hui-Erh Yang, Masaru Narita, Keisuke Miyauchi, Ginro Endo, and Chieh-Chen Huang. 2019. "Identification of A Novel Arsenic Resistance Transposon Nested in A Mercury Resistance Transposon of *Bacillus sp.* MB24." *Microorganisms* 7 (11): 566. https://doi.org/10.3390/microorganisms7110566.

Cullen, William R., and Kenneth J. Reimer. 1989. "Arsenic Speciation in the Environment." *Chemical Reviews* 89 (4): 713–64. https://doi.org/10.1021/cr00094a002.

Dávila-Esqueda, Ma Eugenia, Ma Esther Jiménez-Capdeville, Juan Manuel Delgado, Esperanza De la Cruz, Celia Aradillas-García, Verónica Jiménez-Suárez, Reynaldo Falcón Escobedo, and Joel Rodríguez Llerenas. 2012. "Effects of Arsenic Exposure during the Pre- and Postnatal Development on the Puberty of Female Offspring." *Experimental and Toxicologic Pathology: Official Journal of the GesellschaftFür Toxikologische Pathologie* 64 (1–2): 25–30. https://doi.org/10.1016/j.etp.2010.06.001.

Dey, S., D. Dou, and B. P. Rosen. 1994. "ATP-Dependent Arsenite Transport in Everted Membrane Vesicles of *Escherichia coli*." *The Journal of Biological Chemistry* 269 (41): 25442–46. https://doi.org/10.1016/s0021-9258(18)47270-5.

Dheeman, Dharmendra S., Charles Packianathan, Jitesh K. Pillai, and Barry P. Rosen. 2014. "Pathway of Human AS3MT Arsenic Methylation." *Chemical Research in Toxicology* 27 (11): 1979–89. https://doi.org/10.1021/tx500313k.

Drobná, Zuzana, Felecia S. Walton, Anne W. Harmon, David J. Thomas, and Miroslav Stýblo. 2010. "Interspecies Differences in Metabolism of Arsenic by Cultured Primary Hepatocytes." *Toxicology and Applied Pharmacology* 245 (1): 47–56. https://doi.org/10.1016/j.taap.2010.01.015.

Eisler, R. 1988. "Arsenic Hazards to Fish, Wildlife, and Invertebrates: A Synoptic Review." *Biological Report* 85: 1–12.

European Food Safety Authority (EFSA), Parma, Italy, 2014. "Reasoned opinion on the review of the existing maximum residue levels (MRLs) for thiophanate-methyl and carbendazim according to Article 12 of Regulation (EC) No 396/2005." *EFSA Journal* 12 (12): 3919.

Francisco, Michael JD San, Constance L. Hope, Joshua B. Owolabi, Louis S. Tisa, and Barry P. Rosen. 1990. "Identification of the Metalloregulatory Element of the Plasmid-Encoded Arsenical Resistance Operon." *Nucleic Acids Research* 18 (3): 619–24. https://doi.org/10.1093/nar/18.3.619.

Ghosh, Devanita, Punyasloke Bhadury, and Joyanto Routh. 2014. "Diversity of Arsenite Oxidizing Bacterial Communities in Arsenic-Rich Deltaic Aquifers in West Bengal, India." *Frontiers in Microbiology* 5: 602. https://doi.org/10.3389/fmicb.2014.00602.

Gladysheva, Tatiana B., Kristine L. Oden, and Barry P. Rosen. 1994. "Properties of the Arsenate Reductase of Plasmid R773." *Biochemistry* 33 (23): 7288–93. https://doi.org/10.1021/bi00189a033.

Glick, Bernard R. 2003. "Phytoremediation: Synergistic Use of Plants and Bacteria to Clean up the Environment." *Biotechnology Advances* 21 (5): 383–93. https://doi.org/10.1016/s0734-9750(03)00055-7.

Goswami, Ramansu, Suprabhat Mukherjee, Vipin Singh Rana, Dhira Rani Saha, Rajagopal Raman, Pratap Kumar Padhy, and Shibnath Mazumder. 2015. "Isolation and Characterization of Arsenic-Resistant Bacteria from Contaminated Water-Bodies in West Bengal, India." *Geomicrobiology Journal* 32 (1): 17–26. https://doi.org/10.1080/01490451.2014.920938.

Hagenbuch, Bruno, and Peter J. Meier. 2004. "Organic Anion Transporting Polypeptides of the OATP/ SLC21 Family: Phylogenetic Classification as OATP/SLCO Superfamily, New Nomenclature and Molecular/Functional Properties." *Pflugers Archiv: European Journal of Physiology* 447 (5): 653–65. https://doi.org/10.1007/s00424-003-1168-y.

Huang, Jen-How, Kan-Nian Hu, and Berryinne Decker. 2011. "Organic Arsenic in the Soil Environment: Speciation, Occurrence, Transformation, and Adsorption Behavior." *Water, Air, and Soil Pollution* 219 (1-4): 401–15. https://doi.org/10.1007/s11270-010-0716-2.

Hughes, Michael F. 2002. "Arsenic Toxicity and Potential Mechanisms of Action." *Toxicology Letters* 133 (1): 1–16. https://doi.org/10.1016/s0378-4274(02)00084-x.

Irshad, Sana, Zuoming Xie, Sajid Mehmood, Asad Nawaz, Allah Ditta, and Qaisar Mahmood. 2021. "Insights into Conventional and Recent Technologies for Arsenic Bioremediation: A Systematic Review." *Environmental Science and Pollution Research International* 28 (15): 18870–92. https://doi.org/10.1007/s11356-021-12487-8.

Islam, F. S., R. L. Pederick, A. G. Gault, L. K. Adams, D. A. Polya, J. M. Charnock, and J. R. Lloyd. 2005. "Interactions between the Fe(III)-Reducing Bacterium *Geobactersulfurreducens*and Arsenate, and Capture of the Metalloid by Biogenic Fe(II)." *Applied and Environmental Microbiology* 71 (12): 8642–48. https://doi.org/10.1128/aem.71.12.8642-8648.2005.

Kalia, Kiran, and Dhaval N. Joshi. 2009. "Detoxification of Arsenic." In *Handbook of Toxicology of Chemical Warfare Agents*, Ramesh C. Gupta (Ed.), 1083–1100. Amsterdam: Academic Press.

Kozul, Courtney D., Kenneth H. Ely, Richard I. Enelow, and Joshua W. Hamilton. 2009. "Low-Dose Arsenic Compromises the Immune Response to Influenza A Infection in Vivo." *Environmental Health Perspectives* 117 (9): 1441–47. https://doi.org/10.1289/ehp.0900911.

Kundu, Subrata, Sujit Kumar Ghosh, Madhuri Mandal, Tarasankar Pal, and Anjali Pal. 2002. "Spectrophotometric Determination of Arsenic via Arsine Generation and In-Situ Colour Bleaching of Methylene Blue (MB) in Micellar Medium." *Talanta* 58 (5): 935–42. https://doi.org/10.1016/s0039-9140(02)00434-4.

Lin, Yung-Feng, Jianbo Yang, and Barry P. Rosen. 2007. "ArsD: An As(III) Metallochaperone for the ArsAB As(III)-Translocating ATPase." *Journal of Bioenergetics and Biomembranes* 39 (5-6): 453–58. https://doi.org/10.1007/s10863-007-9113-y.

Liu, J., and B. P. Rosen. 1997. "Ligand Interactions of the ArsC Arsenate Reductase." *The Journal of Biological Chemistry* 272 (34): 21084–89. https://doi.org/10.1074/jbc.272.34.21084.

Liu, Shuang, Fan Zhang, Jian Chen, and Guoxin Sun. 2011. "Arsenic Removal from Contaminated Soil via Biovolatilization by Genetically Engineered Bacteria under Laboratory Conditions." *Journal of Environmental Sciences* 23 (9): 1544–50. https://doi.org/10.1016/s1001-0742(10)60570-0.

Liu, Zijuan, Eckhard Boles, and Barry P. Rosen. 2004. "Arsenic Trioxide Uptake by Hexose Permeases in *Saccharomyces cerevisiae*." *The Journal of Biological Chemistry* 279 (17): 17312–18. https://doi.org/10.1074/jbc.M314006200.

Liu, Zijuan, Miroslav Styblo, and Barry P. Rosen. 2006. "Methylarsonous Acid Transport by Aquaglyceroporins." *Environmental Health Perspectives* 114 (4): 527–31. https://doi.org/10.1289/ehp.8600.

Mandal, Badal Kumar, and Kazuo T. Suzuki. 2002. "Arsenic Round the World: A Review." *Talanta* 58 (1): 201–35. https://doi.org/10.1016/s0039-9140(02)00268-0.

Mateos, Luís M., EfrénOrdóñez, Michal Letek, and José A. Gil. 2006. "*Corynebacterium glutamicum* as a Model Bacterium for the Bioremediation of Arsenic." *International Microbiology: The Official Journal of the Spanish Society for Microbiology* 9 (3): 207–15.

Meng, Yu-Ling, Zijuan Liu, and Barry P. Rosen. 2004. "As(III) and Sb(III) Uptake by GlpF and Efflux by ArsB in *Escherichia coli*." *The Journal of Biological Chemistry* 279 (18): 18334–41. https://doi.org/10.1074/jbc.M400037200.

Molina, María del Carmen, Luis Fernando Bautista, Ignacio Belda, Manuel Carmona, Eduardo Díaz, Gonzalo Durante-Rodríguez, Sara García-Salgado, et al. 2019. "Bioremediation of Soil Contaminated with Arsenic." In *Microorganisms for Sustainability*, A. Kumar and S. Sharma (Eds.), 321–51. Singapore: Springer.

Mukhopadhyay, Rita, Barry P. Rosen, Le T. Phung, and Simon Silver. 2002. "Microbial Arsenic: From Geocycles to Genes and Enzymes." *FEMS Microbiology Reviews* 26 (3): 311–25. https://doi.org/10.1111/j.1574-6976.2002.tb00617.x.

National Research Council. 1977. *Arsenic: Medical and Biological Effects of Environmental Pollutants*. Washington, DC: National Academies Press.

Ordoñez, Omar F., Esteban Lanzarotti, Daniel Kurth, Néstor Cortez, María E. Farías, and Adrian G. Turjanski. 2015. "Genome Comparison of Two *Exiguobacterium* Strains from High Altitude Andean Lakes with Different Arsenic Resistance: Identification and 3D Modeling of the Acr3 Efflux Pump." *Frontiers in Environmental Science* 3. https://doi.org/10.3389/fenvs.2015.00050.

Oremland, Ronald S., and John F. Stolz. 2005. "Arsenic, Microbes and Contaminated Aquifers." *Trends in Microbiology* 13 (2): 45–49. https://doi.org/10.1016/j.tim.2004.12.002.

Oremland, Ronald S., John F. Stolz, and James T. Hollibaugh. 2004. "The Microbial Arsenic Cycle in Mono Lake, California." *FEMS Microbiology Ecology* 48 (1): 15–27. https://doi.org/10.1016/j.femsec.2003.12.016.

Paul, Dhiraj, Sufia K. Kazy, Ashok K. Gupta, Taraknath Pal, and Pinaki Sar. 2015. "Diversity, Metabolic Properties and Arsenic Mobilization Potential of Indigenous Bacteria in Arsenic Contaminated Groundwater of West Bengal, India." *PLoS One* 10 (3): e0118735. https://doi.org/10.1371/journal.pone.0118735.

Porquet, Alain, and Montserrat Filella. 2007. "Structural Evidence of the Similarity of $Sb(OH)_3$ and $As(OH)_3$ with Glycerol: Implications for Their Uptake." *Chemical Research in Toxicology* 20 (9): 1269–76. https://doi.org/10.1021/tx700110m.

Ramaswamy, Sreemathy, and Parjit Kaur. 1998. "Nucleotide Binding to the C-Terminal Nucleotide Binding Domain of ArsA: Studies With An Atp Analogue, 5′-P-Fluorosulfonylbenzoyladenosine (FSBA)." *The Journal of Biological Chemistry* 273 (15): 9243–48. https://doi.org/10.1074/jbc.273.15.9243.

Ramírez-Solís, Alejandro, Rita Mukopadhyay, Barry P. Rosen, and Timothy L. Stemmler. 2004. "Experimental and Theoretical Characterization of Arsenite in Water: Insights into the Coordination Environment of As-O." *Inorganic Chemistry* 43 (9): 2954–59. https://doi.org/10.1021/ic0351592.

Rosenberg, H., R. G. Gerdes, and K. Chegwidden. 1977. "Two Systems for the Uptake of Phosphate in *Escherichia coli*." *Journal of Bacteriology* 131 (2): 505–11. https://doi.org/10.1128/jb.131.2.505-511.1977.

Rosenstein, R., P. Peschel, B. Wieland, and F. Gotz. 1992. "Expression and Regulation of the *Staphylococcus xylosus* Anti- Monite, Arsenite, and Arsenate Resistance Operon." *J Bacteriol* 174: 3676–83.

Sancheti, P., H. Bhonsle, R. H. Patil, M. J. Kulkarni, S. Rapole, and W. N. Gade. 2013. "Arsenomics of *Exiguobacterium*sp. PS (NCIM 5463)3." *RSC Advances*, 1–20. https://nccs.sciencecentral.in/id/eprint/76.

Sanders, Omar I., Christopher Rensing, Masayuki Kuroda, Bharati Mitra, and Barry P. Rosen. 1997. "Antimonite is Accumulated by the Glycerol Facilitator GlpF in *Escherichia coli*." *Journal of Bacteriology* 179 (10): 3365–67. https://doi.org/10.1128/jb.179.10.3365-3367.1997.

Sato, Tsutomu, and Yasuo Kobayashi. 1998. "The *ars* Operon in the *skin*Element of *Bacillus subtilis* Confers Resistance to Arsenate and Arsenite." *Journal of Bacteriology* 180 (7): 1655–61. https://doi.org/10.1128/JB.180.7.1655-1661.1998.

Sengupta, Debanjana. 2021. "Studies on Arsenic Sequestration by Bacteria in Arsenic Contaminated Soil", PhD thesis, St. Xavier's College (Autonomous), Kolkata, affiliated to University of Calcutta.

Sengupta, Debanjana, Siddhartha Chakraborty, Sudeshna Shyam Choudhury, Sayak Ganguli, and Arup Kumar Mitra. 2020. "Isolation and Identification of Unique Arsenotolerant *Exiguobacterium indicum* DSAM62 from Arsenic Rich Environment." *Advances in Zoology and Botany* 8 (4): 298–325. https://doi.org/10.13189/azb.2020.080403.

Sengupta, Debanjana, Arup Kumar Mitra and Sudeshna Shyam Choudhury. 2013. "Identification and Characterization of Effectively Arsenic Tolerant Bacterial Strains from the Potential Arsenic Contaminated Site in 24-Parganas (North) District of West Bengal, India." *Nature Environment and Pollution Technology* 12 (2): 303–8.

Sengupta, Debanjana, Arup Kumar Mitra, Sudeshna Shyam Choudhury, and Ayan Chandra. 2014. "Isotherm Study in Arsenic Tolerant Bacteria Isolated from Arsenic Affected Area in West-Bengal, India." *IOSR Journal of Environmental Science, Toxicology and Food Technology* 8 (1): 8–19. https://doi.org/10.9790/2402-08120819.

Shakoori, Farah R., Iram Aziz, A. Rehman, and A. Shakoori. 2010. "Isolation and Characterization of Arsenic Reducing Bacteria from Industrial Effluents and Their Potential Use in Bioremediation of Wastewater." *Pakistan Journal of Zoology* 42: 331–38.

Shrivastava, Anamika, Devanita Ghosh, Ayusman Dash, and Suatapa Bose. 2015. "Arsenic Contamination in Soil and Sediment in India: Sources, Effects, and Remediation." *Current Pollution Reports* 1 (1): 35–46. https://doi.org/10.1007/s40726-015-0004-2.

Silver, Simon, Guangyong Ji, Stefan Bröer, Saibal Dey, Dexian Dou, and Barry P. Rosen. 1993. "Orphan Enzyme or Patriarch of a New Tribe: The Arsenic Resistance ATPase of Bacterial Plasmids." *Molecular Microbiology* 8 (4): 637–42. https://doi.org/10.1111/j.1365-2958.1993.tb01607.x.

Suresh, K., G. S. N. Reddy, S. Sengupta, and S. Shivaji. 2004. "*Bacillus indicus* Sp. Nov., an Arsenic-Resistant Bacterium from an Aquifer in West Bengal, India." *International Journal of Systematic and Evolutionary Microbiology* 54 (Pt 2): 457–61. https://doi.org/10.1099/ijs.0.02758-0.

Tamaki, S., and W. T. Frankenberger Jr. 1992. "Environmental Biochemistry of Arsenic." *Reviews of Environmental Contamination and Toxicology* 124: 79–110. https://doi.org/10.1007/978-1-4612-2864-6_4.

Vahidnia, A., G. B. van der Voet, and F. A. de Wolff. 2007. "Arsenic Neurotoxicity--A Review." *Human & Experimental Toxicology* 26 (10): 823–32. https://doi.org/10.1177/0960327107084539.

Verma, Shikha, Pankaj Kumar Verma, and Debasis Chakrabarty. 2019. "Arsenic Bio-Volatilization by Engineered Yeast Promotes Rice Growth and Reduces Arsenic Accumulation in Grains." *International Journal of Environmental of Research* 13 (3): 475–85. https://doi.org/10.1007/s41742-019-00188-7.

Verma, Shikha, Pankaj Kumar Verma, Veena Pande, Rudra Deo Tripathi, and Debasis Chakrabarty. 2016. "Transgenic *Arabidopsis thaliana* Expressing Fungal Arsenic Methyltransferase Gene (WaarsM) Showed Enhanced Arsenic Tolerance via Volatilization." *Environmental and Experimental Botany* 132: 113–20. https://doi.org/10.1016/j.envexpbot.2016.08.012.

Villa-Bellosta, Ricardo, and Víctor Sorribas. 2010. "Arsenate Transport by Sodium/Phosphate Cotransporter Type IIb." *Toxicology and Applied Pharmacology* 247 (1): 36–40. https://doi.org/10.1016/j.taap.2010.05.012.

USEPA. 2001. *EPA/Water Quality, Environmental Matters*. Environmental Protection Agency, United States.

WHO. 2018. *Arsenic*. World Health Organization. February 15, 2018. https://www.who.int/news-room/fact-sheets/detail/arsenic.

Wu, Jianhua, and B. P. Rosen. 1993. "Metalloregulated Expression of the Ars Operon." *The Journal of Biological Chemistry* 268 (1): 52–58. https://doi.org/10.1016/s0021-9258(18)54113-2.

Wu, Jilei, Gong Chen, Yilan Liao, Xinming Song, Lijun Pei, Jinfeng Wang, and Xiaoying Zheng. 2011. "Arsenic Levels in the Soil and Risk of Birth Defects: A Population-Based Case-Control Study Using GIS Technology." *Journal of Environmental Health* 74 (4): 20–25.

Wysocki, Robert, Cyrille C. Chéry, Donata Wawrzycka, Marijn Van Hulle, Rita Cornelis, Johan M. Thevelein, and Markus J. Tamás. 2001. "The Glycerol Channel Fps1p Mediates the Uptake of Arsenite and Antimonite in *Saccharomyces cerevisiae*." *Molecular Microbiology* 40 (6): 1391–1401. https://doi.org/10.1046/j.1365-2958.2001.02485.x.

Xu, Chun, and Barry P. Rosen. 1997. "Dimerization is Essential for DNA Binding and Repression by the ArsR Metalloregulatory Protein of *Escherichia coli*." *The Journal of Biological Chemistry* 272 (25): 15734–38. https://doi.org/10.1074/jbc.272.25.15734.

Yuan, Chungang, Xiufen Lu, Jie Qin, Barry P. Rosen, and X. Chris Le. 2008. "Volatile Arsenic Species Released from *Escherichia coli* Expressing the AsIII S-Adenosylmethionine Methyltransferase Gene." *Environmental Science & Technology* 42 (9): 3201–6. https://doi.org/10.1021/es702910g.

Zhang, Jun, Tingting Cao, Zhu Tang, Qirong Shen, Barry P. Rosen, and Fang-Jie Zhao. 2015. "Arsenic Methylation and Volatilization by Arsenite S-Adenosylmethionine Methyltransferase in *Pseudomonas alcaligenes* NBRC14159." *Applied and Environmental Microbiology* 81 (8): 2852–60. https://doi.org/10.1128/AEM.03804-14.

Zhu, Yong-Guan, Masafumi Yoshinaga, Fang-Jie Zhao, and Barry P. Rosen. 2014. "Earth Abides Arsenic Biotransformations." *Annual Review of Earth and Planetary Sciences* 42 (1): 443–67. https://doi.org/10.1146/annurev-earth-060313-054942.

9 Metagenomics for Studying Bioleaching Bioreactor

Sonia Sethi and Harshita Jonwal
Dr. B. Lal Institute of Biotechnology

9.1 INTRODUCTION

Bio-leaching is frequently utilized in the event of wastes containing low quantities of components for which conventional separation procedures are ineffective or impossible. Even in the 21st century, research has looked at additional ways to use bioleaching in practice using diverse bacterial cultures (Conić et al., 2014). Ore resources with high metal concentrations are shrinking, and in the future, rocks with a lower metal percentage or waste will be utilized more frequently (Guezennec et al., 2015). When compared to other procedures, bioleaching has a number of advantages. Inorganic gaseous pollutants are no longer produced. The bioleaching method is particularly cost-effective, not just because it eliminates flue gas.

Several aspects that are linked to the features of microorganisms, chemical reactions (formation of precipitates), or the degree of friction in the suspension impact the process of dissolving metals in solution (Olubambi et al., 2007). pH, particle size, pulp density, stirring frequency (rpm), temperature, nutrient concentration, oxygen content, and total bioleaching duration are only a few of these variables (Amiri et al., 2011). These parameters have a substantial impact on the overall performance of metals transitioning to the liquid phase, and it is critical to explore their impact in order to obtain the best possible metal yield.

The influence of solid substances on the survivability of a combination of bacterial cultures. Particles no longer had a substantial influence on the population of microorganisms, and the particle effect on bacterial decline diminished. Nemati et al. (2000) looked at the impact of various grain sizes on microbe populations and overall efficacy during leaching. The pH value is one of the defining factors of *A. ferrooxidans* cells' metabolic activity. However, several publications (Zhang et al., 2018) differ on the ideal pH for growth and cellular oxidation, as well as the pH cut-off at which the bacteria can execute basic biological tasks. Previous investigations looked into the relationship between pH value and extraction efficiency, but only for different bacterial species, nonstationary systems, or just iron (Mikoda et al., 2019).

Bioleaching is a potential process for recovering metals from low-grade ores and e-waste using microorganisms (Hong and Valix, 2014). Bioleaching is gaining

DOI: 10.1201/9781003451457-9

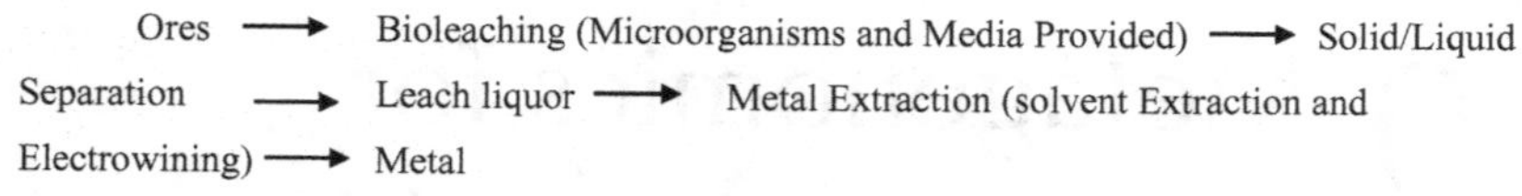

FIGURE 9.1 Generalized flow of the bioleaching process.

popularity due to its decreased operational costs and energy needs in metal recovery. *Acidithiobacillus ferrooxidans, Leptospirillum ferrooxidans, Acidithiobacillus thiooxidans*, and other chemoautotrophic bacteria are routinely utilized for metal recovery from e-waste (Erüst et al., 2013). Bacteria that use CO_2 as a carbon source and inorganic molecules (Fe^{2+} and reduced S) as an energy source use a variety of biooxidants and bioleaching processes to enable metal dissolution (Figure 9.1).

9.2 MICROORGANISMS

9.2.1 *Thiobacillus*

Thiobacillus is the genus of bacteria that is most active in bioleaching. Gram-negative, non-spore-forming rods that develop in aerobic environments. The majority of thiobacilli are chemolithoautotrophic, meaning they use carbon dioxide from the atmosphere to synthesize new cell material. The energy is obtained by oxidizing reduced or partly reduced sulfur compounds such as sulfides, elemental sulfur, and thiosulfate, with sulfate as the ultimate oxidation product.

Bacterial leaching takes place in an acidic environment with a pH range of 1.5–3, where most metal ions stay in solution. *Thiobacillus ferrooxidans and T. thiooxidans*, two acidophilic species, are particularly important. Other thiobacilli may oxidize sulfur and sulfides as well, but they can only develop at higher pH levels where metal ions do not stay in solution. *T. ferrooxidans* is unique among thiobacilli in that it can utilize ferrous iron as an electron donor in addition to obtaining energy through the oxidation of reduced sulfur compounds. *T. ferrooxidans* may grow on reduced inorganic sulfur compounds in the absence of oxygen by utilizing ferric iron as an electron acceptor.

Huber and Stetter have described two new species of acidophilic thiobacilli: *T. cuprinus* is a facultatively chemolithoautotrophic bacterium that oxidizes metal sulfides but not ferrous iron, and *T. prosperus* represents a novel group of halotolerant metal-mobilizing bacteria. This bacterium is said to favor chalcopyrite for copper mobilization. Both strains may have some bioleaching capacity due to their physiological differences.

9.2.2 *Leptospirillum*

Another acidophilic, obligately chemolithotrophic ferrous iron oxidizing bacteria, *Leptospirillum* ferrooxidans, was initially identified from Armenian mine fluids by Markosyan. This bacterium can handle lower pH and greater uranium, molybdenum, and silver concentrations than *T. ferrooxidans*, but it is more susceptible to copper and cannot oxidize sulfur or sulfur compounds. As a result, *L. ferrooxidans* cannot attack mineral sulfides on its own. Only *T. ferrooxidans* or *T. thiooxidans* can be used for this. *T. thiooxidans*, *T. ferrooxidans*, and *L. ferrooxidans* are mesophilic bacteria that thrive at temperatures between 25°C and 35°C.

9.2.3 Thermophilic Bacteria

Th-bacteria, also known as *Thiobacillus*-like bacteria, are moderately thermophilic bacteria that thrive on pyrite, pentlandite, and chalcopyrite at temperatures of about 50°C. The energy source is ferrous iron; however, growth is only visible in the presence of yeast extract. Brierley, Norris, Karavaiko, and their colleagues identified very thermophilic bacteria that thrive at temperatures exceeding 60°C. *Acidianus brierleyi*, formerly known as *Sulfolobus*, is a chemolithoautotrophic, facultatively aerobic, acidophilic Archaeon that thrives on ferrous iron, elemental sulfur, and metal sulfides. Elemental sulfur is employed as an electron acceptor in anaerobic environments and is reduced to H_2S. *Sulfolobus bacteria* are facultatively chemolithotrophic aerobic bacteria that oxidize ferrous iron, elemental sulfur, and sulfide minerals. *Sulfobacillus thermosulfidooxidans*, a spore-forming facultatively autotrophic bacteria, uses the same chemicals as an energy source. However, growth is only possible in the presence of yeast extract.

9.2.4 Heterotrophic Microorganisms

Metal leaching may be caused by heterotrophic bacteria and fungi that require organic inputs for development and energy. Metal solubilization, like manganese leaching, can occur as a result of enzymatic reduction of highly oxidized metal compounds or as a result of the production of organic acids (e.g., lactic acid, oxalic acid, citric acid, gluconic acid) and compounds with at least two hydrophilic reactive groups (e.g., phenol derivatives) that are excreted into the culture medium and dissolve heavy metals by direct displacement of metal ions. The metal leaching provides no advantage to heterotrophic bacteria. Members of the genus Bacillus are the most efficient metal solubilizers among bacteria, whereas the genera *Aspergillus* and *Penicillium* are the most important among fungi (Table 9.1).

TABLE 9.1
Microorganisms Involved in Bioleaching

Microroganisms	Process and Application
Thiobacillus sp	Oxidation of sulfide Metallurgical industries
Leptospirillum sp., *Thermophillic* bacteria *Sulphobacillus*, *Acidophillic* bacteria	Oxidation, desulphurization of coal
Organotrophic microorganisms	Destruction of sulphide minerals, reduction and oxidation Leaching of gold, extraction of metals

9.3 OVERVIEW: MINING, MINE WASTE, AND MICROBIAL SOLUTIONS

The process of gathering usable resources from the ground is known as mining. Coal, gold, and iron ore are just a few examples of mined materials. Iron ore is the raw material used to make iron. Mining has been practiced since prehistoric times.

Large amounts of liquid and solid waste are produced during the extraction of metals and metalloids. Uneconomic materials such as rock, gangue, refuse material, silt, tailings, roasted ore, or processing chemicals are among the mining wastes (Hudson-Edwards and Dold, 2015). Currently, the volume of trash recovered each year is projected to be in the thousands of millions of tonnes, but due to rising demand and the exploitation of low-grade reserves, these numbers are growing rapidly.

These wastes are a source of substantial nuisance in the absence of extremely stringent mining site management, particularly due to the presence of highly poisonous substances such as lead, arsenic, mercury, or radioactive elements such as uranium. Many metals and metalloids, in fact, pose a major hazard to human health and the environment due to their carcinogenicity, cytotoxicity, and mutagenicity. Ingestion of polluted water, inhalation of dust, or eating tainted crops or animals grown on polluted fields pose serious health risks to local communities across the world (Festin et al., 2019).

The creation of acid mine drainage (AMD) and the erosion of tailing wastes by wind or water are two key issues that mine businesses face. AMD are formed when sulfide ores wastes come into contact with oxygen and water, resulting in a low pH and high concentration of sulfate, iron, and hazardous metals and metalloids such as lead and arsenic. AMD can persist for generations after a mine is closed, affecting tens of thousands of kilometers of groundwater, streams, lakes, and estuaries across the world. Tailings contain high numbers of small particles of hazardous metals and metalloids vulnerable to wind and water erosion in the absence of vegetation (Karaca et al., 2018).

The lack of vegetation will have environmental consequences that will be determined by climate conditions. Tailings erosion in arid and semi-arid regions is mostly caused by a mix of wind and water erosion with high gusts and heavy rainfall. In temperate, tropical, and equatorial regions, however, tailing will be more susceptible to leaching, resulting in AMD (Nirola et al., 2016). Furthermore, a vegetative cover might reduce water infiltration into tailings, similar to how evapotranspiration can decrease mine drainage formation in dry and semi-arid regions.

The cost of mining site reclamation is significant, and much study has been conducted on this topic across the world (Aznar-Sánchez et al., 2018). Because metals and metalloids are not biodegradable, they can only be converted to a less toxic and/or less soluble form in soils or precipitated out of solution. Traditional chemical or physical processes (e.g., chemical precipitation or filtration for AMD and soil replacement, solidification/stabilization or soil washing for tailings) are often too expensive or ineffective, especially for abandoned mining sites, given the amount of waste and the need for long-term treatments (Kefeni et al., 2017).

In this environment, creating cost-effective and long-term remediation solutions for AMD and tailings cleanup is both a problem and a requirement, especially for poor nations (Humphries et al., 2017). Microorganisms are vital to life on Earth, despite being invisible to the naked eye. Microbial communities that can tolerate a wide variety of physicochemical pressures have colonized practically every site on Earth over billions of years, including the most severe ecosystems. Microorganisms are geoactive and play an important part in the biosphere because of their diverse metabolic processes. They play an important role in the natural biogeochemical cycles of nutrients, organic matter, metals, and minerals.

Microorganisms, in particular, play a key role in mineral formation and dissolution, redox transformations of metals and metalloids, methylation and demethylation, the production of organometallic complexes, and other processes that might affect metallic element mobility and toxicity. To deal with the presence of metals and metalloids in their environment, prokaryotes and eukaryotes have created a variety of resistance and metabolic strategies. They could also be useful in the cleanup of mining areas. Metals and metalloids have been precipitated in AMD solutions using various techniques based on microbial iron oxidation or sulfate reduction. Phytostabilization also serves to stabilize tailings by creating a vegetative cover that helps to inhibit the spread of potentially hazardous chemicals. Plant development, on the other hand, is restricted in mine waste dumps due to the severe conditions there. Microorganisms can encourage plant development on tailings and contaminated mining soils as well as immobilize contaminants within the soil.

Microbial communities have consequently been the subject of substantial investigation in the hopes of developing efficient, sustainable, and environmentally friendly technology. Rapid advancements in sequencing technology, such as Next Generation Sequencing (NGS), have dramatically advanced microbiological research in recent decades, allowing for the creation of massive volumes of sequences at a fraction of the cost of older sequencing methods.

The use of "omics" methods like metagenomic, metatranscriptomic, metaproteomic, and metabolomic, in combination with metaproteomic and metabolomic, has undeniably increased our understanding of the fine mechanisms at work in prokaryotes and eukaryotic communities in mining contexts (Zuniga et al., 2017). This "omics" approach, when combined with more traditional techniques such as isolation and physiological study, allows researchers to answer questions that were previously unthinkable, such as which microorganisms are present, what they are doing, and what is the relationship between microorganisms and plants (Plewniak et al., 2018).

Mine wastes are undesirable byproducts of mineral extraction, and they can be solid, liquid, or gaseous. Waste rock, poorly extracted ore minerals, gangue minerals, tailings, processing chemicals, and residues account for 20–25 Gt of solid mining waste generated each year worldwide (Hudson-Edwards et al., 2011). Because of their vast distribution and possible toxicity to humans, plants, and wildlife, mine waste pollution is a global environmental hazard.

Heaps and dumps have several advantages, including easy equipment and operation, minimal investment and operating costs, and acceptable returns. On the other hand, it's important to recognize that the operation has some serious limitations: the heaped material is quite heterogeneous, and there's little that can be done to regulate the process except for pH adjustments and the addition of some nutrients. Furthermore, the rates of oxygen and carbon dioxide transfer obtained are modest, necessitating long durations of operation to produce significant conversions.

The intricate network of biological processes involved in bioleaching would be best done in reactors from a process engineering perspective. Reactors would allow for better control of the relevant factors, resulting in improved performance. Volumetric productivity and the degree of extraction may both be greatly improved. The use of reactors in biomining is limited by the massive volumes of run-of-mine ore that must be processed in most circumstances. In 1999, Chile's Chuquicamata copper mine produced 630,000 tonnes of fine copper. The treatment of roughly 6 million tonnes of run-of-mine ore was required to produce that quantity of metal. If this quantity of waste were to be processed in bioreactors, the needed equipment capacity would be on the order of 30 million cubic meters, an inconceivable number. This restricts their use to the treatment of mineral concentrates or the processing of small amounts of ore. Every year, for example, approximately 11,000 tonnes of gold concentrates are biooxidized in reactors. Therefore, the use of bioreactors in biomining, with an emphasis on oxygen and carbon dioxide transfer, maintaining appropriate solids suspensions, and applying bioreactors to commercial applications.

Mixing is one of the most important aspects of bioleaching in agitated tanks. It must be efficient in order to achieve a high oxygen transfer rate and equal distribution of the various components of the slurry throughout the tank in order to comply with oxygen absorption rates of roughly 1,500 mg/L/h. The investment costs of the tanks and impellers, as well as the running expenses of gaseous mass transfer, are the key costs of bioleaching activities. This talk will offer an overview of recent developments made to overcome some of the constraints of bioleaching processes, with a focus on the development of an alternative bioleaching reactor that uses floating agitators to mix and stir the solution.

This innovative idea allows for lower bioleaching costs by operating (1) in lagoons or ponds rather than tanks and (2) with a larger solid loading (>15% w/w) than traditional stirred tank bioreactors. The demand for oxygen is greatly enhanced under these situations of heavy solid loads, and air is replaced by oxygen to provide an adequate oxygen supply. Experiments conducted at lab to pilot sizes (2 L–2 m^3) have proven that this novel device, when operated at solid loads of up to 30%, produces identical leaching results (kinetics and metal yields) as those reported in stirred tank bioreactors with solid loads of less than 20%. The use of oxygen instead of air improves oxygen transfer rate, but it must be done carefully to avoid a high dissolved oxygen (DO) level in the slurry. When the DO concentration exceeded 17 ppm, the bioleach microorganisms' activity decreased, possibly due to oxidative stress caused by the formation of toxic reactive oxygen species.

9.4 BIOMINING AND AMD: GENERATING MICROORGANISMS AND THEIR MODE OF ACTION

The two mechanisms of ore bioleaching are direct and indirect biooxidation. Most researchers generally agree on the indirect biooxidation process (Liu et al., 2017). Contact and noncontact biooxidation methods are included in the indirect biooxidation mechanism. Bacteria would attach to ore and build a biofilm between bacteria and ore in a contact mechanism. Fe^{2+} is oxidized to Fe^{3+} by bacteria inside the biofilm, and Fe^{3+} dissolves the ore. It has been demonstrated that biofilm production is significant in ore bioleaching. The concentration of ferric iron and H^+ in biofilm is substantially greater than in solution, perhaps increasing the rate of ore dissolution. The bacteria do not need to cling to the ore in the noncontact method. Furthermore, the Fe^{3+} generated by bacteria is important in the dissolution of ore.

The indirect contact process is often thought to be the mechanism of Polycholorinated Biphenyls (PCB) bioleaching (Silva et al., 2015). In addition, the mechanism of bioleaching of copper from PCBs is present as follows:

$$Fe^{2+} + O_2 + 4H^+ \longrightarrow \text{Bacteria}Fe^{3+} + 2H_2O \tag{9.1}$$

$$2Fe^{3+} + Cu \rightarrow 2Fe^{2+} + Cu^{2+} \tag{9.2}$$

Copper dissolution from PCBs may be separated into two steps in this process. The ferrous ions are oxidized by bacteria to produce ferric ions in the first phase. The ferric ions mobilize copper from the PCBs in the second phase, and the ferric ions are reduced to ferrous ions. As a result, a Fe^{3+}-Fe^{2+} cycle is produced, and copper is leached away (Mrážiková et al., 2016).

The addition of elemental metal ions (Cu^{2+}, Cr^{3+}, Ni^+, Sn^{2+}, and Zn^{2+}) to bacteria quickly decreased their activity (Bryan et al., 2015). The inclusion of nonmetallic components in the bioleaching process revealed bacterial toxicity. The researchers utilized a mixed culture that outperformed a single bacterium (Jiang et al., 2017). Bacterial tolerance might be enhanced by acclimatization. A two-step procedure was also employed to reduce the harmful impact. Additionally, response surface methodology (RSM) was employed to improve the bioleaching parameters (Arshadi and Mousavi, 2015).

9.4.1 Microbe-Metal Interactions

Through metabolic activities, altering pH or redox conditions, secreting chelating chemicals, and/or passive sorption, microbial activity can solubilize and/or precipitate metals. Microbes can exploit these activities to acquire energy, or they can be energy-intensive and constitute part of metal absorption or resistance mechanisms. Biominerals are microbially precipitated metals, and the process of their creation is called biomineralization. Bioleaching, or biomining if done for industrial reasons, is a term used to describe the process of solubilizing metals.

9.4.2 Oxidation Processes and Their Impact on Metal Mobility

Acidity and AMD are produced as a result of prokaryotes oxidizing Fe(II) and sulfide to get energy. This has an effect on metal mobility because it solubilizes metals linked with reduced minerals like pyrite (FeS_2). Microbial oxidation of Fe(II) and Mn(II) results in the formation of Fe(III) and Mn(IV) oxide minerals, which can be used to sequester metals from solution.

Even in settings where chemical Fe(II) oxidation is very quick, such as in circumneutral oxygenated waters, iron-oxidizing microbes acquire energy by oxidizing Fe(II) to Fe(III). Some prokaryotes, for example, connect Fe(II) oxidation to nitrate reduction and may oxidize Fe(II) under microaerophilic or anoxic circumstances.

Bacteria and archaea do dissimilatory Fe(II) oxidation; some can fix CO_2 via the ribulose-1, 5-biphosphate carboxylaseoxygenase (RuBisCo) enzyme, while others require organic C to thrive (Ilbert & Bonnefoy, 2013).

Chemical Fe(II) oxidation by oxygen happens relatively slowly at low pH, and much more slowly at the extremely low pH levels found in sulfidic mine tailings. However, in such low pH settings, the rate of Fe(II) oxidation actually rises with lowering pH (and increasing Fe[III] solubility), indicating that microbial Fe(II) oxidation plays a substantial role in AMD development (Larson et al., 2014). At low pH, the O_2/H_2O couple's redox potential makes oxygen the more suitable electron acceptor for oxidizing Fe(II). The iron oxidation systems in acidophilic bacteria include several cytochromes that conduct electrons from Fe(II) outside the cell to the electron acceptor inside the cell, a cytochrome oxidase that reduces the electron acceptor (oxygen), a high-potential iron sulfur protein (HiPIP), and a rusticyanin (e.g., *Acidithiobacillus ferrooxidans*). The processes in acidophilic archaea differ from those in acidophilic bacteria, and they contain various cytochromes, HiPIP, copper proteins, and oxidases (Ilbert & Bonnefoy, 2013).

The major cause of AMD is microbial Fe(II) oxidation. Due to low pH values and high concentrations of metals, metalloids, and sulfate, AMD is a global environmental concern that causes substantial contamination of terrestrial and aquatic habitats (Chalkley et al., 2019). Microorganisms derive energy from the oxidation of Fe(II) in iron sulfide minerals (e.g., pyrite), which solubilizes Fe as well as other metals and metalloids, resulting in AMD. At low pH, microbial activity has been demonstrated to boost the iron oxidation rate by five orders of magnitude. This mechanism produces acidity, lowering the pH to dangerously low levels while maintaining metals in suspension.

In the environment, microbial oxidation of reduced sulfur compounds (i.e., H_2S, S[0], sulfite, thiosulfate, and tetrathionate) paired with oxygen reduction is common. S-oxidizers are organisms that are capable of oxidizing reduced sulfur compounds. Phototrophic S oxidizers use anaerobic oxidation to fix CO_2 and generate S(0), which can be stored intracellularly, extracellularly, or sulfate. Other S oxidizers, such as *Thiobacillus denitrificans*, can employ nitrate as an electron acceptor or Fe(III) as an electron acceptor for *A. ferrooxidans*. When reduced sulfur compounds are oxidized, certain S oxidizers create sulfuric acid; this is important in mining wastes when minerals sensitive to acid dissolving (e.g., sphalerite [ZnS], chalcocite [Cu_2S]) are present (Fike et al., 2016).

9.5 REMEDIATION BY SULFATE-REDUCING BACTERIA

Using the oxygen and enhanced carbon supply generated by algae, sulfate-reducing bacteria may be employed to treat acid mine tailings. Sulfate is used as a terminal electron acceptor, converting SO_4^{2-} to H_2S. Sulfate reducing bacteria (SRB) that are acid-tolerant catalyze the reduction of sulfate to sulfide, which converts sulfuric acid to hydrogen sulfide and creates alkalinity. Metal recovery is accomplished by controlling the concentration of the reactant sulfate in the bioreactor via pH control. As a result, toxic metals are successfully removed from mine tailings wastes, and the wastes are converted into valuable products. Bioremediation offers the benefit of minimal maintenance expenses as well as the solid-phase products of water treatment being kept within the marsh sediments.

Metals previously bound within the mineral matrix can be mobilized by SRB present in mine tailings due to sulfide oxidation and reduction of mineral oxides within exposed mine tailings. These SRB are important in the mobilization and removal of hazardous metals through dissolution and precipitation, as well as the recovery of precious metals from low-grade ores. Microorganisms used in microbial leaching of metals from ores/waste come from several genera and include bacteria such as *Acidiphilium cryptum*, *Acidithiobacillus ferrooxidans*, *At. caldus*, *At. thiooxidans*, *Acidianus brierleyi*, *Citrobacter*, *Clostridium*, *Cronobacter*, *Ferribacterium limneticum*, *Ferroplasma acidiphilum*, *Gallionella ferruginea*, *Leptospirillum ferrooxidans*, *L. ferriphilum*, *Ochrobactrum anthropi*, *Sulfolobus* sp., *Sulfobacillus thermosulfidooxidans*, *S. acidophilus*, *Thiobacillus denitrificans*, and *T. thioparus* (Anawar, 2015).

SRB remediation is often based on salt combinations and locally accessible organic substrates for bacteria metabolism, such as manures, sawdust, wasted mushroom compost, sugarcane waste, wood chips, yeast extract, and other carbon sources. In order to obtain maximal metal and sulfate removal, these mixes must be optimized. Organic matter and calcium carbonate are added to lower the initial pH of Acid mine tailings waste (AMTW), which influences SRB development. Sulfate is converted to hydrogen sulfide, and organic waste is converted to hydrogen carbonate as part of the remediation procedure. Heavy metal ions present then react with the hydrogen sulfide gas generated to form insoluble metal sulfide precipitates, allowing the metals to be removed (Kefeni et al., 2017).

By eliminating the carbon (IV) oxide generated in the bioreactor, hydrogen sulfide is continually formed. The reactor is a stirred tank with a long hydraulic retention period. In the chemical process, the H_2S generated is routed through an agitated anaerobic contactor, where targeted metals are precipitated as sulfides. In a standard filtering device, the sulfides can then be dewatered. The filtering machine creates high-quality metal sulfide, which is refined further (Littlejohn et al., 2015). Heavy metals such as As, Cd, Co, Cr, Cu, Hg, Mn, Mo, Ni, Pb, Se, and Zn are commonly targeted.

$$M^{2+} + H_2S \rightarrow MS^{2-} + 2H^+$$

Organisms from the fundamental domains of bacteria, archaea, and eukarya have been discovered at acid mine sites (fungi and algae). The phyla *Acidobacteria*, *Actinobacteria*, *Bacteroidetes*, *Firmicutes*, *Nitrospirae*, the alpha, beta, and gamma classes of the phylum Proteobacteria (the most extensively distributed phylum in acid mine), and several archaea species are among the bacteria found in acid mine. Deltaproteobacteria and Firmicutes are the only organisms that can be found at acid mine sites (Méndez-García et al., 2015). These organisms use solar or chemical energy to perform a wide range of metabolic functions. They develop in aerobic and anaerobic circumstances and use organic or inorganic carbon as their only carbon source (such as ferrous iron, hydrogen, short-chained fatty acids, and reduced sulfur). They combine the oxidation of organic substrates with sulfate reduction to produce hydrogen sulfide as the main end product.

Sulfate reduction is increased when more than one carbon substrate is used. Increases or decreases in substrate mixture have a significant impact on SRB performance and are also crucial in lowering hazardous metals' negative effects through acid buffering and adsorption (Kefeni et al., 2017). SRB oxidizes carboxylic acids (acetate, butyrate, fumarate, malate), amino acids (alanine, glycine, serine), alcohols (ethanol, methanol), hydrogen, methanethiol, some sugars (fructose, glucose), and aromatic compounds (ethylbenzene, benzoate, phenol, and toluene) produced by anaerobic degradation of complex organic compounds. The Calvin-Benson-Bassham cycle is used by these organisms to get cellular carbon, and ribulose bisphosphate carboxylase/oxygenase is a major enzyme in this route.

Sulfate-reducing bacteria may also develop by dividing thiosulfate, sulfite, and sulfur, resulting in sulfate and sulfide production. Sulfur metabolism necessitates the use of ATP in order for two essential enzymes, adenylylsulfate reductase and bisulfite reductase, to decrease sulfate. Sulfate activation and reduction of sulfate to sulfite, sulfite reduction to sulfide, and elemental sulfur reduction are all part of the process. Sulfur-oxidizing acidophiles may also link sulfur oxidation with iron reduction, as demonstrated by *At. thiooxidans* and *At. ferrooxidans.* Sulfide is oxidized by a sulfide/quinone oxidoreductase during the oxidation of reduced inorganic sulfur through sulfite synthesis, which transfers electrons to ubiquinone and creates sulfur. The periplasmic sulfur can be converted to sulfite by the bacteria.

The generated sulfite is subsequently converted to sulfate by the enzyme sulfite oxidoreductase or by adenosine phosphosulfate reductase (Méndez-García et al., 2015). Sulfate is integrated into amino acids like methionine and cysteine, as well as iron-sulfur centers and other metabolites. In the absence of sulfur, SRB can develop and form syntrophic associations with methanogens or other hydrogen scavengers. When electron acceptors get depleted, this helps them survive in the environment.

Between pH 6 and 8, most SRB thrive (Sánchez-Andrea et al., 2015). Beyond this pH range, the microbial sulfate reduction rate and metal removal ability both decrease. Metal sulfide solubility rises when the pH is acidic. Some SRB species may survive pH levels ranging from 5 to 9.5. On solid growth medium at pH 3.6, several acidophilic and acidotolerant SRBs have been identified. The use of acid-tolerant SRB in bioreactor systems has improved performance and led to the creation of novel bioreactor systems for metal recovery. To thrive in this environment, acidophilic organisms must have two physiological characteristics. Even when confronted with a large inward gradient, they must have a remarkable capacity to inhibit proton entrance into the cell cytoplasm. Acidophilic SRB employ membrane-bound ATPases to produce ATP via pH gradients. The capacity to create positive membrane potentials, which is done via active cation inflow, is the second attribute. This provides some protection against negatively charged ions.

9.6 IMPLICATIONS OF MICROORGANISMS IN FORMATION OF ACID MINE DRAINAGE

Sulfide ores are found naturally in several metals (e.g., Pb, Cu, and Zn). When sulfide minerals (particularly pyrite or pyrrhotite) come into contact with oxygen and water, AMD results. Although AMD is a natural occurrence, mining operations like excavation and milling hasten the process by increasing the surface area of sulfide minerals exposed to water, oxygen, and microbes (Simate and Ndlovu, 2014). Low pH increases the mobilization of metals and metalloids, increasing their dispersion into the water, which is characteristic of AMD (Humphries et al., 2017). AMD is one of the most serious sources of water pollution in industrial and post-industrial locations across the world. Several studies have been conducted to define the autochthonous microbial communities (Bacteria, Archaea, and Eucaryota) that thrive in these harsh conditions (Plewniak et al., 2018).

The general mechanism of AMD development is widely understood (Vera et al., 2013). The oxidation of sulfide minerals, such as pyrite (FeS_2), the most prevalent sulfide mineral on Earth, is the first phase. The major oxidant of metal sulfides is ferric iron, according to equation 9.3:

$$FeS_2 + 14\ Fe^{3+} + 8\ H_2O \rightarrow 15\ Fe^{2+} + 2\ SO_4^{2-} + 16\ H^+ \quad (9.3)$$

This reaction results in the formation of ferrous iron, sulfate, and acidity. The rate-determining phase is the regeneration of ferric iron via the oxidation of Fe(II) to Fe(III) (equation 9.4).

$$Fe^{2+} + 0.25\ O_2 + H^+ \rightarrow Fe^{3+} + 0.5\ H_2O \quad (9.4)$$

The physical, chemical, and biological features of the system should be used to identify and construct an appropriate reactor for a biomining process. The complicated composition of the reacting sludge, which is made up of an aqueous liquid, suspended and connected cells, suspended particles, and air bubbles, needs special attention. Bioleaching and biooxidation are best accomplished in a continuous mode of operation with high volumetric productivity and small reactor sizes due to the huge amounts of material to be treated. A continuous stirred tank reactor appears to be the best option when considering the kinetics of microbial development. The autocatalytic nature of the reactor is a key factor when choosing one.

This is true for all fermentation processes, but there is a significant variation in bioleaching. The nutrients used in commercial fermentations are chosen for their high affinity with the microbial population, but the mineral species used in biomining are frequently resistant to microbial activity, signifying a low affinity. Monod's saturation constant, K S, is related to substrate microorganism affinity. As is the case with most sugars, high affinities are reflected in low K S values of a few milligrams per liter. Saturation constants for certain minerals are as high as 3–6 g/L, hundreds of orders of magnitude greater. This condition has an impact on reactor selection. A single agitated tank will require a very large capacity if a high degree of conversion is wanted; hence, an array of reactors will be more suited. It has been demonstrated that a continuous stirred tank reactor followed by a tubular plug flow reactor (PFR) produces the smallest reaction volume required to achieve a given conversion. Because PFRs are impractical due to the necessity for aeration and the presence of solid particles, their performance can be approximated using a sequence of continuous stirred tank reactors. The percolation column, the Pachuca tank, the air-lift column, and certain unusual designs such as rotary reactors have all been researched for their applicability in biomining.

The following are the most important features of bioleaching: (1) It involves groups of acidophilic microorganisms that contain chemolithoautotrophic iron and sulfur oxidizers that fix CO_2 from the air and heterotrophs that scavenge fixed carbon from the chemolithoautotrophs' excretions or dead remnants; (2) it involves oxidative processes that use oxygen from the air as a terminal electron acceptor; (3) several of the key metabolic and chemical reactions involved are exothermic, driving the temperature of the bio-leaching heap from ambient temperatures at the start to as high as 70°C over a period of weeks to months; and (4) the sources of ferrous iron and sulfur compounds that support microbial metabolism are solids, and microbial attack of these substrates results in the cell adhesion and biofilm formation

In this regard, we will now concentrate on the genomes and metagenomics of acidophilic bacteria found in bioleaching wastes or other mining sites. What characterizes a closely similar mining environment is a debatable point. This is especially important to note because a lot of the genomic and metagenomic data used to develop bioleaching models comes from microbes that live in different settings like AMD and acidic (hot) springs. One rationale is that many of the microorganisms identified in the latter habitats are comparable to those found in bioleaching heaps, according to 16S rDNA data. As a result of the lack of data from real bioleaching-derived genomes, related acidophiles may be used to create genetic and metabolic models of specific species as well as to predict ecophysiological interactions that may occur during bioleaching. However, nagging issues remain: how reliable are these models, and how well can ecophysiological interactions be predicted from such (possibly insufficient or even wrong) data? Unfortunately, the closeness of 16S rDNA sequences between two bacteria is insufficient to conclude that they have the same gene complement.

9.7 MOLECULAR METHODS

Traditional molecular techniques used to investigate the structure, function, and dynamics of microbial communities in bio-mining systems include molecular fingerprinting techniques (such as denaturing gradient gel electrophoresis or end-fragment length polymorphism), fluorescence in situ hybridization, and clone library sequencing. The most widely used DNA sequencing technology is next-generation sequencing, which is based on platforms like the Illumina HiSeq, MiSeq, or NovaSeq 6000.

9.7.1 Omics Approaches (Genomics and Bioinformatics)

The genetic and metabolic capacities of acidophiles in bioleaching microorganisms and their near relatives have been revealed thanks to genomics. Bioinformatic analysis of genomic information has been a crucial pathway for acquiring insight into the biology of many of these microbes, which are resistant to genetic modification. We discuss how genome models have been used to predict genes and metabolism, as well as the ecophysiological interactions that are believed to occur during bioleaching, with the caveat that most of this model development has come from examinations of genomes not directly sourced from bioleaching heaps.

In 2000, the first draft genome of a bioleaching-related microorganism (*A. ferrooxidans* ATCC 23270, isolated from a coal waste heap) was published, and the data was used to partially reconstruct its amino acid metabolism and, as a result, to predict sulfur assimilation, regulation, and iron homeostasis, quorum sensing, extracellular polysaccharide formation, CO_2 fixation, iron and sulfur oxidation, and other metabolic aspects. Eight years later, the first complete genome sequence of *A. ferrooxidans* was published, laying the groundwork for comparative genomics-based predictions of its genetic and metabolic potential, which consolidated and extended earlier models in areas such as iron and sulfur oxidation, iron reduction, CO_2 fixation, nitrogen metabolism, ecophysiology, and antisense RNA, as well as studies in heavy metal resistance and quorum sensing (Bevilaqua et al., 2013).

The best-studied model for a bioleaching bacterium is *A. ferrooxidans*. However, it was recognized early on that *A. ferrooxidans* was only one member of a consortium of bioleaching microorganisms, and since 2008, permanent draft genomes from other bioleaching genomes have become available, extending the prediction of the genetic and metabolic potential of bioleaching microorganisms, including models for energy metabolism in *A. caldus* and *A. thiooxidans* as well as models for overall metabolism in *Sulfobacillus*. According to the NCBI database, 55 bacterial and 36 archaeal full genomes from microorganisms found in bioleaching processes are currently publicly available, with a trend toward exponential growth.

The use of omics in biomining has had a substantial influence on our knowledge of the process. The primary goal of genomic research has been to gain a better understanding of metabolism, which is directly involved in important metabolic pathways such as iron and sulfur oxidation, quorum sensing, flagella formation, chemotaxis carbon assimilation, nitrogen fixation, and bioleaching adaptation. The microbial diversity of the bioleaching environment has been identified thanks to genome sequencing. Using existing molecular methods, including quantitative PCR (qPCR), FISH, and DGGE, researchers were able to examine severe leaching settings for new biomining microorganisms with higher leaching potential. Research employed shot-gun sequencing of 16S rRNA genes to assess the microbial diversity at a manganese mining site and their role in manganese solubilization. Proteobacteria lead the microbial population at the manganese mining site with 42.47%, followed by *Actinobacteria* with 23.99%. Two tailing basins from the Panasqueira tungsten mine were researched to determine if there was a relationship between microbial distribution and function and chemical variability in the two basins. By sequencing the 16S rRNA gene, the MiSeq instrument was utilized to identify microbial diversity. The two tailings contain members of the genera *Pseudomonas, Bacillus, Streptococcus, Acinetobacter, Rothia*, and *Cellulomonas*, as well as the *Anaerolineacea* family.

The microbiological distribution of basin 1 was associated with higher potassium and aluminium levels, whereas basin 2's microbiological distribution was linked to higher As-S-Fe levels. The PICRUSt algorithm was used to anticipate the many microbial metabolisms that occur in this harsh environment. The development of a hypothesis for re-creating microbial metabolic pathways in the bioleaching process requires genome sequencing (iron and sulfur oxidation pathways, biofilm formation, etc.). The pet II operon contains a pet II ABC gene cluster (encoding the bc 1 complex) that was co-transcribed with three additional genes: cyc A (encoding cytochrome c), sdr A (encoding a putative dehydrogenase), and the hip gene.

Comparative genomics has been used to investigate the differences between different strains of the same species. There are 2397 genes in common between ATCC 53993 and ATCC 23270T, accounting for 78%–90% of their genomes. Because of this genetic difference, ATCC 53993 has superior copper resistance.

9.7.2 Genome Annotation and Function Analysis, Manual Curation, and Metabolic Pathway Identification

For the genomes of the five biomining bacteria, a pipeline for coding sequence prediction and annotation (CDS) is being applied, as reported before (Travisany et al., 2014). REGANOR, which combines the GLIMMER and CRITICA tools for gene discovery and tRNAscan-SE for tRNA gene identification, was used to do an ab initio CDS and tRNAs prediction. Hidden Markov Models methods were used to predict transmembrane helices TMHMM and the existence of signal peptide cleavage sites using SignalP for every anticipated CDS. BLASTP utilizes PROSITE, Pfam, ProDom, and SMART against the NCBI Non-Redundant, SWISSPROT, TCDB, OMNIOME, the Kyoto Encyclopedia of Genes and Genomes (KEGG), the Cluster of Orthologous Groups database (COG), and InterproScan. METANOR conducted the consensus annotation, while InterPRO hits were utilized to assign Gene Ontology (GO) words to genes. The GenDB platform was used to manually curate automatic annotations. Using Fisher's exact test with Benjamini–Hochberg multiple testing correction, an enrichment analysis of COG categories was performed for the gene sets from each of the five biomining bacteria. Enriched categories have a corrected p-value of less than 0.05.

A collection of matching amino acid sequences was extracted from the SWISSPROT database based on the gene names in this protein list. These sequences were used in a BLASTP against the eggNOG database with an e-value of 1e-10 and a 70% sequence identity to expand the list to additional species. After that, the COG numbers of the high-quality proteins were collected, and clusters of these sequences were extracted to create a database of resistant proteins. Using BLASTP with a threshold e-value of 0.01 and a 35% identity, this database was utilized to look for genes associated with protein resistance in the draught genomes of Chilean biomining bacteria. Finally, each of these possible resistant protein sequences was manually curated using multiple sequence alignments against a reference resistant protein sequence and, when available, particular motivation searches. The Pathway Tools program v13.0 was used to generate pathway/genome databases (PGDB) from annotated genomes of each bacteria to identify metabolic pathways (Mauricio Latorre et al., 2016).

Between the five native bacteria and their matching ATCC reference strains, whole genome alignments revealed significant structural differences (insertions, deletions, and inversions). The consortium had a large number of unique genes that encode unknown proteins and probable transposases. *A. thiooxidans* Licanantay had a total assembly that was 29% greater than public *A. thiooxidans* ATCC 19377, indicating a genome structural difference that might affect the bioleaching process.

The five bacteria share a core of 478 genes, 47 of which are unique to this group (I.e., not detected in the ATCC reference strains). Highlight proteins involved in energy generation and protein biosynthesis (two processes that improve copper extraction efficiency), which were acquired through horizontal gene transfer events within the community (Hemme et al., 2016). The chemolithoautotrophs *A. ferrooxidans* Wenelen, *A. thiooxidans* Licanantay, and *L. ferriphilum* Paiwe make up one clade, whereas *Sb. Thermosulfidooxidans* Cutipay and the heterotroph *A. multivorum* Yenapatur make up the other. The proteins involved in inorganic ion transport and biogenesis are the key functional distinctions between these groups (Mauricio Latorre et al., 2016).

The KEGG database was used for read-based alignments using the BLAST technique with a substantial E-value threshold of 1e10 to obtain greater insight into probable functions in this community. KEGG pathways were related to 148,687 sequences (23.2%) in the PLS sample, including amino acid metabolism (25,218), carbohydrates (19,335), energy (13,843), membrane transport (11,962), and signal transduction (11,887). In the LH sample, 160,203 sequences (24.2%) were linked to KEGG pathways, including amino acid metabolism (28,975), membrane transport (20,780), carbohydrates (19,343), cofactor and vitamin metabolism (12,150), and energy metabolism (12,150). (11,232) (Hu et al., 2015).

The predicted proteins represented by these core gene families are not found in a BLAST search of the NCBI nonredundant database of more than 90 million proteins with a relaxed cut-off of 1.0e5. For any of the five families, there is no clear functional prediction. However, a bioinformatic analysis that combined previously published transcriptomic and proteomic data with pI prediction, motif/domain searches, cellular location predictions, genomic context analyses, and chromosome topology studies suggests that some may be involved in membrane remodeling during cell division, possibly in response to pH stress. Phylogenomic and gene compositional investigations revealed five highly conserved gene families in the core genome of the genus *Acidithiobacilli*. These critical gene groups are missing from the closest extant genus, *Thermithia bacillustepidarius*. Artemis of Sanger (Brettin et al., 2015) and HGTIslandViewer were used to visualize genomic settings in order to foresee genomic islands. With an E-value limit of 1e-5, BLASTN-based scripts were used to discover 16S rRNA sequences from *Acidithiobacillus* genomes. Genes encoding families I–V were mapped onto the genomes of *A. ferrooxidans* ATCC 23270, *A. caldus* ATCC 51756, and *A. carveriadans* SM-1 using a DNA plotter. The origin of replication (Ori) of each genome was predicted between dnaN and dnaA, as previously stated.

An integrated workflow of BLASTP searches against the NCBI NR database was used to annotate protein-coding sequences. Artemis from Sanger was used to display genomic settings. To anticipate genomic islands, Rutherford et al. (2000) employed HGTIslandViewer. Lipo-P Server was used to predict lipoprotein signals (Juncker et al., 2003). BLASTN-based scripts were used to find 16S rRNA sequences from *Acidithiobacillus* genomes with an E-value cutoff of 1e-5.

9.7.3 Metagenomics

Glimmer was used to make gene predictions, and functional gene annotations were processed using cutting-edge approaches including SignalP and TMHMM, as well as local BLAST searches against the NCBI nr, Swissprot, Omniome, PDB, KEGG, COG, and TCDB databases. A GenBank format file was created for each genome after gene annotation and utilized as a general input for metabolic reconstruction using Pathway Tools software v16.0. When a gene encoding for an enzyme related to a metabolic process was found within the genome, we deemed it present and used it to reconstruct a metabolic network. When the five metabolic networks are combined, a network of 2311 reactions is created, with 30% (706 reactions) being unique to specific strains. In contrast, almost 70% (1605 reactions) of the reactions are conserved throughout the five genomes, indicating a core of shared pathways. The synthesis of precursor metabolites, energy, and basal metabolism are all conserved processes. Specific reactions are largely connected to the decomposition of complex carbohydrates, organic acids, and protein synthesis subproducts, demonstrating that secondary metabolism processes are the most important metabolic specificities (Mauricio Latorre et al., 2016).

Sixty-one possible protein coding sequences were sequenced and compared with NCBI-nr, expanded COG (Clusters of Orthologous Groups) (Franceschini et al., 2013), and KEGG databases (Kyoto Encyclopedia of Genes and Genomes). The data was filtered for high-quality (HQ) and cutoff quality using the NGS QC Toolkit v2.3.1 for quality control (FFI). The taxonomy categorization of the protein coding gene sequences was established by MEGAN using the lowest ancestor algorithm (Huson et al., 2011). Allocation of metagenomic sequences to 23 of the 25 COG categories was done, indicating that many functional genes involved in copper bioleaching have yet to be found. In this hostile environment, taxonomic assignment using MEGAN indicated an extremely high microbial richness (over 100 species), which impeded sequence assembly owing to the poor sequencing depth. According to MEGAN analysis, the sulfur- and iron-oxidizing acidophiles, *Acidithiobacillus*-related and *Leptospirillum*-related groups, dominated the microbial community in the mineral surface layer. *A. ferrivorans* has been identified as a significant component of iron- and sulfurous acidophilic bacteria in metal mine-affected settings (Hallberg et al., 2010).

By directly collecting the DNA of all microorganisms from environmental samples, metagenomics examines microbial diversity, population structure, evolutionary links, functional activity, and the interplay between community and environment. It might also lead to new species, genes, enzymes, and compounds being discovered. The first AMD metagenomics study found that *Leptospirillum* group II and *Ferroplasma* type II had nearly full genome reconstructions. Another study revealed the almost complete genome of the dominant bacteria in the microbial community of a gold mine tailings water pit. The inoculated leaching fluid and the leaching heap of a biological heap leaching system include different dominating genera.

The former belongs to the heterotrophic genus *Acidiphilium*, whereas the latter belongs to the autotrophic genus *Acidithiobacillus*. This leaching system had many genes involved in transposition, DNA repair, and heavy metal transport. Microbial communities on the surface of a copper bio-leaching heap were dominated by *Acidithiobacillus*-like, *Thiobacillus*-like, and *Leptospirillum*-like bacteria (Aoshima et al., 2004). They play a crucial function in key metabolic processes (carbon fixation, nitrogen metabolism, ferrous iron oxidation, and sulfur metabolism). High-resolution targeted metagenomics technology has been used for a long time to investigate changes in community structure during adaptation to severe acidity and high gold concentrate pulp contents (Hügler et al., 2007). Functional genes related to heavy metal transport and stress tolerance were found in abundance in the genomes of the community's dominant strains (*A. ferrooxidans*, for example). Similarly, *A. ferrivorans*-like strains were shown to be the most common bacteria in AMD microbial communities at low temperatures (6°C–10°C) (Chen et al., 2015). *Acidithiobacillus* populations' ability to respond to a wide range of hot spring temperatures was mostly due to their high G and proline encoding levels.

The surface layer of low grade copper tailings that were bioleached at the Dexing Copper Mine in China was examined using shotgun metagenome sequencing. The structure and functional properties of a copper bioleaching heap were also studied using bioinformatics. pH, redox potential, sulfur compounds, ferrous iron, and metal ions were all measured at the sample location and tailings.

9.7.4 Transcriptomics and Metatranscriptomics

Transcriptomics is the study of RNA molecules produced by an organism using high-throughput technologies such as microarrays and RNA-seq. Bio-mining microorganisms subjected to organic extractants, quorum sensing super-agonists, severe pH, and other stresses may be revealed using DNA microarrays. When *A. ferrooxidans* ATCC 23270 was exposed to an organic extractant (Lix984n), the expression of genes involved in the pentose phosphate pathway, fatty acid biosynthesis, and glutamate biosynthesis increased.

The specific approach used by *A. ferrooxidans* to deal with short-term stress was to constantly boost the expression of genes producing electron transport proteins such as petI, petII, cyo, and cyd. RNA-seq is becoming increasingly common in transcriptome research due to improvements in sensitivity, accuracy, and resolution (Vikman et al., 2015). This method has been widely utilized to better understand how bio-mining microorganisms adapt to different energy substrates, pH, and heavy metal ions. Comparing *A. ferriphilus* SCUT-1 gene expression patterns in Fe^{2+}, S0, and FeS_2 environments revealed the RISC oxidation model and genes implicated in ferrous oxidation (Fan et al., 2018). Similarly, differentially expressed genes with S0 and S_2O_3 as energy substrates helped to uncover new information on *A. thiooxidans* ATCC 19377's sulfur metabolic model (Camacho et al., 2020). Separate experiments were undertaken in various acidity systems on the variably expressed genes of *A. thioxidans* CCTCC M 2012104 and *A. caldus* CCTCC M 2018054.

High-throughput sequencing of whole RNA or mRNA is used to investigate gene expression and regulation in microbial communities in environmental materials. It can measure changes in bio-mining microorganism community structure and function over time and space, obtaining active expression patterns and community activities in response to changing environments. The environmental transcriptome, for example, revealed metabolic variations between biofilms and planktonic cells of *Leptospirillum* spp. in its natural microbial community. (Moreno-Paz et al., 2010). The metabolism of attached cells was dominated by mixed acid fermentation, whereas the metabolism of planktonic cells was more active. Furthermore, a transcriptome study of biofilms grown in the wild and in a laboratory bioreactor revealed that the species *Leptospirillum* was common in both habitats.

High-throughput sequencing of whole RNA or mRNA is used to investigate gene expression and regulation in microbial communities in environmental materials. It can measure changes in bio-mining microorganism community structure and function over time and space, obtaining active expression patterns and community activities in response to changing environments. The environmental transcriptome of *Leptospirillum* spp. demonstrated metabolic variations between biofilms and planktonic cells in its natural microbial community (Moreno-Paz et al., 2010). The metabolism of attached cells was dominated by mixed acid fermentation, whereas the metabolism of planktonic cells was more active. Furthermore, a transcriptome study of biofilms grown in the wild and in a laboratory bioreactor revealed that the species *Leptospirillum* was common in both habitats. Finally, transcriptomics and metatranscriptomics are less commonly employed in research.

9.7.5 Proteomics and Metaproteomics

Proteomics, like transcriptomics, may examine all proteins produced by biomining microorganisms under a variety of stress situations and offer both qualitative and quantitative data. In biomining microorganisms, comparative proteomics is now the most extensively employed approach. Differentially expressed proteins can be used to anticipate how a strain will react to environmental changes. Proteomics was utilized to investigate *A. ferrooxidans*' stress response and physiological adaptation to pH, phosphate deficiency, temperature, various energy substrates (Mykytczuk et al., 2011), and distinct lifestyles (attached and planktonic cells) (Vargas-Straube et al., 2020). There have also been reports of proteomic alterations in *A. ferrooxidans* in response to heavy metals and potentially hazardous micronutrients, including copper (Oetiker et al., 2018), cadmium, uranium, and potassium. To learn more about the links between the peripheral components of bio-mining bacteria and oxidizable substrates, a proteomics study of the periplasmic area was conducted. The first proteome analysis of *A. ferrooxidans*' periplasmic region, which is the primary site of iron and RISC oxidation, found 131 proteins. Quantitative proteomics is very important for understanding adaptive metabolic pathways within strains. Proteomic analysis of *A. ferrooxidans* ATCC 53993 revealed a significant level of overexpression of proteins encoded by many genes in a unique genomic island at high copper concentrations. By detecting protein changes, proteomics can detect changes in intracellular function and metabolic pathways. The cellular response of *L. ferriphilum* ML-04 to arsenic stress was studied. Proteins involved in phosphate metabolism, protection from reactive oxygen species, glutathione metabolism, DNA synthesis and repair, and protein synthesis, folding, and refolding were all implicated in the cellular response of *L. ferriphilum* ML-04 to arsenic stress.

At any given moment, metaproteomics examines the whole protein pool retrieved directly from complex environmental microbial populations (Ramos-Zúñiga et al., 2019). Metaproteomics may infer genetic information, metabolic capacity, microbial diversity, and niche distribution by recognizing active proteins. AMD has supplied the bulk of the biomining microbial communities examined by metaproteomics technology so far. In 2005, the first proteomics investigation of an AMD microbial biofilm community was published, focusing on the community's metabolic function division.

9.7.6 Current Status of Genome Projects from Bioleaching Heaps and Related Mining Environments

As of March 2016, there were 157 genomes of acidophiles deposited in public databases. Of these, 29 (20%) are derived from microorganisms associated with bioleaching heaps or related biomining environments. A list of these genomes is provided in Table 9.1. Three metagenomic studies have been carried out on bioleaching heaps, whereas ten metagenomic studies of other acidic environments have been published (Table 9.2).

TABLE 9.2
List of Genomes Studied

Organism	NCBI Accession	Source
Acidiplasma cupricumulans BH2	LKBH00000000	Mineral sulfide ore, Myanmar
Acidiplasma cupricumulans JCM 13668	BBDK00000000	Industrial-scale chalcocite bioleach heap, Myanmar
Acidiplasma sp. MBA-1	JYHS00000000	Bioleaching bioreactor pulp, Russia
Sulfolobus acidocaldarius Ron12/I	NC_020247	Uranium mine heaps, Germany
Acidiphilium angustum ATCC 35903^T	JNJH00000000	Waste coal mine waters, USA
Acidiphilium cryptum JF-5	NC_009484	Acidic coal mine lake sediment, Germany
Acidiphilium sp. JA12-A1	JFHO00000000	Pilot treatment plant water, Germany
Acidithiobacillus caldus ATCC 51756^T	CP005986	Coal spoil enrichment culture, UK
Acidithiobacillus caldus SM-1	NC_015850	Pilot bioleaching reactor, China
Acidithiobacillus ferrivorans CF27	CCCS000000000	Abandoned copper/cobalt mine drainage, USA
Acidithiobacillus ferrivorans SS3	NC_015942	Enrichment culture from mine-impacted soil samples, Russia
Acidithiobacillus ferrooxidans ATCC 23270^T	NC_011761	Acid, bituminous coal mine effluent, USA

(*Continued*)

TABLE 9.2 (*Continued*)
List of Genomes Studied

Organism	NCBI Accession	Source
Acidithiobacillus ferrooxidans ATCC 53993	NC_011206	Copper deposits, Armenia
Acidithiobacillus sp. GGI-221	AEFB00000000	Mine water, India
Acidithiobacillus thiooxidans A01	AZMO00000000	Wastewater of coal dump, China
Acidithiobacillus thiooxidans ATCC 19377^T	AFOH00000000	Kimmeridge clay, UK
Acidithiobacillus thiooxidans Licanantay	JMEB00000000	Copper mine, Chile
Acidithrix ferrooxidans DSM 28176^T	JXYS00000000	acidic stream draining in abandoned copper mine, UK
Ferrimicrobium acidiphilum DSM 19497^T	JQKF00000000	Mine water, UK
Leptospirillum ferriphilum DSM 14647^T	JPGK00000000	Enrichment culture, Peru
Leptospirillum sp. Sp-Cl	LGSH00000000	Industrial bioleaching solution, Chile
"*Ferrovum myxofaciens*" P3G^T	JPOQ00000000	Stream draining an abandoned copper mine, UK
Ferrovum sp. JA12	LJWX00000000	Pilot treatment plant water, Germany
Ferrovum sp. Z-31	LRRD00000000	AMD water, Germany
Ferrovum sp. PN-J185	LQZA00000000	AMD water, Germany
"*Acidibacillus ferrooxidans*" DSM 5130^T	LPVJ00000000	Neutral drainage from copper mine, Brazil
Sulfobacillus acidophilus DSM 10332^T	NC_016884	Coal spoil heap, UK
Sulfobacillus thermosulfidooxidans CBAR13	LGRO00000000	Percolate solution of a bioleaching heap in copper mine, Chile
Sulfobacillus thermosulfidooxidans Cutipay	ALWJ00000000	Naturally mining environment, Chile
Sulfobacillus thermosulfidooxidans DSM 9293^T	(2506210005)	Spontaneously heated ore deposit, Kazakhstan
Bioleaching heap surface Metagenome	(4664533.3)	Dexing Copper Mine, China
Bioleaching heap PLS sample Metagenome	(4554868.3)	Dexing Copper Mine, China
Bioleaching heap sample Metagenome	(4554867.3)	Dexing Copper Mine, China

9.8 CONCLUSION

The bulk of acidophile genomes come from AMD and other low pH settings, and they've been utilized to build models of bioleaching microorganisms' genetic and metabolic capabilities, as well as bioleaching ecophysiology. Despite the fact that many of these nonbioleaching genomes are related to bioleaching microorganisms at the 16S rDNA level, there is a concern that surrogate model building will likely portray an incomplete picture of the metabolic potential and interaction dynamics operating within a bioleaching heap. Clearly, additional genomes from bioleaching heaps, as well as from different phases of bioleaching and places within a heap, are needed. However, with a few exceptions, a major issue has been the difficulty of acquiring access to commercial bioleaching heaps by the scientific community.

9.9 FUTURE PROSPECTS

Genomes and metagenomes can only be used to analyze genetic and metabolic potential forecasts. To really identify what part of this potential is manifested during bioleaching and how it changes in time and place, transcriptomic, proteomic, and metabolic data are required. Given the difficulty in separating RNA and proteins from bioleaching heaps, such information will be tough to come by. There is very little information that connects physico-chemical characteristics in the bioleaching heap, such as mineral composition, Eh, pH, temperature, and CO_2 and O_2 levels, to microorganism distribution and function. Even if such data is obtained, it is frequently unavailable to researchers.

REFERENCES

Amiri, F., Mousavi, S.M. and Yaghmaei, S. Enhancement of bioleaching of a spent Ni/Mo hydroprocessing catalyst by *Penicillium simplicissimum*. *Sep Purif Techno*, 80(2011): 566–576.

Anawar, H.M. Sustainable rehabilitation of mining waste and acid mine drainage using geochemistry, mine type, mineralogy, texture, ore extraction and climate knowledge. *J Environ Manage*, 158(2015): 111–121.

Aoshima, M., Ishii, M. and Igarashi, Y. A novel enzyme, cityl-CoA lyase, catalysing the second step of the citrate cleavage reaction in *Hydrogenobacter thermophilus* TK-6. *Mol Microbiol*, 52(3)(2004): 763–70.

Arshadi, M. and Mousavi, S.M. Multi-objective optimization of heavy metals bioleaching from discarded mobile phone PCBs: simultaneous Cu and Ni recovery using *Acidithiobacillus ferrooxidans*. *Sep Purif Technol*, 147(2015): 210–219

Aznar-Sánchez, J., García-Gómez, J., Velasco-Muñoz, J. and Carretero-Gómez, A. Mining waste and its sustainable management: advances in worldwide research. *Minerals*, 8(7) (2018): 284.

Bevilaqua, D., Lahti, H., Suegama, P.H., Garcia, O.J., Benedetti, A.V., Puhakka, J.A. and Tuovinen, O.H. Effect of Na-chloride on the bioleaching of a chalcopyrite concentrate in shake flasks and stirred tank bioreactors. *Hydrometallurgy*, 138(2013): 1–13.

Brettin, T., Davis, J.J., Disz, T., Edwards, R.A., Gerdes, S., Olsen, G.J., et al. *RASTtk*: a modular and extensible implementation of the RAST algorithm for building custom annotation pipelines and annotating batches of genomes. *Sci Rep*, 5(2015): 8365.

Bryan, C.G., Watkin, E.L., McCredden, T.J., Wong, Z.R., Harrison, S.T.L. and Kaksonen, A.H. The use of pyrite as a source of lixiviant in the bioleaching of electronic waste. *Hydrometallurgy*, 152(2015): 33–43

Camacho, D., Frazao, R., Fouillen, A., Nanci, A., Lang, B.F., Apte, S.C., et al. New insights Into *Acidithiobacillus thiooxidans* sulfur metabolism through coupled gene expression, solution chemistry, microscopy, and spectroscopy analyses. *Front Microbiol*, 11(2020): 411.

Chalkley, R., Child, F., Al-Thaqafi, K., Dean, A.P., White, K.N. and Pittman, J.K. Macroalgae as spatial and temporal bioindicators of coastal metal pollution following remediation and diversion of acid mine drainage. *Ecotoxicol Environ Saf*, 182(2019): 109458.

Chen, L.X., Hu, M., Huang, L.N., Hua, Z.S., Kuang, J.L., Li, S.J. and Shu, W.S. Comparative metagenomic and metatranscriptomic analyses of microbial communities in acid mine drainage. *ISME J*, 9(7)(2015): 1579–1592.

Conić, V.T., Rajčić Vujasinović, M.M., Trujić, V.K. and Cvetkovski, V.B. Copper, zinc, and iron bioleaching from polymetallic sulphide concentrate. *Trans Nonferrous Met Soc China,* 24(2014): 3688–3695.

Erüst, C., Akcil, A., Gahan, C.S., Tuncuk, A. and Deveci, H. Biohydrometallurgy of secondary metal resources: a potential alternative approach for metal recovery. *J Chem Technol Biotechnol*, 88(2013): 2115–2132

Fan, W., Peng, Y., Meng, Y., Zhang, W., Zhu, N., Wang, J, et al. Transcriptomic analysis reveals reduced inorganic sulfur compound oxidation mechanism in *Acidithiobacillus ferriphilus. Microbiology,* 87(2018): 486–501.

Festin, E.S., Tigabu, M., Chileshe, M.N., Syampungani, S. and Oden, P.C. Progresses in restoration of post-mining landscape in Africa. *J Forest Res*, 30(2019): 381–396.

Fike, D.A., Bradley, A.S. and Leavitt, W.D. Geomicrobiology of sulfur. In Ehrlich H. L., Newman D. K., &Kappler A. (Eds.), *Ehrlich's Geomicrobiology* (2016): 37. CRC Press.

Franceschini, A., Szklarczyk, D., Frankild, S., Kuhn, M., Simonovic, M., Roth, A., et al. STRING v9. 1: protein–protein interaction networks, with increased coverage and integration. *Nucleic Acids Res*, 41(D1)(2013): D808–D815.

Guezennec, A.G., Bru, K., Jacob, J. and d'Hugues, P. Co-processing of sulfidic mining wastes and metal-rich post-consumer wastes by biohydrometallurgy. *Miner Eng*, 75(2015): 45–53.

Hallberg, K.B., González-Toril, E. and Johnson, D.B. *Acidithiobacillus ferrivor*ans, sp. nov.; facultatively anaerobic, psychrotolerant iron-, and sulfur-oxidizing acidophiles isolated from metal mine-impacted environments. *Extremophiles,* 14(1)(2010): 9–19.

Hemme, C.L., Green, S.J., Rishishwar, L., Prakash, O., Pettenato, A., Chakraborty, R., et al. Lateral gene transfer in a heavy metal-contaminated-groundwater microbial community. *MBio*, 7(2)(2016): 10–1128.

Hong, Y. and Valix, M. Bioleaching of electronic waste using acidophilic sulfur oxidising bacteria. *J Clean Prod*, 65(2014): 465–472

Hu, Q., Guo, X., Liang, Y., Hao, X., Ma, L., Yin, H. and Liu, X. Comparative metagenomics reveals microbial community differentiation in a biological heap leaching system. *Res Microbiol*, 166(6)(2015): 525–534.

Hudson-Edwards, K. and Dold, B. Mine waste characterization, management and remediation. *Minerals*, 5(1)(2015): 82–85.

Hudson-Edwards, K.A., Jamieson, H.E., and Lottermoser, B.G. Mine wastes: past, present, future. *Elements*, 7(2011): 375–380.

Hügler, M., Huber, H., Molyneaux, S.J., Vetriani, C. and Sievert, S.M. Autotrophic CO_2 fixation via the reductive tricarboxylic acid cycle in different lineages within the phylum Aquificae: evidence for two ways of citrate cleavage. *Environ Microbiol*, 9(1)(2007): 81–92.

Humphries, M.S., McCarthy, T.S. and Pillay, L. Attenuation of pollution arising from acid mine drainage by a natural wetland on the Witwatersrand. *S Afr J Sci*, 113(1–2)(2017): 1–9.

Huson, D.H., Mitra, S., Ruscheweyh, H.J., Weber, N. and Schuster, S.C. Integrative analysis of environmental sequences using MEGAN4. *Genome Res*, 21(9) (2011): 1552–1560.

Ilbert, M. and Bonnefoy, V. Insight into the evolution of the iron oxidation pathways. *Biochim Biophys*, 1827(2013): 161–175.

Jiang, L.L., Zhou, J.J., Quan, C.S. and Xiu, Z.L. Advances in industrial microbiome based on microbial consortium for biorefinery. *Bioresour Bioprocess*, 4(2017): 11

Juncker, A. S., Willenbrock, H., Von Heijne, G., Brunak, S., Nielsen, H., and Krogh, A. Prediction of lipoprotein signal peptides in Gram-negative bacteria. *Protein Sci,* 12(2003): 1652–1662.

Karaca, O., Cameselle, C. and Reddy, K.R. Mine tailing disposal sites: contamination problems, remedial options and phytocaps for sustainable remediation. *Rev Environ Sci Bio/Technol*, 17(2018): 205–228.

Kefeni, K.K., Msagati, T.A.M. and Mamba, B.B. Acid mine drainage: prevention, treatment options, and resource recovery: a review. *J Clean Prod*, 151(2017): 475–493.

Larson, L.N., Sánchez-España, J., Kaley, B., Sheng, Y., Burgos, W.D. and Burgos, W.D. Thermodynamic controls on the kinetics of microbial low-pH Fe(II) oxidation. *Environ Sci Technol*, 48(2014): 9246–9254.

Latorre, M., Cortés, M.P., Travisany, D., Di Genova, A., Budinich, M., Reyes-Jara, A., et al. The bioleaching potential of a bacterial consortium. *Bioresour Technol*, 218(2016): 659–666.

Littlejohn, P., Kratochvil, D., and Consigny, A. *Using Novel Technology for Residue Management and Sustainable Mine Closure.* Vancouver, BC: Mine Closure, (2015): 1–10.

Liu, J., Wu, W., Zhang, X., Zhu, M. and Tan, W. Adhesion properties of and factors influencing *Leptospirillumferriphilum* in the biooxidation of refractory gold-bearing pyrite. *Int J Miner Process*, 160(2017): 39–46

Méndez-García, C., Peláez, A. I., Mesa, V., Sánchez, J., Golyshina, O. V., and Ferrer, M. Microbial diversity and metabolic networks in acid mine drainage habitats. *Front Microbiol*, 6(2015): 475.

Mikoda, B., Potysz, A. and Kmiecik, E. Bacterial leaching of critical metal values from Polish copper metallurgical slags using *Acidithiobacillusthiooxidans*. *J Environ Manag*, 236(2019): 436–445.

Moreno-Paz, M., Gómez, M.J., Arcas, A. and Parro, V. Environmental transcriptome analysis reveals physiological differences between biofilm and planktonic modes of life of the iron oxidizing bacteria *Leptospirillum* spp. in their natural microbial community. *BMC Genomics*, 11(2010): 404.

Mrážiková, A., Kaduková, J., Marcinčáková, R., Velgosová, O., Willner, J., Fornalczyk, A. and Saternus, M. The effect of specific conditions on Cu, Ni, Zn and Al recovery from PCBS waste using acidophilic bacterial strains. *Arch Metall Mater*, 61(2016): 261–264

Mykytczuk, N.C., Trevors, J.T., Foote, S.J., Leduc, L.G., Ferroni, G.D. and Twine, S.M. Proteomic insights into cold adaptation of psychrotrophic and mesophilic *Acidithiobacillus ferrooxidans* strains. *Antonie Van Leeuwenhoek*, 100(2011): 259–277.

Nemati, M., Lowenadler, J. and Harrison, S.T.L. Particle size effects in bioleaching of pyrite by acidophilic thermophile *Sulfolobusmetallicus* (BC). *Appl Microbiol Biotechnol*, 53(2000): 173–179.

Nirola, R., Megharaj, M., Beecham, S., Aryal, R., Thavamani, P., Vankateswarlu, K., and Saint, C. Remediation of metalliferous mines, revegetation challenges and emerging prospects in semi-arid and arid conditions. *Environ Sci Pollut Res*, 23(20)(2016): 20131–20150

Oetiker, N., Norambuena, R., Martínez-Bussenius, C., Navarro, C.A., Amaya, F., Álvarez, S.A, et al. Possible role of envelope components in the extreme copper resistance of the biomining *Acidithiobacillus ferrooxidans*. *Genes,* 9(2018): 347.

Olubambi, P.A., Ndlovu, S., Potgieter, J.H. and Borode, J.O. Effects of ore mineralogy on the microbial leaching of low grade complex sulphide ores. *Hydrometallurgy*, 86(2007): 96–104.

Plewniak, F., Crognale, S., Rossetti, S. and Bertin, P.N. A genomic outlook on bioremediation: the case of arsenic removal. *Front Microbiol*, 9(2018): 820.

Ramos-Zúñiga, J., Gallardo, S., Martínez-Bussenius, C., Norambuena, R., Navarro, C.A., Paradela, A., et al. Response of the biomining *Acidithiobacillus ferrooxidans* to high cadmium concentrations. *J Proteomics,* 198(2019): 132–144.

Rutherford, K., Parkhill, J., Crook, J., Horsnell, T., Rice, P., Rajandream, M.A., et al. Artemis: sequence visualization and annotation. *Bioinformatics*, 16(2000): 944–945.

Sánchez-Andrea, I., Stams, A.J., Hedrich, S., Ňancucheo, I. and Johnson, D.B. *Desulfosporosinus acididurans* sp. nov.: an acidophilic sulfate-reducing bacterium isolated from acidic sediments. *Extremophiles*, 19(2015): 39–47.

Silva, R.A., Park, J., Lee, E., Park, J., Choi, S. Q. and Kim, H. Influence of bacterial adhesion on copper extraction from printed circuit boards. *Sep Purif Technol*, 143(2015): 169–176

Simate, G. S. and Ndlovu, S. Acid mine drainage: challenges and opportunities. *J Environ Chem Eng*, 2(3)(2014): 1785–1803.

Travisany, D., Cortes, M.P., Latorre, M., Di Genova, A., Budinich, M., Bobadilla-Fazzini, R.A., et al. A new genome of *Acidithiobacillusthiooxidans* provides insights into adaptation to a bioleaching environment. *Res Microbiol*, 165(9)(2014): 743–752.

Vargas-Straube, M.J., Beard, S., Norambuena, R., Paradela, A., Vera, M. and Jerez, C.A. High copper concentration reduces biofilm formation in *Acidithiobacillus ferrooxidans* by decreasing production of extracellular polymeric substances and its adherence to elemental sulfur. *J Proteomics,* 225(2020): 103874.

Vera, M., Schippers, A. and Sand, W. Progress in bioleaching: fundamentals and mechanisms of bacterial metal sulfide oxidation - part A. *Appl Microbiol Biotechnol*, 97(17)(2013): 7529–7541.

Vikman, P., Fadista, J. and Oskolkov, N. RNA sequencing: current and prospective uses in metabolic research. *J Mol Endocrinol*, 53(2014): 93–101.

Zhang, S., Yan, L., Xing, W., Chen, P., Zhang, Y. and Wang, W. *Acidithiobacillusferrooxidans* and its potential application. *Extremophiles*, 22(2018): 563–579.

Zuniga, C., Zaramela, L. and Zengler, K. Elucidation of complexity and prediction of interactions in microbial communities. *Microb Biotechnol*, 10(6)(2017): 1500–1522.

10 Bioleaching
A Method of Microbial Recovery of Metal Values from Industrial Wastes

Satarupa Dey
Shyampur Siddheswari Mahavidyalaya
(affiliated to University of Calcutta)

10.1 INTRODUCTION

Industrialization and urbanization have led to the extensive production of industrial wastes of diverse types. A huge amount of these wastes comprises, different types of E wastes, such as computer hard drives, permanent magnets, mobile parts, lamp phosphors, catalysts, and rechargeable batteries. Apart from them, waste from other industries such as the thermal power industry, spent catalyst, and alumina factories contribute considerably to the pollution. Most of this industrial waste consists of several valuable rare earth elements and heavy metals such as Ni, Co, Te, In, Mo, Ga, and W (Hennebel et al., 2015). Most of these metals are considered important as they are scarce and have low substitutability.

Several traditional methods have been reported to extract metal values from industrial waste. The processes involved in chemical extraction of metal values include extraction using a suitable solvent, exchange of ions, coprecipitation, and crystallization; however, most of these methods are not considered cost-effective or environmentally friendly (Yin et al., 2017). Another drawback of the chemical extraction process lies in the fact that most of these methods are not efficient when the metal is present in low concentrations (Lo et al., 2014) and need the aid of acids or bases, which consequently generate secondary waste (Dodson et al., 2015). Biological techniques have several advantages, such as being low-cost and eco-friendly. A few applications of bioleaching technology and its advantages over conventional pyrometallurgical and hydrometallurgical methods are depicted in Figure 10.1.

Strategies including biosorbents, bioleaching, and biomineralization have been used to recover metals from wastes for a very long time (Gu et al., 2018). Among all these processes, bioleaching has been considered a very suitable method for the recovery of metals from metal waste particles having lower concentrations of metals. In most cases, bioleaching is considered when other chemical methods are not cost-effective. However, until now, large-scale use of bioleaching is still in its infancy, and further research is necessary for large-scale industrial use. Various chemolithotrophic,

DOI: 10.1201/9781003451457-10

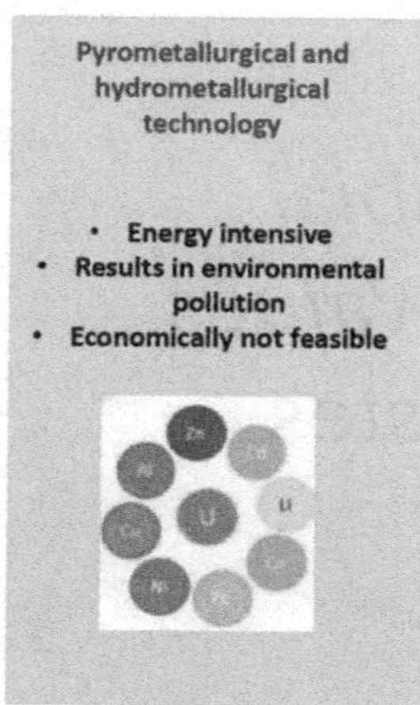

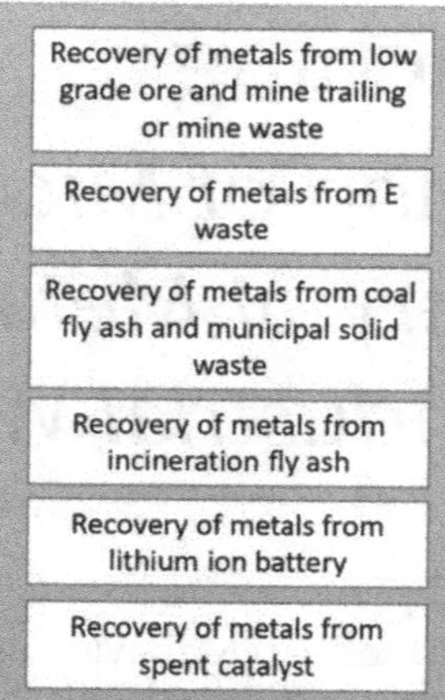

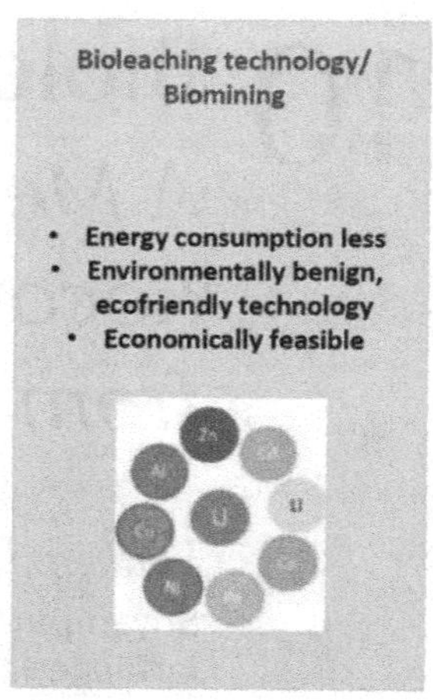

FIGURE 10.1 Benefits of bioleaching over conventional technology and application of bioleaching. (Bioleaching has certain advantages over pyrometallurgical and hydrological technology, which can be used for the recovery of metals from low grade ore, mine tailing, mine waste, E-waste, fly ash, municipal solid waste, spent catalyst, and lithium ion batteries).

heterotrophic, and cyanogenic bacteria and fungi, such as *Acidithiobacillus thiooxidans*, *A. ferrooxidans*, *Chromobacterium violaceum*, *Pseudomonas* sp., *Bacillus megaterium*, *Aspergillus niger*, and *Penicillium simplicissimum*, have been extensively used for bioleaching of metal values (Gu et al., 2018). These microbes produce inorganic, organic, and cyanide to perform bioleaching, which can be effectively used to obtain metals from industrial wastes at extremely low concentrations (Pollmann et al., 2018). In bioleaching, the extracted metals are usually recovered following methods like precipitation, solvent extraction, concentration, and ion exchange, or by techniques like membrane technology and dialysis, and by processes like bioadsorption, bioreduction, bioflotation, and microbial electrochemistry (Hsu et al., 2019).

This chapter mainly deals with the different factors that enhance bioleaching and also gives a detailed account of the microbes used in the bioleaching of metal values from industrial wastes of different types.

10.2 FACTORS ENHANCING THE BIOLEACHING PROCESS

It has been suggested by several researchers that pH plays a crucial role in the process of bioleaching of metal values (Xin et al., 2016). Wu et al. (2019) reported an enhancement in the recovery rate of Co and Ni to 96% and 97%, respectively, under a suitable pH, which was largely associated with an increase in biomass production. Both sulfur-oxidizing bacteria (SOB) and iron-oxidizing bacteria (IOB) can be used for the extraction of Co and Ni, where the efficiency of extraction was reported to be 100% and 99.3% after pH adjustment, respectively. Other than pH, application of biochar additionally leads to better bioleaching performance through regulation of electron transfer from Fe-mediated microbial communities to metals (Wang et al. 2016).

The use of suitable microbes is a crucial factor that leads to the efficient leaching of metals from industrial waste. Cyanogenic microbes are widely used for the extraction of precious metals from industrial wastes, as cyanide production is directly associated with the recovery of metals. Indigenous isolates obtained from waste land have shown further increased efficiency of bioleaching. Isolates like *Pseudomonas balearica* SAE1 obtained from E waste showed a leaching efficiency of around 68.5% for Au and 33.8% for Ag under optimal conditions when the pulp density was kept at 10 g/L (Kumar et al., 2018). Another strain of *Pseudomonas fluorescens* showed very high cyanide production and associated bioleaching when tryptone and yeast extract media were used (Yuan et al., 2018). Maneesuwannarat et al. (2016) reported an isolate of *Cellulosimicrobium funkei* from metal-contaminated soils that showed a leaching efficiency of 70% and recovered Gallium from GaAs solar cell waste. In most cases, the isolates that can leach valuable metals from industrial waste were reported to show a high degree of tolerance to different groups of metals (Natarajan & Ting, 2015; Zotti et al., 2014). In most cases, fungal bioleaching is more convenient than bacterial bioleaching as they can function at a much lower pH in a much shorter time and they secrete organic acids such as citric and oxalic acids, which makes it almost comparable to acid leaching (Kim et al., 2016). Kim et al. (2016) also stated an *Aspergillus* genus that exhibited a high recovery of metals like Zn, Mn, Ni, and Cd (90–95%) from Zn–Mn batteries and Ni–Cd batteries through the formation of citric and oxalic acids. A similar study was also conducted by Di Piazza et al. (2017), where it was reported that *Penicillium expansum* was able to bioconcentrate Lanthanum (390 ppm) and Terbium (1520 ppm) from waste from electrical and electronic equipments (WEEE).

In most cases, the bioleaching microbes are adapted to high concentrations of metals caused by the dissolution of metals (Ma et al., 2019). Those microbes are armored with the aid of numerous membrane transporter proteins and extra cellular polysaccharides (EPS), which alleviate the toxicity of metals via adsorbption at the cellular surface and via extracellular efflux. *Acidithiobacillus ferrooxidans* has been reported to be treated with different concentrations of light emitting diode (LED) powder to increase its adaptation to higher Fe (III) levels, cell amount, lower pH, and oxidation-reduction potential (Pourhossein and Mousavi, 2018). This adapted strain was able to increase the bioleaching efficiency by 1.6 fold, and it was able to recover up to 100% of indium (Jowkar et al., 2018). A similar adaptation was also seen in *Aspergillus niger*, which improved organic acid production and resulted in high metal recovery (Bahaloo-Horeh et al., 2018). By following three vital steps, such as consortium production, directed evolution, and chemostat selection, the efficiency of bioleaching can be improved. In a study by Pathak et al. (2017), the addition of catalysts can enhance the dissolution of metal waste and improve the bioleaching rate. Citric acid, Fe (II), and elemental sulfur increase the recovery of cobalt and copper by means of moderate thermophiles along with *Leptospirillum ferriphilum* DX, *Ferroplasma thermophilum* L1, and *Sulfobacillus acidophiles* TPY at a pulp density of 50 g/L (Liu et al., 2020). Starky medium,

9 K medium, or a mix of both can be used to induce bioleaching of bacterial communities through the S-mediated pathway, the Fe-mediated pathway, or a combined S- and Fe-mediated pathway (Xie et al., 2019). The two-step method also helps in efficient metal mobilization, which is extensively used to recover gold from waste printed circuit board (PCB) using *Chromobacterium violaceum* (Li et al., 2015). Even though Cu and Au concentrations were high, *Acidithiobacillus ferrooxidans* was able to remove 80% of Cu and other base metals at optimum conditions. Pretreatment, oxygen supplementation, and the addition of nutrition were reported to enhance the leaching efficiency (Tran et al., 2011). The use of *Acidithiobacillus ferrivorans* and *Acidithiobacillus thiooxidans*, or *P. fluorescens* and *P. putida*, increased gold recovery from waste PCBs (Işıldar et al., 2016). Indirect bioleaching occurs when the media extract is used to recover nearly 83%, 97%, and 84% of copper, nickel, and gallium, respectively, within 30 to 15 days when the pulp density is maintained at 20 g/L (Pourhossein and Mousavi, 2019).

Commercial metal extraction from low-grade metal ores and industrial wastes has been done by bioleaching. Nevertheless, low metal concentrations and multiple metal toxicity may hinder bioleaching from industrial waste, so different techniques have been developed to enhance bioleaching. It is also necessary to explore novel strains of bacteria to improve the efficiency of bioleaching.

10.3 MICROBIAL RECOVERY OF METAL VALUES FROM DIFFERENT INDUSTRIAL WASTES

10.3.1 Microbial Recovery of Metal Values from Fly Ash

Fly ash is known to contain a high amount of toxic metals and other toxic soluble components, including Si, Al, Fe, Cd, Mn, B, As, and Hg, which are deposited in specialized landfills (Tiwary et al., 2008). Fly ash is known to cause extensive water and air pollution; moreover, it contains enough Al and Zn and can be extracted economically. Fly ash can therefore be considered an artificial ore for metal extraction. Many chemical treatments, such as EDTA, ammonium nitrate, ammonium chloride, and several organic acids, have several disadvantages, including the release of hazardous chemicals and high energy costs. Fungal isolates of sulfur-oxidizing and IOB are known to produce gluconic acid or citric acid and sulfuric acid, which help in oxidizing a reduced metal compound (Ishigaki et al., 2005). By combining SOB with IOB, Brombacher et al. (1998) created a laboratory-scale leaching plant (LSLP) for the treatment of fly ash, where nearly 81% of Zn and 52% of Al were removed. The leaching performance of Cu, Ni, and Cr had been found to be almost 89%, 64%, and 12%, respectively. Ishigaki et al. (2005) also employed a consortium of SOB and IOB for the bioleaching of fly ash. Park and Liang (2019) reported the use of *Candida bombicola*, *Phanerochaete chrysosporium*, and *Cryptococcus curvatus* for leaching trace elements and REEs from fly ash. Similar effective bioleaching of Nd (76%), Pb (59%), and Co (55%), was also reported by using a mixed culture of acidophilic bacteria (Funari et al., 2017). The detailed account of different types of microbes used in the bioleaching of metals from fly ash is listed in Table 10.1.

TABLE 10.1
Recovery of Metal Residues by Microbes from Fly Ash

Element	Approach	Microbes	Conditions	References
Co, others		*Thiobacillus thiooxidans*	Not known	US Pat. No. 5278069 (1994)
Fe, Zn, Al, Mn, Zn, Pb, Cu	One step, Two step	*Aspergillus niger*	1% (w/v) FA; 100 mL sucrose medium; 30°C; 120 rpm, 40 days	Wu and Ting (2006)
Zn, Cu, Pb, Fe	One step, Two step	*Aspergillus niger*	1% (w/v) FA; 100 mL sucrose medium; 30°C; 120 rpm, 27 days	Xu et al. (2014)
Fe; Zn, Al	Two step	*Aspergillus niger*	2.7% (w/v) FA; 100 mL sucrose medium; 30°C; 120 rpm, 26 days	Xu and Ting (2004)
Cadmium (Cd) (99); Cu (56); Pb (28); Zn (62)	Water-washing, One step, Two step	*Aspergillus niger*	1% (w/v) FA; 100 ml sucrose medium; 30°C; 140 rpm	Wang et al. (2009)
Cu, Zn, Al, Fe, Ni, Cd	Two step	Sulfur-oxidizing bacteria (SOB)	1% (w/v) FA; 100 mL inorganic medium; 30°C; 110 rpm	Krebs et al. (2001)
Cu, Pb, Zn	Two step	SOB and iron-oxidizing bacteria (IOB)	10%(v/v) FA; 30°C; 150 rpm; 48 days	Funari et al. (2019)
Mg, Zn, Mn, Al	Two step	SOB and IOB	5% (v/v) FA; 30°C; 150 rpm	Funari et al. (2017)
Cu		*Alkalibacterium* sp. TRTYP6	1% (v/v) of bacteria culture; 20% (w/v) FA; 100 mL; sterile medium, 30 days	Ramanathan and Ting (2016)

10.3.2 Microbial Recovery of Metal Values from Spent Catalyst

Spent hydro-processing catalyst is the solid waste produced from petrochemical industries. It is said to incorporate numerous toxic metals consisting of Al, V, Mo, Co, Ni, and Fe, as well as elemental sulfur, carbon and oils (Mishra et al., 2008). Several different types of approaches using pyrometallurgical or hydrometallurgical routes can be used to recover metals from spent hydro-processing catalyst. The bioleaching route can also be used to leach metals from spent catalysts. Briand et al. (1996) used *Acidithiobacillus thiooxidans* to extract metals from spent vanadium-phosphorus catalysts. Later, Bosio et al. (2008) used *Acidithiobacillus thiooxidans* for the extraction of metals from spent nickel catalysts. Gholami et al. (2011) conducted a study on the effects of pH, temperature, pulp density, rotation speed, and toxicity of heavy metals present in waste catalyst on bacteria. Aung and Ting (2005) and Anjum et al. (2010) reported the use of *Aspergillus niger* for the bioleaching of

TABLE 10.2
Recovery of Metal Residues by Microbes from Spent Catalyst

Element	Catalyst Source	Microbes	Metal Recovery	References
Ni, V, Mo	Spent refinery LC-Fining catalysts	*Acidithiobacillus ferrooxidans*, *Acidithiobacillus thiooxidans*, *Leptospirillum ferrooxidans*	30–40% Mo(VI), 83% Ni(II), 90% V(V)	Beolchini et al. (2010)
Al, Mo, Ni, V	Spent petroleum catalyst	*Acidithiobacillus ferrooxidans*, *Acidithiobacillus thiooxidans*	26% Mo(VI) 85% Ni(II), and 92% V(V)	Kim et al. (2008)
V, Mo, Ni, Co, Al	Vanadium rich spent refinery catalysts	*Acidithiobacillus thiooxidans*	94.8% V, 88.3% Ni, and 46.3% Mo	Mishra et al. (2007)
Mo, V, Ni, Al	Spent petroleum refinery catalyst	*Sulfolobus metallicus*	94–97% Ni, 54–59% Al, 2–11 V, and 4–27% Mo	Kim et al. (2012)
Mo, V, Ni, Al, Fe	Spent petroleum refinery catalyst	*Acidithiobacillus ferrooxidans*, *Acidithiobacillus thiooxidans*		Pradhan et al. (2010)
Mo, Ni, Co, Al	Spent Mo–Co–Ni refinery catalyst	*Acidithiobacillus ferrooxidans*, *Acidithiobacillus thiooxidans*	63% Al, 96% Co, 84% Mo and 99% Ni	Gholami et al. (2011)

spent refinery processing catalysts; it was also reported to produce citric, oxalic, malic, and gluconic acids, which helped in the bioleaching of metals. The detailed account of different types of microbes used in the bioleaching of metals from spent catalyst listed in detail in Table 10.2.

10.3.3 Microbial Recovery of Metal Values from Button Cell Batteries

Recycling of small silver oxide primary cells obtained from hearing aids, digital thermometers, insulin pumps, portable medical monitors, hospital pagers, watches, toys, and calculators (Aktas, 2010). These silver oxide cells are mostly discarded after the life of the instrument is over. Bioleaching can be employed for the recycling of these cells to recover valuable metals from the scrap (Aktas, 2010). In a study by Aktas (2010), it was reported that silver can be recovered from spent silver oxide button cells by using hydrometallurgical methods. The entire process is affected by the concentration of acid, reaction temperature, and shaking rate. They extracted the dissolved silver by using potassium chloride solution and zinc powder.

10.3.4 Microbial Recovery of Metal Values from Spent Batteries

With the increased use of mobile and cordless devices,the number of spent batteries has increased. These batteries contain huge amounts of Co, Mn, Ni, Cd, and Pb and several organic components such as diethylcarbonate and ethylcarbonate, which are extremely hazardous and pose environmental threats. To recover metals from spent batteries, usually two different methods are followed: pyrometallurgical and hydrometallurgical strategies. As pyrometallurgical strategies are costly, energy intensive, and Li cannot be obtained from the treatment, hydrometallurgical strategies are preferred. Usually, hydrometallurgical strategies are done in several processes, like one-step bioleaching, two-step bioleaching, and spent medium leaching, where inorganic and organic acids are used, which are followed solvent extraction, precipitation, or electrochemical deposition (Asadi Dalini et al., 2021).

For bioleaching, numerous acidophilic sulfer and iron oxidizing microbes can be used where Fe(III) acts as a chemical oxidant and reacts with sulfide minerals where Fe(III) is converted to Fe(II) (Ghassa et al., 2015). Under suitable abiotic conditions and low pH, bioleaching by acidophilic microbes is conducted. On the other hand, heterotrophic leaching has also been widely tested, in which *Aspergillus* and *Penicillium* spp. have been used (Pathak et al., 2021). Here, the bioleaching is done mainly by acidolysis of minerals, metal sequestration, and complex formation, followed by metal redox reactions. Usually, the target metals and efficiency vary with the type of spent batteries used. Isolate *Acidithiobacillus ferrooxidans* and *Acidithiobacillus thiooxidans* have been widely used to extract metals like Co, Li, Ni, and Cu from Li, Zn, and Ni spent batteries when the reaction condition is maintained at low pH (Zeng et al., 2012; Zeng et al., 2013; Heydarian et al., 2018). However, with heterotrophic isolates like *Aspergillus niger*, the spent medium is used to extract Ci, Li, Ni, Mn, and Al from Li spent batteries (Bahaloo-Horeh et al., 2017). The detailed account of different types of microbes used in the bioleaching of metals from spent batteries is listed in Table 10.3.

10.3.5 Microbial Recovery of Metal Values from E-Waste

Diverse microbial groups, including autotrophic bacteria, heterotrophic bacteria, and heterotrophic fungi, can be utilized for extracting metal values from E-waste (Awasthi et al., 2016). Metal values from E-wastes can be extracted using one-step, two-step, and spent medium bioleaching methods. The details of the processing of E waste are represented in Figure 10.2.

Both iron- and SOB can be used to extract metals from the waste. Microorganism such as *Acidithiobacillus ferrooxidans*, *Acidithiobacillus thiooxidans*, *Leptospirillum ferrooxidans*, and *Sulfolobus* sp. are acidophilic in nature and are reported to be most efficient in the extraction of metals. Microbes such as *Acidianus brierleyi*, *Sulfobacillus thermosulfidooxidans*, and *Metallosphaera sedula* are thermophilic and chemolithoautotrophic in nature and are also used for bioleaching purposes (Plessis et al., 2007). Oxidation of iron and sulfide produces ferric ions and sulfuric acid, which in turn convert insoluble sulfides to soluble sulfate forms. Heterotrophic microbes, on the other hand, can tolerate a wide range of pH and produce organic

TABLE 10.3
Recovery of Metal Residues by Microbes from Spent Batteries

Types of Battery	Microbes	Conditions	Metal recovery	References
Li spent battery	*Acidithiobacillus ferrooxidans*	pH 2.5 and 0.5% pulp density, 25 days	65% Co and 10% Li	Mishra et al. (2008)
	Acidithiobacillus ferrooxidans	1% pulp density, 10 days	43% of Co	Zeng et al. (2012)
	Acidithiobacillus ferrooxidans	20 mg/ L Ag+, and 1% pulp density, 7 days	98% Co	Zeng et al. (2013)
	A. ferrooxidans and *Acidithiobacillus thiooxidans*	4% pulp density and pH of 1.5, 16 days	99% Li, 50% Co, and 89% Ni	Heydarian et al. (2018)
	A. ferrooxidans and *A. thiooxidans*	Indirect leaching, initial pH 1.1–2.5 final pH 2 to 4 25°C, 2 h.	82% Mn, 74% Cu, 60% Li, 49% Ni, and 53% Co	Boxall et al. (2018)
	Acidithiobacillus caldus, *L. ferriphilum*, *Sulfobacillus* spp., and *Ferroplasma* spp.	S, Fe2+, and Fe scrap, 2 days	Co and Ni	Ghassa et al. (2020a, b)
	Aspergillus niger	Spent media leaching, 8 days	100% Cu, 95% Li, 45% Co, 38% Ni, 70% Mn, and 65% Al	Bahaloo-Horeh et al. (2017)
Zn spent battery	Mixed culture of *Alicyclobacillus* spp. and *Sulfobacillus* spp.	1% pulp density	100% Zn and 94% Mn	Xin et al. (2012)
	Alicyclobacillus-Sulfobacillus mixed culture	NA	48% to 62% Zn and 31% to 62% Mn	Niu et al. (2015)
	Alicyclobacillus and *Sulfobacillus* spp.	Indirect leaching, 1 day	96% Zn	Xin et al. (2012)
	Aspergillus niger	Spent media leaching, 8 days	Zn, Mn	Kim et al. (2016)
Ni spent battery	*Aspergillus niger*	Spent media leaching, 8 days	95% Ni and Cd	Kim et al. (2016)

acids and complexants that can be used to treat moderately alkaline wastes (Natarajan et al., 2015). Isolates such as *Bacillus*, *Pseudomonas*, *Chromobacterium*, *Aspergillus*, and *Penicillium* have been used for heterotrophic bioleaching. Cyanogenic and heterotrophic isolates such as *Pseudomonas aeruginosa*, *Pseudomonas fluorescens*, and *Pseudomonas putida* produce several types of metabolites that can be used to

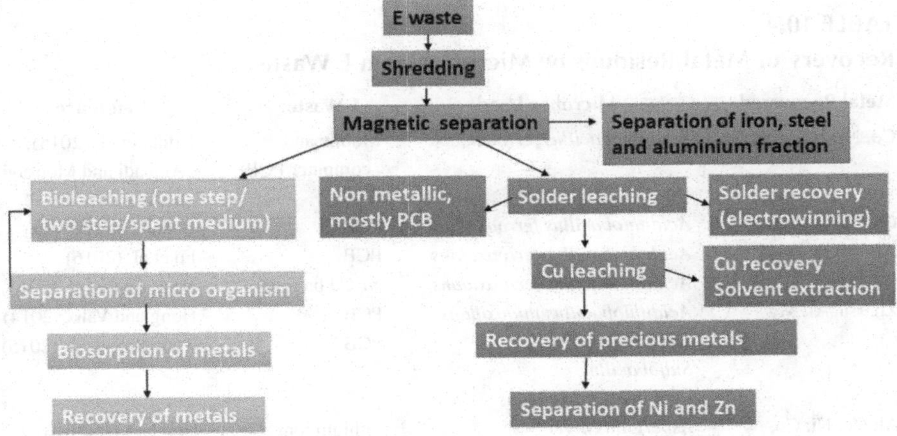

FIGURE 10.2 Processes involved in the recovery of precious metals from E-waste. (Processing and recovery of precious metals from E-waste using Solder leaching, Cu leaching, and bioleaching).

leach metals. Researchers like Pradhan and Kumar (2012) and Işıldar et al. (2016) reported the recovery of gold by *Pseudomonas* using crushed WEEE. Later, Sahni et al. (2016) used *Chromobacterium violaceum* to extract 13.79% of copper, 2.55% of silver, and 0.44% of gold from an obsolete SIM card. Both *Acidithiobacillus thiooxidans* and *Leptospirillum ferriphilum* were capable of extracting Cu, Mn, and Zn from powdered spent batteries (Niu et al., 2015). Maneesuwannarat et al. (2016) used *Cellulosimicrobium funkei* for the leaching of gallium arsenide (GaAs) from semiconductors. Marra et al. (2018) reported the use of *Acidithiobacillus thiooxidans* (DSM 9463) and *Pseudomonas putida* (WSC361) to extract nearly 99% of cerium, europium, and neodymium and 48% of Au, respectively, from shredded E wastes. The list of microbes playing a vital role in the bioleaching of E-waste is tabulated in Table 10.4.

10.3.6 Microbial Recovery of Metal Values from Bauxite Residue

Bauxite ores are used for the extraction of alumina and are usually heterogeneous in composition, comprising aluminium hydroxide, silica, iron oxides, titania, and aluminium silicates. Alumina is usually extracted by the Bayers process, where the ore is crushed and subsequently digested with sodium hydroxide. The solid residue formed after the Bayers process, also named as red mud, is extremely alkaline (pH 9–12), highly saline, and polymetallic in nature, containing residual minerals, trace elements, and rare earth elements (Samal et al., 2013). The bauxite residues also contain oxides of titanium and silicon, along with scandium, yttrium, lanthanum, neodymium, gadolinium, and radionuclides such as radium and thorium (Qu et al., 2013), and are referred to as "polymetallic raw materials". Bioleaching can be effectively used to recover metal values from low-grade ore waste, and can play an effective

TABLE 10.4
Recovery of Metal Residues by Microbes from E Waste

Metal Recovered	Microbes Used	E Waste	References
Cd, Ni	*Acidithiobacillus ferrooxidans*	Mobile and computer PCB	Isildar et al. (2015); Arshadi and Mousavi (2014)
Cu	*Acidithiobacillus ferrooxidans*	PCB	Chen et al. (2015)
Al	*Acidithiobacillus ferrooxidans*	PCB	Fu et al. (2016)
Cd	*Acidithiobacillus ferrooxidans*	Ni-Cd batteries	Velgosová et al. (2013)
Zn, Cu, Al, Mg	*Acidithiobacillus thiooxidans*	PCB	Hong and Valix (2014)
Cu	Thermophiles *Sulfobacillus thermosulfidooxidans*	PCB	Rodrigues et al. (2015)
Al, Co, Ni, Co, Cu, Li, Mn	*Aspergillus niger*	Lithium ion battery	Horeh et al. (2016, 2018)
Cd, Co, Ni, Zn	Mixture of six *Aspergillus* species: *A. fumigatus*, *A. flavipes*, *A. japonicus*, *A. tubingensis*, *A. versicolor*, *A. niger*	ZMB and Ni-Cd batteries	Kim et al. (2016)
Au, Cu	*Chromobacterium violaceum*	PCB	Chi et al. (2011)
Au	*Chromobacterium violaceum*	Electronic scrap	Natarajan and Ting (2014)
Au, Ag, Cu	*Chromobacterium violaceum*	Sim card	Sahni et al. (2016)
Au, Cu	*Bacillus megaterium*	PCB	Arshadi and Mousavi (2015)
Au, Ag, Cu	*Pseudomonas* sp.	PCB	Ruan et al. 2014)
Au	*Pseudomonas aeruginosa*, *Pseudomonas fluorescens*, *Chromobacterium violaceum*	PCB	Natarajan and Ting (2015)
Ni, Au, Cu	*Chromobacterium violaceum*, *Pseudomonas fluorescens*, *Bacillus megaterium*	Electronic scrap	Faramarzi et al. (2004)

role in the treatment of bauxite residues, and can be tagged as an environmentally benign, low-energy-requiring, cost effective green process. Both autotrophic and heterotrophic organisms can be employed in the leaching of rare earths and trace metals from industrial wastes. However, for effective bioleaching from alkaline waste, mostly heterotrophic bacteria and fungi are used, as they are capable of rapid growth and adaptability (Santini et al., 2015). In most cases, three different options, such as the one-step, two-step, and spent medium bioleaching processes are employed in the treatment of bauxite residue. The mechanism of metal leaching using heterotrophic microorganisms has been well documented and is largely based on solid-phase dissolution or transformation mediated by the production of organic acids. It was Vachon et al. (1994) who first reported the bioleaching of aluminium from the red mud using heterotrophic microorganisms such as sewage sludge bacteria (indigenous thiobacilli) and fungal strains such as *Aspergillus niger, Penicillium simplicissimum, Penicillium notaturn,* and *Trichoderma viride. Penicillium simplicissimum* was reported to be the most suitable fungal strain capable of leaching nearly 75% of Al when red mud was used at a 10% v/v concentration. Later, Ghorbani et al. (2008) used indigenous fungi, *Aspergillus niger* and *Penicillium simplicissimum,* and solubilized nearly 2,082 mg of Al_2O_3/L at 15% pulp density. Similarly, Urík et al. (2015) also used *Penicillium, Aspergillus, Eurotium,* and *Emericella* in batch culture for the bioleaching of Al from red mud. In the same study, a strain of *Aspergillus niger* was reported to solubilize approximately 141 mg/L of aluminium from 0.2 g of red mud following acidification of the culture medium. It was also reported that this acidification was mainly due to the production of citric, oxalic, and gluconic acids by the isolate. Several different conditions, such as one-step, two-step, and spent medium bioleaching, were used in the bioleaching of various heavy metals from red mud samples using *A. niger* by Qu et al. (2013). In each case, the pulp density was maintained at 1%. Among the different metals, Pb and Zn were released most efficiently (80%), which was followed by Cu (67%), Ni (50%), As (44%), Ba (31%), Cr (26%), and Fe and Zr (11%). Bioleaching of REE from bauxite residue was first reported by Qu and Lian (2013). They used an indigenous fungal strain, *Penicillium tricolor* RM-10, which was reported to produce oxalic and citric acids and was identified as the main lixiviants for the leaching process. The highest extraction of REE was achieved in a two-step process. Among chemolithotrophic bacteria, *Acetobacter* sp. isolated from red mud impoundments was explored under one-step, two-step, and spent medium bioleaching processes (Qu et al., 2019). *Acetobacter* sp. was reported to produce high amounts of acetic, oxalic, and malic acids but the high metal leaching performance of red mud was recorded under a one-step processes, and the leaching ratios of Al, Lu, Y, Sc, and Th were 55%, 53%, 61%, 52%, and 53%, respectively, at 2% pulp density. Later, Zhang et al. (2020) used *Acidianus manzaensis* for the extraction of Al and REs from bauxite residues using both aerobic and anaerobic bi-stage bioleaching methods. It was reported that 85.1%, 82.4%, 86.8%, 85.3%, and 78.6% leaching of Al, Ce, Gd, Y, and Sc, respectively, occurred after 22 days. Table 10.5 represents a comprehensive list of different microorganisms and their efficiency in leaching metal values from bauxite residues.

TABLE 10.5
Bioleaching of Metal Values and Rare Earth Elements from Bauxite Residue

Metal Recovered	Microbes Used	Bioleaching Process	Pulp Density	References
Al, Lu, Y, Sc, and Th	*Acetobacter*	One step	2%	Qu et al. (2019)
Al, Ce, Gd, Y, and Sc	*Acidianus manzaensis*	Aerobic and anaerobic bi-stage bioleaching	0.6 g	Zhang et al. (2020)
Al	*Aspergillus niger*	Spent medium	3%	Vachon et al. (1994)

(Continued)

TABLE 10.5 (*Continued*)
Bioleaching of Metal Values and Rare Earth Elements from Bauxite Residue

Metal Recovered	Microbes Used	Bioleaching Process	Pulp Density	References
Al and Ti	*Aspergillus niger*	Spent medium	15%	Ghorbani et al. (2008)
Pb and Zn, Cu, As, Ni, Cr, Fe, and Zr	*Aspergillus niger*	One-step Two-step and spent medium	1%	Qu et al. (2013)
Al	*Aspergillus niger*	One step	0.2 g	Urík et al. (2015)
Al	*Penicillium notatum*	Spent medium	3%	Vachon et al. (1994)
Al andTi	*Penicillium notatum*	Spent medium	15%	Ghorbani et al. (2008)
Al	*Penicillium simplicissimum*	One-step Spent medium	3%	Vachon et al. (1994)
Ra, Th, K	*Penicillium tricolor* RM 10	One-step Two-step and Spent medium	10%	Qu and Lian (2013)
Al	*Penicillium crustosum* G-140	One step	0.2 g	Urík et al. (2015)

10.4 CONCLUSION AND FUTURE PERSPECTIVE

Treatment and sustainable disposal of industrial waste have become global problems, and it has become necessary to develop environmentally friendly treatment processes. Biotechnological approaches can act as a cost-effective and ecofriendly alternative for the treatment and extraction of precious metal from industrial scraps. Moreover, a detailed knowledge of the interactions between microbes and various types of industrial waste is necessary for the time for successful application of microbes in the extraction of valuable metals from industrial waste.To improve the yield of metal value after treatment, it has become necessary to find new microbial strains that can tolerate high concentrations of toxic metals. A detailed knowledge of the community of microbes thriving in polluted areas and their variation both spatially and temporally is required to enhance the economic viability of the process and for large-scale applications.

REFERENCES

Aktas, Serdar. "Silver recovery from spent silver oxide button cells." *Hydrometallurgy* 104, no. 1 (2010): 106–111.

Anjum, Fozia, Haq Nawaz Bhatti, Muhammad Asgher, and Muhammad Shahid. "Leaching of metal ions from black shale by organic acids produced by *Aspergillus niger*." *Applied Clay Science* 47, no. 3–4 (2010): 356–361.

Arshadi, M., and S. M. Mousavi. "Enhancement of simultaneous gold and copper extraction from computer printed circuit boards using *Bacillus megaterium*." *Bioresource Technology* 175 (2015): 315–324.

Arshadi, M., and S. M. Mousavi. "Simultaneous recovery of Ni and Cu from computer-printed circuit boards using bioleaching: Statistical evaluation and optimization." *Bioresource Technology* 174 (2014): 233–242.

Asadi Dalini, E., Gh Karimi, S. Zandevakili, and M. Goodarzi. "A review on environmental, economic and hydrometallurgical processes of recycling spent lithium-ion batteries." *Mineral Processing and Extractive Metallurgy Review* 42, no. 7 (2021): 451–472.

Aung, Khin Moh Moh, and Yen-Peng Ting. "Bioleaching of spent fluid catalytic cracking catalyst using *Aspergillus niger*." *Journal of Biotechnology* 116, no. 2 (2005): 159–170.

Awasthi, Abhishek Kumar, Xianlai Zeng, and Jinhui Li. "Comparative examining and analysis of e-waste recycling in typical developing and developed countries." *Procedia Environmental Sciences* 35 (2016): 676–680.

Horeh Bahaloo, Nazanin, and Seyyed Mohammad Mousavi. "Enhanced recovery of valuable metals from spent lithium-ion batteries through optimization of organic acids produced by *Aspergillus niger*." *Waste Management* 60 (2017): 666–679.

Horeh Bahaloo, Nazanin, Seyyed Mohammad Mousavi, and Mahsa Baniasadi. "Use of adapted metal tolerant *Aspergillus niger* to enhance bioleaching efficiency of valuable metals from spent lithium-ion mobile phone batteries." *Journal of Cleaner Production* 197 (2018): 1546–1557.

Beolchini, Francesca, V. Fonti, F. Ferella, and Francesco Vegliò. "Metal recovery from spent refinery catalysts by means of biotechnological strategies." *Journal of Hazardous Materials* 178, no. 1–3 (2010): 529–534.

Bosio, V., M. Viera, and E. Donati. "Integrated bacterial process for the treatment of a spent nickel catalyst." *Journal of Hazardous Materials* 154, no. 1–3 (2008): 804–810.

Boxall, Naomi J., Ka Yu Cheng, Warren Bruckard, and Anna H. Kaksonen. "Application of indirect non-contact bioleaching for extracting metals from waste lithium-ion batteries." *Journal of Hazardous Materials* 360 (2018): 504–511.

Briand, Laura, Horacio Thomas, and Edgardo Donati. "Vanadium (V) reduction in *Thiobacillus thiooxidans* cultures on elemental sulfur." *Biotechnology Letters* 18, no. 5 (1996): 505–508.

Brombacher, Christoph, Reinhard Bachofen, and Helmut Brandl. "Development of a laboratory-scale leaching plant for metal extraction from fly ash by *Thiobacillus* strains." *Applied and Environmental Microbiology* 64, no. 4 (1998): 1237–1241.

Chen, Shu, Yuankun Yang, Congqiang Liu, Faqin Dong, and Bijun Liu. "Column bioleaching copper and its kinetics of waste printed circuit boards (WPCBs) by *Acidithiobacillus ferrooxidans*." *Chemosphere* 141 (2015): 162–168.

Chi, Tran D., Jae-chun Lee, B. D. Pandey, KyoungkeunYoo, and JinkiJeong. "Bioleaching of gold and copper from waste mobile phone PCBs by using a cyanogenic bacterium." *Minerals Engineering* 24, no. 11 (2011): 1219–1222.

Di Piazza, Simone, Grazia Cecchi, Anna Maria Cardinale, Cristina Carbone, Mauro Giorgio Mariotti, Marco Giovine, and MircaZotti. "*Penicillium expansum* Link strain for a biometallurgical method to recover REEs from WEEE." *Waste Management* 60 (2017): 596–600.

Dodson, Jennifer R., Helen L. Parker, Andrea Muñoz García, Alexandra Hicken, Kaana Asemave, Thomas J. Farmer, He He, James H. Clark, and Andrew J. Hunt. "Bio-derived materials as a green route for precious & critical metal recovery and re-use." *Green Chemistry* 17, no. 4 (2015): 1951–1965.

Faramarzi, Mohammad A., Marion Stagars, Enrico Pensini, Walter Krebs, and Helmut Brandl. "Metal solubilization from metal-containing solid materials by cyanogenic *Chromobacterium violaceum*." *Journal of Biotechnology* 113, no. 1–3 (2004): 321–326.

Fu, Kaibin, Bin Wang, Haiyan Chen, Mengjun Chen, and Shu Chen. "Bioleaching of Al from coarse-grained waste printed circuit boards in a stirred tank reactor." *Procedia Environmental Sciences* 31 (2016): 897–902.

Funari, V., H. I. Gomes, M. Cappelletti, S. Fedi, E. Dinelli, Michael Rogerson, W. M. Mayes, and M. Rovere. "Optimization routes for the bioleaching of MSWI fly and bottom ashes using microorganisms collected from a natural system." *Waste and Biomass Valorization* 10, no. 12 (2019): 3833–3842.

Funari, Valerio, Jarno Mäkinen, Justin Salminen, Roberto Braga, Enrico Dinelli, and Hannu Revitzer. "Metal removal from municipal solid waste incineration fly ash: A comparison between chemical leaching and bioleaching." *Waste Management* 60 (2017): 397–406.

Ghassa, Sina, Zohreh Boruomand, MarzieMoradian, HadiAbdollahi, and Ata Akcil. "Microbial dissolution of Zn-Pb sulfide minerals using mesophilic iron and sulfur-oxidizing acidophiles." *Mineral Processing and Extractive Metallurgy Review* 36, no. 2 (2015): 112–122.

Ghassa, Sina, Akbar Farzanegan, Mahdi Gharabaghi, and Hadi Abdollahi. "Novel bioleaching of waste lithium ion batteries by mixed moderate thermophilic microorganisms, using iron scrap as energy source and reducing agent." *Hydrometallurgy* 197 (2020a): 105465.

Ghassa, Sina, Akbar Farzanegan, Mahdi Gharabaghi, and Hadi Abdollahi. "The reductive leaching of waste lithium ion batteries in presence of iron ions: Process optimization and kinetics modelling." *Journal of Cleaner Production* 262 (2020b): 121312.

Gholami, Roya Mafi, Seyed Mehdi Borghei, and Seyyed Mohammad Mousavi. "Bacterial leaching of a spent Mo-Co-Ni refinery catalyst using *Acidithiobacillus ferrooxidans* and *Acidithiobacillus thiooxidans*." *Hydrometallurgy* 106, no. 1-2 (2011): 26–31.

Ghorbani, Y., M. Oliazadeh, and A. Shahvedi. "Aluminum solubilization from red mud by some indigenous fungi in Iran." *Journal of Applied Biosciences* 7 (2008): 207–213.

Gu, Tingyue, Seyed Omid Rastegar, Seyyed Mohammad Mousavi, Ming Li, and Minghua Zhou. "Advances in bioleaching for recovery of metals and bioremediation of fuel ash and sewage sludge." *Bioresource Technology* 261 (2018): 428–440.

Hennebel, Tom, Nico Boon, Synthia Maes, and Markus Lenz. "Biotechnologies for critical raw material recovery from primary and secondary sources: R&D priorities and future perspectives." *New Biotechnology* 32, no. 1 (2015): 121–127.

Heydarian, Ahmad, Seyyed Mohammad Mousavi, Farzane Vakilchap, and Mahsa Baniasadi. "Application of a mixed culture of adapted acidophilic bacteria in two-step bioleaching of spent lithium-ion laptop batteries." *Journal of Power Sources* 378 (2018): 19–30.

Hong, Y., and M. Valix. "Bioleaching of electronic waste using acidophilic sulfur oxidising bacteria." *Journal of Cleaner Production* 65 (2014): 465–472.

Horeh, N. B., Mousavi, S. M., and Shojaosadati, S. A. "Bioleaching of valuable metals from spent lithium-ion mobile phone batteries using Aspergillus niger." *Journal of power sources*, 320 (2016): 257–266.

Horeh Bahaloo, Nazanin, and Seyyed Mohammad Mousavi. "Enhanced recovery of valuable metals from spent lithium-ion batteries through optimization of organic acids produced by *Aspergillus niger*." *Waste Management* 60 (2017): 666–679.

Horeh Bahaloo, Nazanin, Seyyed Mohammad Mousavi, and Mahsa Baniasadi. "Use of adapted metal tolerant *Aspergillus niger* to enhance bioleaching efficiency of valuable metals from spent lithium-ion mobile phone batteries." *Journal of Cleaner Production* 197 (2018): 1546–1557.

Hsu, Emily, Katayun Barmak, Alan C. West, and Ah-Hyung A. Park. "Advancements in the treatment and processing of electronic waste with sustainability: A review of metal extraction and recovery technologies." *Green Chemistry* 21, no. 5 (2019): 919–936.

Ishigaki, Tomonori, Akane Nakanishi, Masafumi Tateda, Michihiko Ike, and Masanori Fujita. "Bioleaching of metal from municipal waste incineration fly ash using a mixed culture of sulfur-oxidizing and iron-oxidizing bacteria." *Chemosphere* 60, no. 8 (2005): 1087–1094.

Işıldar, A., van de Vossenberg, J., Rene, E. R., van Hullebusch, E. D., and Lens, P. N. "Two-step bioleaching of copper and gold from discarded printed circuit boards (PCB)." *Waste Management* 57 (2016): 149–157. doi: 10.1016/j.wasman.2015.11.033.

Işıldar, Arda, Jack van de Vossenberg, Eldon R. Rene, Eric D. van Hullebusch, and Piet N. L. Lens. "Two-step bioleaching of copper and gold from discarded printed circuit boards (PCB)." *Waste Management* 57 (2016): 149–157.

Jowkar, Mohammad Javad, Nazanin Bahaloo-Horeh, Seyyed Mohammad Mousavi, and Fatemeh Pourhossein. "Bioleaching of indium from discarded liquid crystal displays." *Journal of Cleaner Production* 180 (2018): 417–429.

Kim, Dong J., D. Mishra, Park KH, and Ralph DE. "Metal leaching from spent petroleum catalyst by acidophilic bacteria in presence of pyrite." *Materials Transactions* 49, no. 10 (2008): 2383–2388.

Kim, D. J., H. Srichandan, C. S. Gahan, and S. W. Lee. "Thermophilic bioleaching of spent petroleum refinery catalyst using *Sulfolobus metallicus*." *Canadian Metallurgical Quarterly* 51, no. 4 (2012): 403–412.

Kim, Min-Ji, Ja-Yeon Seo, Yong-Seok Choi, and Gyu-Hyeok Kim. "Bioleaching of spent Zn-Mn or Ni-Cd batteries by *Aspergillus* species." *Waste Management* 51 (2016): 168–173.

Krebs, W., R. Bachofen, and H. Brandl. "Growth stimulation of sulfur oxidizing bacteria for optimization of metal leaching efficiency of fly ash from municipal solid waste incineration." *Hydrometallurgy* 59, no. 2–3 (2001): 283–290.

Kumar, Anil, Harvinder Singh Saini, and Sudhir Kumar. "Bioleaching of gold and silver from waste printed circuit boards by *Pseudomonas balearica* SAE1 isolated from an e-waste recycling facility." *Current microbiology* 75, no. 2 (2018): 194–201.

Li, Jingying, Changjin Liang, and Chuanjing Ma. "Bioleaching of gold from waste printed circuit boards by *Chromobacterium violaceum*." *Journal of Material Cycles and Waste Management* 17, no. 3 (2015): 529–539.

Liu, Ronghui, Zhenhua Mao, Wenxian Liu, Yuguang Wang, Haina Cheng, Hongbo Zhou, and Kaifang Zhao. "Selective removal of cobalt and copper from Fe (III)-enriched high-pressure acid leach residue using the hybrid bioleaching technique." *Journal of Hazardous Materials* 384 (2020): 121462.

Lo, Yung-Chung, Chieh-Lun Cheng, Yin-Lung Han, Bor-Yann Chen, and Jo-Shu Chang. "Recovery of high-value metals from geothermal sites by biosorption and bioaccumulation." *Bioresource Technology* 160 (2014): 182–190.

Ma, Liyuan, Hongmei Wang, Jiangjun Wu, Yuguang Wang, Du Zhang, and Xueduan Liu. "Metatranscriptomics reveals microbial adaptation and resistance to extreme environment coupling with bioleaching performance." *Bioresource Technology* 280 (2019): 9–17.

Maneesuwannarat, Sirikan, Alisa S. Vangnai, Mitsuo Yamashita, and PaitipThiravetyan. "Bioleaching of gallium from gallium arsenide by *Cellulosimicrobium funkei* and its application to semiconductor/electronic wastes." *Process Safety and Environmental Protection* 99 (2016): 80–87.

Marra, Alessandra, Alessandra Cesaro, Eldon R. Rene, Vincenzo Belgiorno, and Piet N. L. Lens. "Bioleaching of metals from WEEE shredding dust." *Journal of Environmental Management* 210 (2018): 180–190.

Mishra, D., D. J. Kim, D. E. Ralph, J. G. Ahn, and Y. H. Rhee. "Bioleaching of vanadium rich spent refinery catalysts using sulfur oxidizing lithotrophs." *Hydrometallurgy* 1, no. 88 (2007): 202–209.

Mishra, Debaraj, Dong-Jin Kim, D. E. Ralph, Jong-Gwan Ahn, and Young-Ha Rhee. "Bioleaching of metals from spent lithium ion secondary batteries using Acidithiobacillusferrooxidans." *Waste Management* 28, no. 2 (2008): 333–338.

Natarajan, Gayathri, and Yen-Peng Ting. "Gold biorecovery from e-waste: An improved strategy through spent medium leaching with pH modification." *Chemosphere* 136 (2015): 232–238.

Natarajan, Gayathri, and Yen-Peng Ting. "Pretreatment of e-waste and mutation of alkali-tolerant cyanogenic bacteria promote gold biorecovery." *Bioresource Technology* 152 (2014): 80–85.

Niu, Zhirui, Qifei Huang, Jia Wang, Yiran Yang, Baoping Xin, and Shi Chen. "Metallic ions catalysis for improving bioleaching yield of Zn and Mn from spent Zn-Mn batteries at high pulp density of 10%." *Journal of Hazardous Materials* 298 (2015): 170–177.

Park, Stephen, and Yanna Liang. "Bioleaching of trace elements and rare earth elements from coal fly ash." *International Journal of Coal Science & Technology* 6, no. 1 (2019): 74–83.

Pathak, Ashish, Richa Kothari, Mari Vinoba, Nazima Habibi, and V. V. Tyagi. "Fungal bioleaching of metals from refinery spent catalysts: A critical review of current research, challenges, and future directions." *Journal of Environmental Management* 280 (2021): 111789.

Pathak, Ashish, Liam Morrison, and Mark Gerard Healy. "Catalytic potential of selected metal ions for bioleaching, and potential techno-economic and environmental issues: A critical review." *Bioresource Technology* 229 (2017): 211–221.

Plessis, Chris A. du, John D. Batty, and David W. Dew. "Commercial applications of thermophile bioleaching." In *Biomining*, Rawlings, D.E., Johnson, D.B. (eds.), pp. 57–80. Springer, Berlin, Heidelberg, 2007.

Pollmann, Katrin, Sabine Kutschke, Sabine Matys, Johannes Raff, Gregor Hlawacek, and Franziska L. Lederer. "Bio-recycling of metals: Recycling of technical products using biological applications." *Biotechnology Advances* 36, no. 4 (2018): 1048–1062.

Pourhossein, Fatemeh, and Seyyed Mohammad Mousavi. "Enhancement of copper, nickel, and gallium recovery from LED waste by adaptation of *Acidithiobacillus ferrooxidans*." *Waste Management* 79 (2018): 98–108.

Pourhossein, Fatemeh, and Seyyed Mohammad Mousavi. "A novel step-wise indirect bioleaching using biogenic ferric agent for enhancement recovery of valuable metals from waste light emitting diode (WLED)." *Journal of Hazardous Materials* 378 (2019): 120648.

Pradhan, Debabrata, Debaraj Mishra, Dong J. Kim, Jong G. Ahn, G. Roy Chaudhury, and Seoung W. Lee. "Bioleaching kinetics and multivariate analysis of spent petroleum catalyst dissolution using two acidophiles." *Journal of Hazardous Materials* 175, no. 1–3 (2010): 267–273.

Pradhan, Jatindra Kumar, and Sudhir Kumar. "Metals bioleaching from electronic waste by *Chromobacterium violaceum* and *Pseudomonads* sp." *Waste Management & Research* 30, no. 11 (2012): 1151–1159.

Qu, Yang, Hui Li, Xiaoqing Wang, Wenjie Tian, Ben Shi, Minjie Yao, and Ying Zhang. "Bioleaching of major, rare earth, and radioactive elements from red mud by using indigenous chemoheterotrophic bacterium *Acetobacter* sp." *Minerals* 9, no. 2 (2019): 67.

Qu, Yang, and Bin Lian. "Bioleaching of rare earth and radioactive elements from red mud using *Penicillium tricolor* RM-10." *Bioresource Technology* 136 (2013): 16–23.

Qu, Yang, Bin Lian, Binbin Mo, and Congqiang Liu. "Bioleaching of heavy metals from red mud using *Aspergillus niger*." *Hydrometallurgy* 136 (2013): 71–77.

Ramanathan, Thulasya, and Yen-Peng Ting. "Alkaline bioleaching of municipal solid waste incineration fly ash by autochthonous extremophiles." *Chemosphere* 160 (2016): 54–61.

Rodrigues, Michael L. M., Versiane A. Leão, Otavio Gomes, Fanny Lambert, David Bastin, and Stoyan Gaydardzhiev. "Copper extraction from coarsely ground printed circuit boards using moderate thermophilic bacteria in a rotating-drum reactor." *Waste Management* 41 (2015): 148–158.

Ruan, Jujun, Xingjiong Zhu, Yiming Qian, and Jian Hu. "A new strain for recovering precious metalsfrom waste printed circuit boards." *Waste Management* 34, no. 5 (2014): 901–907.

Sahni, Aditya, Anil Kumar, and Sudhir Kumar. "Chemo-biohydrometallurgy-a hybrid technology to recover metals from obsolete mobile SIM cards." *Environmental Nanotechnology, Monitoring & Management* 6 (2016): 130–133.

Samal, Sneha, Ajoy K. Ray, and Amitava Bandopadhyay. "Proposal for resources, utilization and processes of red mud in India-a review." *International Journal of Mineral Processing* 118 (2013): 43–55.

Santini, Talitha C., Janice L. Kerr, and Lesley A. Warren. "Microbially-driven strategies for bioremediation of bauxite residue." *Journal of Hazardous Materials* 293 (2015): 131–157.

Tiwari, S., Kumari, B., and Singh, S. N. (2008). "Evaluation of metal mobility/immobility in fly ash induced by bacterial strains isolated from the rhizospheric zone of Typha latifolia growing on fly ash dumps." *Bioresource Technology* 99 (5): 1305–1310.

Tran, Chi Dac, Jae-Chun Lee, Banshi Dhar Pandey, JinkiJeong, KyoungkeunYoo, and Trung Hai Huynh. "Bacterial cyanide generation in presence of metal ions (Na^+, Mg^{2+}, Fe^{2+}, Pb^{2+}) and gold bioleaching from waste PCBs." *Journal of Chemical Engineering of Japan* 44 (2011): 692–700.

Urík, Martin, Marek Bujdoš, Barbora Milová-Žiaková, Petra Mikušová, Marek Slovák, and Peter Matúš. "Aluminium leaching from red mud by filamentous fungi." *Journal of Inorganic Biochemistry* 152 (2015): 154–159.

Vachon, Pascale, Rajeshwar D. Tyagi, Jean Christian Auclair, and Kevin J. Wilkinson. "Chemical and biological leaching of aluminum from red mud." *Environmental Science & Technology* 28, no. 1 (1994): 26–30.

Velgosová, Oksana, Jana Kaduková, Renáta Marcinčáková, Pavol Palfy, and Jarmila Trpčevská. "Influence of H_2SO_4 and ferric iron on Cd bioleaching from spent Ni-Cd batteries." *Waste Management* 33, no. 2 (2013): 456–461.

Wang, Qunhui, Jie Yang, Qi Wang, and Tingji Wu. "Effects of water-washing pretreatment on bioleaching of heavy metals from municipal solid waste incinerator fly ash." *Journal of Hazardous Materials* 162, no. 2–3 (2009): 812–818.

Wang, Shuhua, Yue Zheng, Weifu Yan, Lixiang Chen, Gurumurthy Dummi Mahadevan, and Feng Zhao. "Enhanced bioleaching efficiency of metals from E-wastes driven by biochar." *Journal of Hazardous Materials* 320 (2016): 393–400.

Wu, Hung-Yee, and Yen-Peng Ting. "Metal extraction from municipal solid waste (MSW) incinerator fly ash-chemical leaching and fungal bioleaching." *Enzyme and Microbial Technology* 38, no. 6 (2006): 839–847.

Wu, Shijuan, Tengfei Li, Xian Xia, Zijie Zhou, Shixue Zheng, and Gejiao Wang. "Reduction of tellurite in *Shinella* sp. WSJ-2 and adsorption removal of multiple dyes and metals by biogenic tellurium nanorods." *International Biodeterioration & Biodegradation* 144 (2019): 104751.

Xie, Yaling, Shuhua Wang, Xiaochun Tian, Liming Che, Xuee Wu, and Feng Zhao. "Leaching of indium from end-of-life LCD panels via catalysis by synergistic microbial communities." *Science of the Total Environment* 655 (2019): 781–786.

Xin, Baoping, Wenfeng Jiang, Hina Aslam, Kai Zhang, Changhao Liu, Renqing Wang, and Yutao Wang. "Bioleaching of zinc and manganese from spent Zn-Mn batteries and mechanism exploration." *Bioresource Technology* 106 (2012): 147–153.

Xin, Yayun, Xingming Guo, Shi Chen, Jing Wang, Feng Wu, and Baoping Xin. "Bioleaching of valuable metals Li, Co, Ni and Mn from spent electric vehicle Li-ion batteries for the purpose of recovery." *Journal of Cleaner Production* 116 (2016): 249–258.

Xu, Tong-Jiang, Thulasya Ramanathan, and Yen-Peng Ting. "Bioleaching of incineration fly ash by *Aspergillus niger*-precipitation of metallic salt crystals and morphological alteration of the fungus." *Biotechnology Reports* 3 (2014): 8–14.

Xu, Tong-Jiang, and Yen-Peng Ting. "Optimisation on bioleaching of incinerator fly ash by *Aspergillus niger*-use of central composite design." *Enzyme and Microbial Technology* 35, no. 5 (2004): 444–454.

Yin, Xuemiao, Yaxing Wang, Xiaojing Bai, Yumin Wang, Lanhua Chen, Chengliang Xiao, Juan Diwu et al. "Rare earth separations by selective borate crystallization." *Nature Communications* 8, no. 1 (2017): 1–8.

Yuan, Zhihui, Jujun Ruan, Yaying Li, and Rongliang Qiu. "A new model for simulating microbial cyanide production and optimizing the medium parameters for recovering precious metals from waste printed circuit boards." *Journal of Hazardous Materials* 353 (2018): 135–141.

Zeng, Guisheng, Xiaorong Deng, Shenglian Luo, Xubiao Luo, and Jianping Zou. "A copper-catalyzed bioleaching process for enhancement of cobalt dissolution from spent lithium-ion batteries." *Journal of Hazardous Materials* 199 (2012): 164–169.

Zeng, Guisheng, Shenglian Luo, Xiaorong Deng, Lei Li, and Chaktong Au. "Influence of silver ions on bioleaching of cobalt from spent lithium batteries." *Minerals Engineering* 49 (2013): 40–44.

Zhang, Duo-rui, Hong-rui Chen, Zhen-yuan Nie, Jin-lan Xia, Er-ping Li, Xiao-lu Fan, and Lei Zheng. "Extraction of Al and rare earths (Ce, Gd, Sc, Y) from red mud by aerobic and anaerobic bi-stage bioleaching." *Chemical Engineering Journal* 401 (2020): 125914.

Zotti, Mirca, Simone Di Piazza, Enrica Roccotiello, Gabriella Lucchetti, Mauro Giorgio Mariotti, and Pietro Marescotti. "Microfungi in highly copper-contaminated soils from an abandoned Fe-Cu sulphide mine: Growth responses, tolerance and bioaccumulation." *Chemosphere* 117 (2014): 471–476.

11 Treasure Hunt
Converting E-Waste to E-Wealth

Deeksha Dave
IGNOU

Vartika Jain
Govt. Meera Girls College

11.1 INTRODUCTION

Human society has changed enormously from the origin of *Homo sapiens* to the present. Technological advancements has led humanity to a materialistic era in which humans' desire to own a gadget or anything else outweighs their need. The results are the development of rapidly changing fashion with use and throw culture. The use of electronic items has proliferated in recent decades, with electronic and electrical gadgets becoming an integral part of the modern-day lifestyle (Widmer et al., 2005). This includes television, cell phone, laptop, desktop, printer, scanner, camera, i-pad, i-pod, audio/video disc, microwave, washing machine, refrigerator, digital music recorders, air conditioners, irons, routers, toasters, dishwashers, toys, vacuum cleaners, medical devices, etc. (Ghimire and Ariya, 2020; Purchase et al., 2020). All such gadgets are advancing day by day with new features and upgraded technology, and the shelf life of these gadgets has also decreased (Montalvo et al., 2016). Hence, consumers are bound to change these gadgets with passing time, resulting in a huge mass of outdated gadgets, which is called "E-waste". Damaged electronic items also pile up as E-waste.

According to a study, 53.6 million metric tonnes of E-waste were generated globally in 2019, amounting to 7.3 kg per capita (UN 2021). In 2019, 24.9 million tonnes of E-waste were generated in the Asia-Pacific region (Andeobu et al., 2021). Europe tops the list on the grounds of the amount generated per capita (16.2 kg), and Asia records the highest E-waste generation of about 24.9 Mt in 2019 (Forti et al., 2020). Among the Asian countries, China (10.1 Mt), India (3.23 Mt), Japan (2.57 Mt), and Indonesia (1.62 Mt) are the major producers, contributing to about 70.36% of total E-waste generation (Baldé et al., 2020). China is anticipated to be the largest generator of approximately 28.4 million tonnes of E-waste in 2030, replacing the USA as the largest global E-waste producer (Fu et al., 2018). Unfortunately, both China and India lag far behind in the collection and recycling

DOI: 10.1201/9781003451457-11

of E-waste (Ghimire and Ariya, 2020). As per Forti et al. (2020), India is the third largest generator of E-waste with 3.2 million tons generated each year, and this is going to rise further as we move to digitalization as well as increasing consumer demand owing to continuous technology updates and the shortening life span of electronic items (Borthakur and Govind, 2018).

However, the temporary comfort provided by these gadgets is creating permanent pollution and health hazards (Purchase et al., 2020; Dutta et al., 2021). The majority of E-waste is dumped in open landfills, bodies of water, and agricultural lands, and some of it is inflamed. All these methods to get rid of E-waste are very dangerous for the environment. E-waste not only creates water and soil pollution but also air pollution (Cesaro et al., 2019). Humans are affected by this pollution in several ways, such as direct inhalation, drinking contaminated water, consuming grains, vegetables, or fruits grown in polluted soils, etc. Cancer (Yang et al., 2020) is one of the many diseases that can result from this, as are headaches (Siddiqi et al., 2003), diarrhea (Ankit et al., 2021), insomnia (Dai et al., 2020), fatigue (Adanu et al., 2020), muscular pain (Ankit et al., 2021), and reproductive disorders (Parvez et al., 2021). Not only humans but also birds and animals are affected by hazardous E-waste. It is reported that hexabromocyclododecane was detected in the muscles of birds living near an E-waste recycling region in Qingyuan, China, demonstrating the hazards associated with the dismantling of E-waste (Sun et al., 2012). To avoid such a situation, proper management of E-waste is the need of the hour.

Substantial work has been done covering the various aspects of E-waste generation, such as health hazards related to it, circular economy potential, and environmental impact (Sengupta et al., 2022; Pajunen and Holuszko, 2022). Several issues related to E-waste have been addressed by various scientists around the world, such as global E-waste generation and management (Awasthi and Li, 2017), recycling processes and methods (Islam et al., 2020), E-waste as a source of circular economy (Pini et al., 2019), environmental and health-related aspects (Awasthi et al., 2016; Grant et al., 2013; Vaccari et al., 2019), and urban mining and sustainable management of E-waste (Khaiwal and Mor, 2019; Ghosh, 2020).

The management of E-waste could be economically beneficial as E-waste is rich in rare, precious, and strategic metals. For example, a cell phone contains various metals such as Pd, Cd, Ag, Al, Cu, Co, Sn, Ni, Li, and Sr (Ribeiro et al., 2022). Appropriate recycling technologies are required to tap the hidden potential of E-waste, which in turn can help to maintain the sustainability of resources, thereby creating less pressure on the environment. Further, this may also create livelihood opportunities, leading to socioeconomic growth. There are still certain drawbacks associated with urban mining, such as high costs, poor efficiency, the production of more toxic waste, etc. These are being resolved as technology advances (Zeng et al., 2018; Rai et al., 2021). The present chapter describes the types of E-waste, best practices being employed for urban mining of E-waste and as well as challenges and opportunities in this upcoming sector in global and particularly Indian contexts.

11.2 E-WASTE: NATURE, CHARACTERISTICS, AND COMPOSITION

All kinds of electric and electronic equipment and their components or sub-assemblies, such as printed circuit boards (PCBs), printed wiring boards, cathode ray tubes, semiconductor chips, batteries, data tapes, light bulbs, hard disc drives, speakers, fluorescent lights, capacitors, power supply boxes, vibrators, liquid crystal displays, etc., create a heterogeneous composition of E-waste (Sethurajan et al., 2019; Ghimire and Ariya, 2020). These items contain various types of materials, some of which are hazardous for human health and the environment, and some of which are highly valuable, having the potential to be tapped for economic benefits and sustainability (Rai et al., 2021). E-waste mainly contains metals (61%) and plastic (20%), and its recovery requires special technology (Kaya 2018). Table 11.1 gives the details of the different categories of metals present in E-waste.

Precious metals present in various E-waste could be classified as precious metals: Au, Ag; platinum group metals: Pd, Pt, Rh, Ir, and Ru; base metals: Cu, Al, Ni, Sn, Zn, and Fe; metals of concern (hazardous): Hg, Be, In, Pb, Cd, As, and Sb; and scarce elements: Te, Ga, Se, Ta, and Ge (Khaliq et al., 2014). Table 11.2 shows the different metals present in various electronic gadgets. E-waste plastic parts are recycled to produce plastic products and to generate heat, oil, and gas (Mourshed et al., 2017).

TABLE 11.1
Categories of Metals Present in E-Waste

Categories of E-Waste	Type	References
Nonhazardous metals	Iron and steel	Dutta et al. (2021)
Hazardous metals	Lead, cadmium, chromium, mercury, arsenic, nickel, barium, beryllium, copper, cobalt, lithium	Wath et al. (2010); Dutta et al. (2021)
Precious metals	Gallium, gold, germanium, silver, platinum, palladium, selenium	Wath et al. (2010); Dasgupta et al. (2014)
Rare earth metals	Neodymium, praseodymium, tantalum, indium	Dutta et al. (2021); Dasgupta et al. (2014)
Organic and inorganic chemicals	Polycyclic aromatic hydrocarbon, poly vinyl chloride, halogens (bromine, fluorine, chlorine), polychlorinated dibenzo-dioxins, polychlorinated biphenyls	Pinto (2008)

TABLE 11.2
Metals to be Tapped from Electronic Gadgets

Electronic Gadgets	Metals Present	References
Fluorescent lamps	Yttrium, europium, cerium, lanthanum, terbium	Rai et al. (2021)
Electric gas lamps	Thorium	Rai et al. (2021)
Liquid crystal displays panels	Indium	Matsumiya et al. (2018)
Light-emitting diodes	Gallium arsenide and gallium phosphide	Zhou et al. (2019)
Batteries	Lithium, nickel, cobalt, zinc, and manganese	Sethurajan et al. (2019)
Spent printed circuit boards	Gold, silver, palladium, copper, nickel, and iron	Li et al. (2018)

Electronic waste is categorized into various types based on differences in size, composition, life span, and weight. The amount generated in each category depends on socioeconomic factors, income levels, and consumption patterns. Therefore, the collection logistics and disposal mechanisms also vary by place. According to the European Commission, E-waste is divided into the following chief categories (Balde et al., 2015):

- **Temperature Exchange Equipment**: All temperature exchange equipment such as refrigerators, air conditioners, and geysers
- **Screens and Monitors**: Televisions, monitors, laptops, notebooks, and tablets
- **Lamps**: fluorescent lights and LED lamps.
- **Large Equipment**: Dishwashers, washing machines, large photocopying machines, etc.
- **Small Equipment**: Electric mixers, juicers, microwaves, tape recorders, toasters, vacuum cleaners, video cameras, printers, electronic toys, etc.
- **Small IT and Telecommunications Equipment**: Mobile phones, calculators, telephones, etc.

According to the Global E-waste Monitoring Report of 2020 (Forti et al., 2020), E-waste in 2019 mainly consisted of small equipment (17.4 Mt), followed by large equipment (13.1 Mt), temperature exchange equipment (10.8 Mt), screens and monitors (6.7 Mt), lamps (4.7 Mt), and small IT and telecommunication equipment (0.9 Mt). In India, computer devices account for nearly 70% of E-waste. The rest, 12%, is generated by the telecom sector, 8% from medical equipment, and 7% from electrical items. India is the second largest generator of mobile waste in the world (Aneja, 2018). It is noteworthy that a major share of electronic waste comes from public and private companies, whereas individual households are responsible for a meager 16% of the electronic waste (Mitra and Maiti, 2021).

E-waste is also categorized based on its use (Forti et al., 2020). For example:

- **Home Appliances**: Wide-ranging electronic items are used in homes that later become waste when broken or become nonfunctional. Examples include TVs, washing machines, heaters, geysers, microwaves, refrigerators, voltage stabilizers, etc.
- **Communications and Information Technology Devices**: Devices used for communication fall under this category. They are used everywhere, from home to offices, public places, malls, shopping centers, etc. Examples are smartphones, personal computers, laptops, printers, compact discs, routers, etc.
- **Entertainment Devices**: A major share of E-waste is produced by the entertainment industry, for example, stereos, TVs, radios, i-pods, video games, etc.
- **Electronic Utilities**: The appliances that help transform the way we live, work, and enjoy life, namely, heating pads, remote controls, massage chairs, lamps, treadmills, CCTV cameras, smart watches, etc.

- **Office Equipment**: The government and private offices usually produce E-waste comprising fax machines, printers, cartridges, IT servers, modems, cords, cables, etc.
- **Medicine and Health Care**: Commonly used products in this sector are dialysis machines, testing equipment, blood sugar testing instruments, blood pressure recorders, imaging equipment, and so on.

11.3 DEALING WITH E-WASTE

Electronic items contain various elements; some are hazardous, some are valuable, and some are both. Because E-waste is an alternative source for obtaining various metals, recycling is proposed as the ideal solution to deal with the increasing amounts of E-waste. The valuable elements can be recycled through physical recycling or by utilizing thermal, chemical, or metallurgical methods (Debnath et al., 2018) in formal or informal settings (Perkins et al., 2014). It is important for environmental reasons as well as for valuation reasons. The vast treasure that E-waste possesses can be converted into wealth for the benefit of the planet as well as humanity (Figure 11.1). The E-waste sector has huge potential in terms of job creation. The sector is expected to generate 4.5 lakh direct ventures by 2025 (Singh et al., 2021).

For recycling, collection of E-waste is followed by dismantling of the items, preprocessing, and end-processing to recover the metals (Okwu and Onyeje, 2014). According to estimates (Szcs and Szentannai, 2021), less than 20% of E-waste is recycled and documented, and the other 80% is dumped in landfills without further processing. Moreover, poor disposal, unscientific storage, and outdated practices for processing E-waste are also caused by ineffective recycling techniques (Wakuma and Woldemikael, 2021).

Still, recycling remains the most successful method of managing E-waste (Debnath et al., 2018). Unfortunately, only 15% of E-waste is being recycled at the moment (Sahajwalla and Gaikwad, 2018). According to Forti et al. (2020), of these,

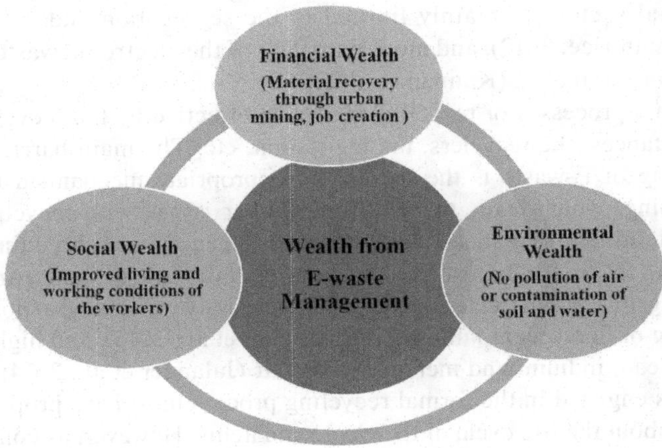

FIGURE 11.1 Wealth from E-waste management.

Europe has the highest recycling rate at 42.5%, followed by Asia (11.7%), America (9.4%), Oceania (8.8%), and Africa (0.9%). India has a low collection and recycling rate (0.93% in 2016) compared to other countries that produce E-waste around the world (such as 52% in Germany in 2017). This shows that E-waste management in India is still in its early stages (Mohan and Amit, 2020). Recycling could be further categorized into the following two categories.

11.3.1 Formal Recycling

Formal recycling facilities are substantial, licensed, and controlled units with closer attention to environmental and human health safety. These facilities disassemble, recycle, refurbish, and shred electronic equipment in order to recover recyclable components like plastic, glass, metals, etc. A thorough collection, sorting, pre-treatment, and metal recovery procedure are all part of a formal, methodical approach for recycling E-waste. The electronic components are sorted either manually or automatically (Ceballos et al., 2014a,b, 2015). In the pre-treatment process, E-waste is physically dismantled followed by de-soldering, size reduction, and fractionation processes for obtaining metallic and nonmetallic components. Then, the metallic fractions of the E-waste are subjected to the recovery process, which includes pyro-metallurgical, hydro-metallurgical, and bio-hydrometallurgical methods for the recovery of metals (Debnath et al., 2018; Zhang and Zhang, 2018; Annamalai and Gurumurthy, 2019).

Manufacturers and regulatory organizations are working to establish proper E-waste recycling mechanisms all around the world. Formal E-recycling facilities are the standard in developed nations like the US, Canada, and Sweden (Perkins et al., 2014). These nations serve as the norm or reference point for creating suitable E-waste management regulations in other parts of the world as well (Andeobu et al., 2021). Recently, formal E-recycling facilities have begun to be established in underdeveloped nations. China recently stopped performing impromptu recycling in Guiyu in order to transform the area into centralized facilities in an industrial park (Standaert, 2015). In India, the processes followed in the formal sector are mainly limited to the segregation and dismantling of E-waste (Chatterjee, 2012), and more than 95% of the electronic waste is recycled through informal means (Kannan et al., 2016).

The formal processes of recycling are made effective by the recycling system, the circumstances, the recyclers, the legislation, etc. The main barrier to the formal recycling of E-waste is the lack of an appropriate mechanism for disposal and processing, which leads to landfilling. Other issues with recycling E-waste include contamination by metal dust, improper cleaning and shredding (Ceballos et al., 2014a), excessive exposure to barium, cadmium, lead, and yttrium during the recycling of CRTs (Lecler et al., 2015), excessive mercury exposure during the recycling of fluorescent lamps (Zimmermann et al., 2014), and higher levels of chromium, lead, indium, and mercury in the air (Julander et al., 2014). Therefore, the recyclers engaged in the formal recycling process must have proper skills and knowledge about the life cycle of the electronic items. However, as compared with informal recycling, the formal recycling sector poses less exposure risk for both workers and the environment (Ádám et al., 2021).

Due to operational and regulatory issues, the formal recycling industry also confronts significant challenges. Polymers, including polystyrene, polyvinyl chloride, and polycarbonate, are typically used in the construction of electrical and electronic equipment. More polymers and fewer ferrous metals are used by the producers since there is a demand for lightweight electrical products. This situation makes it more difficult to recycle E-waste. Additionally, it might be challenging to distinguish mixed polymers, which presents a threat to the recycling and recovery of usable elements from E-waste (Debnath et al., 2020). In addition, recyclers have typically been selective, which implies that recycling businesses are picky about the type of E-waste they pick up. Instead of taking larger, less expensive products like washing machines and refrigerators, they only take profitable tablets, smartphones, and computers (Turaga et al., 2019).

11.3.2 Informal Recycling

Electronic garbage is frequently recycled illegally, unregulatedly, chaotically, and in toxic environments through a method known as "informal recycling" (Chi et al., 2011). It is characterized by practices such as open burning, acid baths, and crude handling of chemicals carried out in unventilated spaces and in a decentralized manner. It is common to find manual sorting, dismantling, and de-soldering of PCBs over coal grills to release valuable chips (Chan and Wong, 2013) or open burning of wires to retrieve copper (Matsukami et al., 2015).

In the past, informal scrap networks have made money by gathering, extracting, and selling recyclable items, contributing significantly to solid waste management (Davis and Garb, 2015). The best markets for informal E-waste recycling are in developing nations in Asia and Africa because of their low labor costs and lax environmental laws (Tansel, 2017; Feng, 2021). It gives migrants from rural areas a source of income and meets the demand for workers in the field of recycling E-waste (Borthakur, 2022). According to the United Nations Global E-waste Monitor (2020), the informal sector continues to play a significant role in recycling E-waste in low- and middle-income nations (Kyere et al., 2016; Forti et al., 2020). According to Baldé et al. (2017) and Sthiannopkao and Wong (2013), small-sized E-waste is illegally exported (as used equipment) from high-income nations like the United States, United Kingdom, Canada, and Germany to low-income Asian and African nations, frequently under the guise of charitable giving or donations. This 'toxic trade' of E-waste complicates the problem of sustainable E-waste management (Borthakur, 2020).

E-waste is a dual concern for developing nations like India and China because of large domestic production and illegal importation from wealthy nations. (Thakur and Kumar, 2022). According to Dutta et al. (2021), the informal sector recycles 95% of the country's electronic garbage, with Dharavi in Mumbai and Seelmapur in Delhi being the two biggest marketplaces. In India and many other developing countries, the entire value chain of E-waste is hampered by issues such as a lack of infrastructure, lax enforcement of E-waste laws and regulations, and a weakened framework for end-of-life items (Arya and Kumar, 2020).

The practices followed by the informal sector are neither safe for the workers nor for the environment (Widmer et al., 2005), particularly for the vulnerable population

(Heacock et al., 2016). One of the main issues in the unregulated recycling industry is high exposure to toxic compounds (Ádám et al., 2021). Studies have revealed elevated levels of Pd, Cu, Mn, and Zn in the blood samples of Nigerian E-waste pickers (Popoola et al., 2019). Additionally, precious components in E-waste are not effectively removed for reuse and recycling but rather dumped on open ground (Song and Li, 2014; Awasthi and Li, 2017). Moreover, the E-waste streams also contain harmful materials such as heavy metals, brominated flame retardants, polychlorinated biphenyls, persistent organic pollutants, polycyclic aromatic hydrocarbons, and polychlorinated dibenzofurans (Ahirwar and Tripathi, 2021). Some hazardous compounds, including polychlorinated dibenzofurans and polycyclic aromatic hydrocarbons, are released into the atmosphere during recycling, further contributing to environmental contamination (Orlins and Guan, 2016). Besides, workers typically heat PCBs to remove solder before dipping them in cyanide and other potent chemicals to extract gold and other metals. These are potentially fatal procedures (Hao et al., 2020). Lethal working conditions coupled with primitive, outdated technology present a challenge in the form of a lack of monitoring and control over the emission of hazardous chemicals (Pradhan and Kumar, 2014). Besides, child labor is another issue associated with informal recycling. For example, 35,000–45,000 kids between the ages of 10 and 14 handle E-waste in Delhi alone, and The International Labor Organization cautions that if E-waste recycling is left to the unorganized sector, India will face significant environmental and human health issues (Thakur, 2015).

11.4 BEST PRACTICES FOR E-WASTE MANAGEMENT

Not only the environmental and health impacts arising out of informal recycling are important, but the latest ideas in E-waste management, concerns, upcoming opportunities, and constraints also need research and exploration. A tail-end strategy involves framing the rules and putting them into practice. Electronics waste management can be dealt with better while we work on best practices at the level of technology, business, and policy interventions and management.

Upstream and downstream actions must be the focus of any potential remedies to the E-waste problem (Figure 11.2). Upstream actions include educating users,

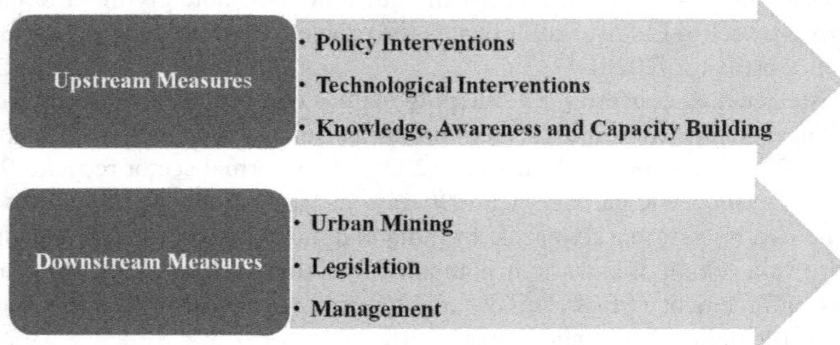

FIGURE 11.2 Best practices in E-waste management.

sharing equipment, finding ways to correctly recycle garbage, and encouraging environmentally friendly user behavior. Downstream efforts, including international cooperation, improved domestic law enforcement, monitoring of imports and exports, involvement of the informal sector, and expansion of formal E-waste collection and recycling activities, can serve to reduce the illegal traffic in E-waste (Bakhiyi et al., 2018).

11.4.1 Urban Mining: Leveraging the Anatomy of E-Waste

The removal of metals from the soil for use in various ways is known as virgin mining. Mineral overexploitation has put tremendous strain on natural resources, altered the earth's geology, and upset the ecological balance. Urban mining, in contrast, is promoted as a sustainable method for recovering the metals present in the anthroposphere and ensuring their availability in the future. In the context of managing E-waste, urban mining refers to recovering usable electronic devices and their components from the E-waste stream in order to retrieve the valuable materials and lessen reliance on mining for those resources. Compared to virgin mining, it is not only more affordable but also more environmentally benign (Zeng et al., 2021).

Diverse materials, including hazardous ones like lead, polychlorinated biphenyls, polybrominated biphenyls, mercury, polybrominated biphenyl ethers, and brominated flame retardants, as well as valuable ones like iron, steel, copper, aluminium, and plastics, are present in the E-waste stream. Many precious metals that may be utilized for other things are wasted when E-waste is disposed of in landfills. Rare Earth elements, albeit employed in very small amounts in the production of parts for electronic products, could potentially be found in E-waste. Electrical and electronic waste and mine tailings are prospective secondary Rare Earth elements resources because of the overwhelming difficulties of conventional Rare Earth elements exploration and mining (Balaram, 2022). It is advised to collect these separately because they need specific handling and cannot be disposed of in landfills.

Urban mining entails either collecting used electronics from the dump or buying them from businesses, recyclers, or garbage collectors. It can be done in a variety of ways. Secondary mining, which is the extraction of resources that are left over from mining virgin materials, is the first category of urban mining. Mine tailings are these leftover sediments, which include harmful compounds and may harm the environment if disposed of improperly. Heavy metals are removed from these tailings by a thermochemical process. The extraction of precious metals from industrial waste dumps is the second sort of urban mining. These sites, along with abandoned factories, contain copper, iron, aluminum, etc., which can act as useful reservoirs. The third type of urban mining is landfill mining, which is the collection of waste found in landfills or even the recovery of metals found in leachate (Park et al., 2017).

The limits of recovery techniques including pyrometallurgy, hydrometallurgy, biometallurgy, and bio-hydrometallurgy include the production of toxic compounds, high costs, challenges with large-scale production, and high selectivity (Hsu et al., 2019). To recover both valuable and rare-earth metals, processes based on electrochemistry that cause little environmental harm and are commercially viable could be used (Rai et al., 2021). For example, effective capturing of precious metals from

PCB leachate has been reported using a porphyrin-based, stable, nonporous, covalent organic polymer (COP-180) with economic viability (Hong et al., 2020). Additionally, it has been demonstrated that using microwave heating can produce a variety of metallic and nonmetallic phases and recover metals from used PCBs (Mahapatra et al., 2019). Recently, a solvent-free and sustainable process, namely Flash Joule heating, has been proposed in which >80% recovery yields are obtained for Rh, Pd, Ag, and >60% for Au through evaporative separation at high temperature. It is a comparatively less energy-consuming process that also removes toxic substances within a fraction of a second (Deng et al., 2021).

Despite the fact that only 17.4% of E-waste generated globally is processed in formal sectors, over 15% of demand is satisfied for practically all valuable metals (Murthy and Ramakrishna, 2022). Envisioning a future where 100% of E-waste is processed formally, around 85%–95% of the demand or more would be met. Urban mining is the new normal. The practice is crucial to achieving a sustainable and circular economy. Virgin mining should not be considered an option in order to achieve a sustainable and decarbonized environment. Urban mining is extensively exploited in China and is highly appreciated for providing alternative solutions to meet the carrying capacity of the country's resources in manufacturing Electrical and Electronic Equipment (EEE) at large (Wen et al., 2015). According to Zeng et al. (2018), urban mining is more cost-effective than virgin mining. Taking the example of obtaining copper and gold from E-waste streams, there are possibilities in the direction wherein urban mining can contribute positively toward a circular economy and replace linear economic models. However, it is difficult to implement urban mining due to various barriers, such as a lack of know-how and needed technology (Kazançoglu et al., 2020).

Urban conflicts because of E-waste are also evident (Minter, 2016; Daum et al., 2017) that occur between residents, waste pickers, and recyclers and lead to social problems (Petricca et al., 2020). In this context, urban mining not only reduces climate change by recycling products, but it also helps to reduce conflict in various zones. In countries like the Congo, mining of various minerals causes conflicts in the region. By lowering the demand for conflict minerals, urban mining can support regional peacekeeping efforts.

11.4.2 Policy Interventions

Devising suitable region-specific policies is required to mitigate the growing concerns about E-waste (Borthakur, 2022). For example, Nigeria, Ghana, and Tanzania, with Kenya, Senegal, and Egypt, are major countries in Africa receiving transboundary E-waste and creating conditions of environmental distress (Maes and Preston-Whyte, 2022). Hence, E-waste policy intervention seeks sectorial measures like regulating imports and exports, capacity building, institutional mechanisms and coordination, waste collection and disposal, resource mobilization, monitoring, evaluation, and reporting, etc. It is imperative to have a local specific, systematic, organized, and stepwise approach toward formulating and implementing the E-waste policies in the respective countries (Borthakur, 2020).

A change in policy-level solutions and support from corporate and business organizations can also help in a great manner. E-waste policies in India are mostly influenced by European Union guidelines, and in 2016, the E-waste management Rules were expanded to include "Producer Responsibility Organization" (PRO) to facilitate collection and recycling of E-waste and introduced buy-back and exchange schemes along with deposit refunds under the ambit of Extended Producer Responsibility (EPR), which is a policy tool that compels manufacturers of electronic items to manage the disposal of their products after their useful lives. EPR states that the manufacturer's responsibility extends beyond simply creating the electronic product and selling it to the customer and that the producer is also accountable for any carelessness or mishandling (Anonymous, 2016). Moreover, just like filling out an income tax return, the producers of electronic goods have to fill in the details of the electronic waste that was managed by them in the previous year so that the government is aware that environmental considerations have been addressed.

The ambitious sustainable development goals (SDGs), particularly SDG-12, pertaining to Responsible Production and Consumption and SDG-13, fighting climate change, are crucial in terms of E-waste management and proper recycling. These objectives clarify how the economy, SDGs, and people are interconnected. Moreover, urban mining of the precious metals in dumped electronic waste could help in achieving the Agenda for Sustainable Development Goals 2030 due to its profitable prospects, extended business opportunities and potential for employment (Arya and Kumar, 2020).

11.4.3 Technological Interventions

This is an era of technological change where it is obvious that large volumes of E-waste will be generated and discarded. "Designed to landfill" must give way to "Designed to Refurbish" and "Designed to Repair," and the circular model must replace the linear model. Technological innovations with better methods and long-lasting, recyclable, and affordable materials must be brought in. However, sometimes the recyclers are lacking in funds and find difficulty meeting their ends. Conditional selling by the tech companies can compel the customers to buy new technology only after exchanging old electronic products for the new ones, and discounts or incentives such as exchange offers given by companies and hardware stores could be of much help to customers who give away their old electronic devices (Manish and Chakraborty, 2019).

Moreover, working with E-waste disposal firms can help sellers make the best use of the Internet of Things (IoT), tracking technology, and Radio Frequency Identification. For instance, Microsoft India encourages customers to drop off discarded electronics, batteries, and packaging materials at designated drop-off locations. E Parisara Pvt. Ltd. has been chosen by IBM to oversee the collection and recycling of old IBM products. An E-waste policy has also been implemented by HCL Infosystems Ltd. for the online processing of recycling requests from both individuals and businesses, and a helpdesk has been established to address questions about waste electrical and electronic equipment (Mitra and Maiti, 2021).

11.4.4 Management

Managing E-waste is done in different ways at different places. According to Chi et al. (2014), the most common method is to dump it into waste bins. This is done for the electronic items, which are small in size. Another method is through formalized regulators, which are run by government bodies, and with the increase in E-waste generation today, individual waste dealers and companies have also come up to aid the government in managing the electronic waste. The third option is to give it to the local junk collectors.

It is increasingly clear that top-down regulatory approaches are not effective for international E-waste management, which is of high value and not suitable for inefficient and unsafe basic recycling in emerging market economies. The management approach should involve stakeholders at various levels and in various capacities. Engaging the private sector, including manufacturers, retailers, and labor investors, is also necessary (Awasthi et al., 2019). EPR and Producer Responsibility schemes are taken up by companies as corporate social responsibility actions. Effective EPR instruments such as deposit refund, producer takeback, material tax, and labeling the products and their components also help in this regard (Murthy and Ramakrishna, 2022).

Some of the E-waste management measures could include setting up collection centers, proper deposit return schemes, take-back centers, etc. Other measures such as collection, pre-processing, and shipment of electronic items also need revamping in the context of threats looming improper disposal of E-waste. To cite an example, initiatives such as *EcoATM kiosks* in supermarkets in America, wherein people drop off electronics, assess their value, and dispense cash or coupons in return, are worth considering. The service providers in E-waste management must create enabling conditions so as to make the collection and recycling processes smooth. Suitable and appropriate technology, along with machinery and infrastructure, should be adopted so that the valuables in E-waste will not be wasted in the long run (Thorat, 2021). Thus, better and innovative waste management practices like Resource recovery, the linear model, and design for end-of-life must be prioritized with incentives and regulations for better management of the forthcoming digital tsunami.

11.4.5 Legislation

The Global E-waste Monitor report (2017) by The United Nations University (UNU) states that only 66% of the world's population spread over 67 countries is covered under E-waste legislation (Baldé et al., 2017). However, more countries are planning and trying to adopt E-waste legislation. Asia is the largest producer of E-waste, and most Asian countries are covered under E-waste legislation (Baldé et al., 2017). In contrast, Africa generates the least amount of E-waste, and most African countries do not have E-waste legislation (Baldé et al., 2017).

The Basel Convention (1992) is a global regulatory framework regarding E-waste that was amended in 2019 (Maes and Preston-Whyte, 2022). Though the European Union has been a pioneer in framing the E-waste directives for its member nations, it is still trying to amend its regulations to effectively manage E-waste. The major concern is weak and inefficient law enforcement in most of the countries with E-waste legislation. As there is no uniformity in legislation across all countries, it is difficult to monitor E-waste recycling on a global scale (Patil and Ramakrishna, 2020). International trade, as well as factors such as resource depletion, have compelled several countries, including emerging economies, to change their E-waste policies and legislation (Zlamparet et al., 2017).

The period of economic liberalization in India led to an increasing penetration of electrical and electronic goods into the market due to the evolution of the IT and electronics sectors. During these years, E-waste grew silently but largely remained unnoticed. In 2008, the Ministry of Environment and Forests, Government of India, issued the Guidelines for Environmentally Sound Management of E-waste (2008) to provide guidance for the identification of various sources of E-waste and prescribe procedures for handling E-waste in an environmentally sound manner. After that, in 2011, India became the first South Asian country to frame a law regarding E-waste. Then after, under the E-Waste (Management) Rules (2016), environmentally sound E-waste management guidelines have been presented for recycling norms, collection and transportation, dismantling of E-waste, testing of hazardous substances, etc. These rules bring in EPR, which makes producers more responsible for the proper collection and management of electronic waste. The Deposit Refund Scheme' has been introduced as an additional economic instrument within the purview of this legislation, whereby the producers charge an additional amount as a deposit at the time of sale of the electrical and electronic equipment and return it to the consumer along with interest when the electrical and electronic equipment is returned (Anonymous, 2016).

Recently, the Ministry of Environment, Forest, and Climate Change, Government of India, published the Battery Waste Management Rules, 2022, to ensure environmentally sound management of waste batteries. The rules cover all types of batteries, *viz*., electric vehicle batteries, portable batteries, automotive batteries, and industrial batteries. The rules are based on the concept of EPR, where the producers of batteries are responsible for the collection and recycling/refurbishment of waste batteries and the use of recovered materials from waste into new batteries. EPR makes it mandatory that all waste batteries be collected and sent for recycling and does not allow their disposal in landfills or incineration (Website 1, 2022).

E-waste in India is an issue that has yet to be adequately addressed by researchers, the scientific community, academicians, policymakers, the government, and nongovernmental organizations (Borthakur and Govind, 2017). It is required that India mandatorily introduce "right to repair" laws. It must get easier for people to get the devices repaired rather than discard them altogether. Despite these rules, in the absence of enough collection centers, the manufacturers are not following the EPR rules mentioned under the E-waste Management Rules, 2016 (Patil and Ramakrishna, 2020).

11.5 CHALLENGES IN THE TREASURE HUNT

Understanding the design and development of the present policies is crucial since sustainable management of E-waste is still a long way from being a reality (Borthakur, 2022). The generation of electronic items is only going to increase along with the consumerist lifestyle, and unless measures are taken, human civilization will be drowned in the tsunami of E-waste. Technically qualified human resources, the absence of laws particularly addressing E-waste, and an inadequate explanation of the roles and duties of the institutions and stakeholders engaged in managing E-waste are some major constraints. The cost of employing various recycling methods in relation to their advantages for the environment should also be taken into account (Rene et al., 2021). The risk of heavy metal contamination of food and vegetables growing near E-waste recycling units is another important issue to be addressed during urban mining (Fu et al., 2013). Soil and water pollution near E-waste recycling units also needs attention, as indicated in some studies (Singh et al., 2018; Amphalop et al., 2020). On the other hand, developing nations struggle with inadequate infrastructure, a lack of financial resources, inadequate manpower, the complex nature of waste, political unwillingness, corruption, etc. (Robaina et al., 2020). Due to diverse socioeconomic, cultural, political, and technological conditions, it is difficult to frame an E-waste management framework in the Indian scenario (Borthakur and Govind, 2017; Borthakur, 2022). The challenges also lie in generating a common legal framework, licensing, regularization of the informal sector, benchmark technologies, responsibilities of various stakeholders, and entrepreneurial opportunities to enhance the formal capacity.

11.6 DISCUSSION

E-waste is a significant source of cash for local recyclers, scrap dealers, and dismantlers due to the abundance of precious metals in it (Deubzer et al., 2019). Urban mining is like treating the hidden, untapped reserves of the city as the source of raw materials. According to a scientific calculation, 50 million tons of E-waste remain to be recovered worldwide. To leverage this, an interdisciplinary framework and models are prerequisites to working upon several aspects of E-waste management (Borthakur, 2022), along with appropriately addressing the concerns of EPR and social responsibility (Murthy and Ramakrishna, 2022). Producers may be required by regulations and performance standards to use a minimum amount of recycled materials to encourage the return and recycling of end-of-life products. Through EPR initiatives, they will have to assume either the financial responsibility or the operational and administrative responsibilities of the collection/recycling process (Patil and Ramakrishna, 2020).

Moreover, consumers should be made aware of the hazards posed by E-waste so that they can adopt better means to recycle their old items through outsourcing to E-waste disposal companies. Reducing the lifespan of electronic gadgets to promote the "buy new, purchase frequently" mentality is against the circular economy idea and requires stringent implementation of regulations. For example, in 2020, regulators in France fined Apple €25 million for deliberately introducing an update that slowed down older iPhones (Malinauskaite and Erdem, 2021). Additionally, problems with software compatibility, device upgrades, and repairs lead to a severe E-waste problem (Parajuly et al., 2019; Zeng et al., 2018; Bridgens et al., 2019). 'Design for Environment' is a design approach that must be adopted to foresee the impact of a product on the environment and human health.

Peng et al. (2018) suggest that the efficacious execution of a well-operating E-waste management system necessitates close coordination, collaboration, information sharing, knowledge transfer, and conviction among all the workers in the formal and informal sectors engaged in the E-waste supply chain structure. Currently, the success of informal E-waste collectors over formal collectors in emerging economies is due to the fact that informal collectors frequently provide more flexibility, convenience, accessibility, and even financial incentives for the E-waste collected (Otto et al., 2018). However, the existing informal sectors should be considered in new formal E-waste recycling systems, and more policies need to be implemented to increase recycling rates, working conditions, and the effectiveness of the associated informal players (Chi et al., 2011).

11.7 CONCLUSION AND RECOMMENDATIONS

E-waste is one of the major domestic wastes in the current era of digitalization. Mere dumping it into landfills or burning it is a huge threat to the environment and, consequently, to human health. Recycling is the best choice for its proper disposal, which is an effective means to recover the highly valuable precious and rare earth metals from E-waste. This is referred to as "urban mining," and with the advancement of tools and techniques, this process should become more economically viable as well as environmentally friendly. The market for recycling E-waste is stimulated by the economic value of E-waste, and the E-waste recycling market could be projected as an upcoming opportunity for young entrepreneurs to invest in and convert E-waste into eco-E-wealth. Though E-waste recycling seems like a great hope, there are many issues to be addressed. The generation of toxic waste in this process should also be kept in mind. Awareness and capacity enhancement regarding the use of E-waste for the circular economy, along with a sustainable funding source, whether through public or private partnerships, are required. To stop creating massive amounts of electronic waste, the consumerist mindset must also be changed. E-waste collection facilities should be readily accessible. Moreover, framing rules and regulations and strict enforcement of existing laws related to E-waste management are also warranted. Urban mining of E-waste can alter the culture of resource conservation through a multifaceted strategy and an all-encompassing solution to the current issue. Further, it is expected that by overcoming these challenges, E-waste could help in the green economy sector and, thereby, foster rapidly developing digital technology without bias.

REFERENCES

Ádám, B., T. Göen, and P. T. J. Scheepe. 2021. From inequitable to sustainable e-waste processing for reduction of impact on human health and the environment. *Environmental Research* 194:110728. https://doi.org/10.1016/j.envres.2021.110728

Adanu, S. K., S. F. Gbedemah, and M. K. Attah. 2020. Challenges of adopting sustainable technologies in e-waste management at Agbogbloshie, Ghana. *Heliyon* 6(8):e04548. https://doi.org/10.1016/j.heliyon.2020.e04548.

Ahirwar, R., and A. K. Tripathi. 2021. E-waste management: A review of recycling process, environmental and occupational health hazards, and potential solutions. *Environmental Nanotechnology, Monitoring & Management* 15:100409.

Amphalop, N., N. Suwantarat, T. Prueksasit, C. Yachusr, and S. Srithongouthai. 2020. Ecological risk assessment of arsenic, cadmium, copper, and lead contamination in soil in e-waste separating household area, Buriram province, Thailand. *Environmental Science and Pollution Research* 27: 44396–44411. https://doi.org/10.1007/s11356-020-10325-x

Andeobu, L., S. Wibowo, and S. Grandhi. 2021. A systematic review of e-waste generation and environmental management of Asia Pacific countries. *International Journal of Environmental Research and Public Health* 18(17):9051.

Aneja, U. 2018. *A Digital India Must Embrace Circular Economy*. Chatham House. Available online at: https://www.chathamhouse.org/2018/01/digital-india-must-embrace-circular-economy. (Accessed on: 11 March 2022).

Ankit, L., V. K. Saha, J. Tiwari, S. Rawat, J. Singh, and K. Bauddh. 2021. Electronic waste and their leachates impact on human health and environment: Global ecological threat and management. *Environmental Technology & Innovation* 24: 102049. https://doi.org/10.1016/j.eti.2021.102049

Annamalai, M., and K. Gurumurthy. 2019. Enhanced bioleaching of copper from circuit boards of computer waste by *Acidithiobacillus ferrooxidans*. *Environmental Chemistry Letters* 17(4):1873–1879.

Anonymous. 2016. Salient features of the e-waste (management) rules, 2016 and its likely implication. Available online at: https://archive.pib.gov.in/documents/rlink/2016/mar/p201632302.pdf. (Accessed on: 18 October 2022).

Arya, S., and S. Kumar. 2020. E-waste in India at a glance: Current trends, regulations, challenges and management strategies. *Journal of Cleaner Production* 271:122707.

Awasthi, A. K., X. Zeng, and J. Li. 2016. Relationship between e-waste recycling and human health risk in India: A critical review. *Environmental Science and Pollution Research* 23(12):11509–11532.

Awasthi, A. K., and J. Li. 2017. Management of electrical and electronic waste: A comparative evaluation of China and India. *Renewable and Sustainable Energy Reviews* 76: 434–447.

Awasthi, A. K., Li, Jinhui, L. Koh, and A. O. Ogunseitan. 2019. Circular economy and electronic waste. *Nature Electronics* 2(3):86–89.

Bakhiyi, B., S. Gravel, D. Ceballos, M. A. Flynn, and J. Zayed. 2018. Has the question of E-waste opened a Pandora's box? An overview of unpredictable issues and challenges. *Environment International* 110:173–192.

Balaram, V. 2022. Sources and applications of rare earth elements. In *Environmental Technologies to Treat Rare Earth Elements Pollution: Principles and Engineering*, ed. A. Sinharoy, and P. N. L. Lens. IWA Publishing. https://doi.org/10.2166/9781789062236

Baldé, C. P., V. Forti, V. Gray, R. Kuehr, and P. Stegmann. 2017. The Global E-Waste Monitor. 2017. *Quantities, Flows and Resources*. United Nations University, International Telecommunication Union, and International Solid Waste Association, Bonn, Geneva, Vienna.

Balde, C.P., R. Kuehr, K. Blumenthal, S. Fondeur Gill, M. Kern, P. Micheli, E. Magpantay, and J. Huisman. 2015. *E-Waste Statistics: Guidelines on Classifications, Reporting and Indicators*. United Nations University, IAS - SCYCLE, Bonn, Germany.

Baldé, C.P., S. van den Brink, V. Forti, A. van der Schalk, and F. Hopstaken. 2020. *The Dutch WEEE Flows 2020: What Happened between 2010 and 2018?* United Nations University (UNU)/United Nations Institute for Training and Research (UNITAR), Tokyo, Japan.

Borthakur, A. 2020. Policy approaches on e-waste in the emerging economies: A review of the existing governance with special reference to India and South Africa. *Journal of Cleaner Production* 252: 119885.

Borthakur, A. 2022. Design, adoption and implementation of electronic waste policies in India. *Environmental Science and Pollution Research* 2022: 1–10.

Borthakur, A., and M. Govind. 2017. How well are we managing e-waste in India: Evidences from the city of Bangalore. *Energy, Ecology and Environment* 2(4):225–235.

Borthakur, A., and M. Govind. 2018. Public understandings of e-waste and its disposal in urban India: From a review towards a conceptual framework. *Journal of Cleaner Production* 172: 1053–1066.

Bridgens, B., K. Hobson, D. Lilley, J. Lee, J.L. Scott, and G. T. Wilson. 2019. Closing the loop on e-waste: A multidisciplinary perspective. *Journal of Industrial Ecology* 23(1):169–181.

Ceballos, D. M., W. Gong, and E. Page. 2015. A pilot assessment of occupational health hazards in the US electronic scrap recycling industry. *Journal of Occupational and Environmental Hygiene* 12(7):482–488.

Ceballos, D., W. Gong, and E. Page. 2014a. Hazard Evaluation Report: A Pilot Assessment of Occupational Health Hazards in the US Electronic Waste Recycling Industry, Cincinnati, OH: Department of Health and Human Services, Centers for Disease Control and Prevention. National Institute for Occupational Safety and Health. E-scrap Survey Report, 1–19. Available online at: https://www.cdc.gov/niosh/hhe/reports/pdfs/e-scrap_survey_report.pdf (Accessed on: 28 Februry 2022)

Ceballos, D., L. Chen, E. Page, A. Echt, C. A. Oza, and J. Ramsey. 2014b. Evaluation of occupational exposures at an electronic scrap recycling facility. Health Hazard Evaluation Program, Report no. 2012-0100-3217. Available online at: https://www.cdc.gov/niosh/hhe/reports/pdfs/2012-0100-3217.pdf (Accessed on: 2 March 2022).

Cesaro, A., V. Belgiorno, G. Gorrasi, G. Viscusi, M. Vaccari, G. Vinti, and S. Salhofer. 2019. A relative risk assessment of the open burning of WEEE. *Environmental Science and Pollution Research* 26(11):11042–11052.

Chan, J. K. Y. and M. H. Wong. 2013. A review of environmental fate, body burdens, and human health risk assessment of PCDD/Fs at two typical electronic waste recycling sites in China. *Science of the Total Environment* 463:1111–1123.

Chatterjee, S. 2012. Sustainable electronic waste management and recycling process. *American Journal of Environmental Engineering* 2(1):23–33.

Chi, X., M. Streicher-Porte, M. Y. Wang, and M. A. Reuter. 2011. Informal electronic waste recycling: A sector review with special focus on China. *Waste Management* 31(4):731–742.

Chi, X., M. Y. L. Wang, and M. A. Reuter. 2014. E-waste collection channels and household recycling behaviors in Taizhou of China. *Journal of Cleaner Production* 80:87–95.

Dai, Q, X. Xu, B. Eskenazi, et al. 2020. Severe dioxin-like compound (DLC) contamination in e-waste recycling areas: An under-recognized threat to local health. *Environment International* 139:105731. https://doi.org/10.1016/j.envint.2020.105731

Dasgupta, D., A. Debsarkar, D. Chatterjee, and A. Gangopadhyay. 2014. E-waste management in India: Issues and concern. *Journal of International Environmental Application and Science* 9(2):257–261.

Daum, K., J. Stoler, and R. J. Grant. 2017. Toward a more sustainable trajectory for e-waste policy: A review of a decade of e-waste research in Accra, Ghana. *International Journal of Environmental Research and Public Health* 14(2):135.

Davis, J. M. and Y. Garb. 2015. A model for partnering with the informal e-waste industry: Rationale, principles and a case study. *Resources, Conservation and Recycling* 105:73–83.

Debnath, B., R. Chowdhury, and S. K. Ghosh. 2018. Sustainability of metal recovery from e-waste. *Frontiers of Environmental Science Engineering* 12(6):1–12.

Debnath, B., R. Chowdhury, and S. K. Ghosh. 2020. Recycling of polymers from WEEE: Issues, challenges and opportunities. In *Urban Mining and Sustainable Waste Management*, ed. S. K. Ghosh, pp. 69–80. Springer, Singapore.

Deng, B., D. X. Luong, Z. Wang, C. Kittrell, E. A. McHugh, and J. M. Tour. 2021. Urban mining by flash Joule heating. *Nature Communications* 12:5794. https://doi.org/10.1038/s41467-021-26038-9

Deubzer, O. 2019. Reduction of hazardous materials in electrical and electronic equipment. In *Waste Electrical and Electronic Equipment (WEEE) Handbook*, eds. V. Goodship, A. Stevels, and J. Huisman, pp. 207–230. Woodhead Publishing Series, Elsevier. https://doi.org/10.1016/C2016-0-03853-6

Dutta, D., S. Arya, and S. K. E. Lichtfouse. 2021. Electronic waste pollution and the COVID-19 pandemic. *Environmental Chemistry Letters*. https://doi.org/10.1007/s10311-021-01286-9

Feng, H. 2021. Making and Unmaking of Guiyu: The Global Center of E-waste (Doctoral dissertation, Duke University).

Forti, V., C. P. Baldé, R. Kuehr, and G. Bel. 2020. *The Global E-Waste Monitor 2020*, p. 120. United Nations University (UNU), International Telecommunication Union (ITU) & International Solid Waste Association (ISWA), Bonn/Geneva/Rotterdam.

Fu, J., A. Zhang, T. Wang, et al. 2013. Influence of e-waste dismantling and its regulations: Temporal trend, spatial distribution of heavy metals in rice grains, and its potential health risk. *Environmental Science & Technology* 47:7437–7445. https://doi.org/10.1021/es304903b

Fu, J., H. Zhang, A. Zhang, and G. Jiang. 2018. E-waste recycling in China: A challenging field. *Environmental Science & Technology* 52(12):6727–6728.

Ghimire, H. and P. A. Ariya. 2020. E-wastes: Bridging the knowledge gaps in global production budgets, composition, recycling and sustainability implications. *Sustainable Chemistry* 1:154–182. https://doi.org/10.3390/suschem1020012

Ghosh, S. K. ed. 2020. *Urban Mining and Sustainable Waste Management*. Springer, Berlin, Heidelberg.

Grant, K., F. C. Goldizen, P. D. Sly, et al. 2013. Health consequences of exposure to e-waste: A systematic review. *The Lancet Global Health* 1(6): e350–e361.

Hao, J., Y. Wang, Y. Wu, and F. Guo. 2020. Metal recovery from waste printed circuit boards: A review for current status and perspectives. *Resources, Conservation and Recycling* 157:104787.

Heacock, M., C. B. Kelly, K. A. Asante, et al. 2016. E-waste and harm to vulnerable populations: A growing global problem. *Environmental Health Perspectives* 124(5):550–555.

Hong, Y., D. Thirionb, S. Subramanian, et al. 2020. Precious metal recovery from electronic waste by a porous porphyrin polymer. *Proceedings of the National Academy of Sciences of the United States of America* 117(28):16174–16180. www.pnas.org/cgi/doi/10.1073/pnas.2000606117

Hsu, E., K. Barmak, A. C. West, and A.-H. A. Park. 2019. Advancements in the treatment and processing of electronic waste with sustainability: A review of metal extraction and recovery technologies. *Green Chemistry* 21:919–936. https://doi.org/10.1039/C8GC03688H

Islam, A., T. Ahmed, M. R. Awual, et al. 2020. Advances in sustainable approaches to recover metals from e-waste-A review. *Journal of Cleaner Production* 244:118815.

Julander, A., L. Lundgren, L. Skare, M. Grandér, B. Palm, M. Vahter, and C. Lidén. 2014. Formal recycling of e-waste leads to increased exposure to toxic metals: An occupational exposure study from Sweden. *Environment International* 73:243–251.

Kannan, D., K. Govindan, and M. Shankar. 2016. Formalize recycling of electronic waste. *Nature* 530(7590):281.

Kaya, M. 2018. Current WEEE recycling solutions. In *Waste Electrical and Electronic Equipment Recycling*, ed. F. Vegliò, and I. Birloaga, pp. 33–93. Elsevier Ltd., Amsterdam, the Netherlands.

Kazançoglu, Y., E. Ada, Y. Ozturkoglu, and M. Ozbiltekin. 2020. Analysis of the barriers to urban mining for resource melioration in emerging economies. *Resources Policy* 68:101768.

Khaiwal, R., and S. Mor. 2019. E-waste generation and management practices in Chandigarh, India and economic evaluation for sustainable recycling. *Journal of Cleaner Production* 221:286–294.

Khaliq, A., M. A. Rhamdhani, G. Brooks, and S. Masood. 2014. Metal extraction processes for electronic waste and existing industrial routes: A review and Australian Perspective. *Resources* 3:152–179. https://doi.org/10.3390/resources3010152

Kyere, V. N., K. Greve, and S. M. Atiemo. 2016. Spatial assessment of soil contamination by heavy metals from informal electronic waste recycling in Agbogbloshie, Ghana. *Environmental Health and Toxicology*, 31:e2016006.

Lecler, M. T., F. Zimmermann, E. Silvente, F. Clerc, A. Chollot, and J. Grosjean. 2015. Exposure to hazardous substances in Cathode Ray Tube (CRT) recycling sites in France. *Waste Management* 39:226–235.

Li, H., J. Eksteen, and E. Oraby. 2018. Hydrometallurgical recovery of metals from waste printed circuit boards (WPCBs): Current status and perspectives-A review. *Resources, Conservation and Recycling* 139:122–139.

Maes, T., and F. Preston-Whyte. 2022. E-waste it wisely: Lessons from Africa. *SN Applied Sciences* 4:72. https://doi.org/10.1007/s42452-022-04962-9

Mahapatra, R. P., S. S. Srikant, R. B. Rao, and B. Mohan. 2019. Recovery of basic valuable metals and alloys from e-waste using microwave heating followed by leaching and cementation process. *Sådhanå* 44:209. https://doi.org/10.1007/s12046-019-1193-y

Malinauskaite, J., and F. B. Erdem. 2021. Planned obsolescence in the context of a holistic legal sphere and the circular economy. *Oxford Journal of Legal Studies* 41(3):719–749.

Manish, A., and P. Chakraborty. 2019. E-Waste Management in India: Challenges and Opportunities. Available online at: https://www.teriin.org/article/e-waste-management-india-challenges-and-opportunities. (Accessed on: 26 April 2022).

Matsukami, H., N. M. Tue, G. Suzuki, et al. 2015. Flame retardant emission from e-waste recycling operation in northern Vietnam: Environmental occurrence of emerging organophosphorus esters used as alternatives for PBDEs. *Science of the Total Environment* 514:492–499.

Matsumiya, M., M. Sumi, Y. Uchino, and I. Yanagi. 2018. Recovery of indium based on the combined methods of ionic liquid extraction and electrodeposition. *Separation and Purification Technology* 201:25–29.

Minter, A. 2016. The burning truth behind an e-waste dump in Africa. *Smithsonian Magazine.* https://www.smithsonianmag.com/science-nature/burning-truth-behind-e-waste-dump-africa-180957597/. (Accessed on: 15 April 2022).

Mitra, S., and D. K. Maiti. 2021. Environmental problems and management aspects of waste electrical and electronic equipment and use of clean energy for sustainable development. In *Environmental Management of Waste Electrical and Electronic Equipment*, ed. C. Hussain, pp. 3–21. Elsevier.

Mohan, T. V., and R. K. Amit. 2020. Dismantlers' dilemma in end-of-life vehicle recycling markets: A system dynamics model. *Annals of Operations Research* 290(1):591–619.

Montalvo, C., D. Peck, and E. Rietveld. 2016. *A Longer Lifetime for Products: Benefits for Consumers and Companies*. European Parliament, Brussels. Available online at: https://www.europarl.europa.eu/studies (Accessed on: 12.3.2022)

Mourshed, M., M. H. Masud, F. Rashid, and M. U. H. Joardder. 2017. Towards the effective plastic waste management in Bangladesh: A review. *Environmental Science and Pollution Research* 24(35):27021–27046.

Murthy, V., and S. Ramakrishna. 2022. A review on global e-waste management: Urban mining towards a sustainable future and circular economy. *Sustainability* 14(2):647.

Okwu, P. I., and I. N. Onyeje. 2014. Extraction of valuable substances from e-waste. *American Journal of Engineering Research* 3(1):299–304.

Orlins, S., and D. Guan. 2016. China's toxic informal e-waste recycling: Local approaches to a global environmental problem. *Journal of Cleaner Production* 114:71–80.

Otto, S., A. Kibbe, L. Henn, L. Hentschke, and F. G. Kaiser. 2018. The economy of e-waste collection at the individual level: A practice oriented approach of categorizing determinants of E-waste collection into behavioral costs and motivation. *Journal of Cleaner Production*, 204:33–40.

Pajunen, N., and M. E. Holuszko. 2022. Circular economy in electronics and the future of e-waste. In *Electronic Waste: Recycling and Reprocessing for a Sustainable Future*, ed. M. E. Holuszko, A. Kumar, D.C.R. Espinosa, pp. 299–314. Wiley-VCH GmbH, Weinheim. DOI:10.1002/9783527816392

Parajuly, K., R. Kuehr, A. K. Awasthi, et al. 2019. *Future E-Waste Scenarios*. StEP Initiative, UNU VIE-SCYCLE, UNEP IETC, Bonn and Osaka.

Park, J. K., T. Clark, N. Krueger, and J. Mahoney. 2017. A review of urban mining in the past, present and future. *Advances in Recycling & Waste Management* 2:127. https://doi.org/10.4172/2475-7675.1000127

Parvez, S. M., F. Jahan, M.-N. Brune, J. F. Gorman, M. J. Rahman, and D. Carpenter. 2021. Health consequences of exposure to e-waste: An updated systematic review. *The Lancet. Planetary Health* 5(12):E905–E920. https://doi.org/10.1016/S2542-5196(21)00263-1

Patil, R., and S. Ramakrishna. 2020. A comprehensive analysis of e-waste legislation worldwide. *Environmental Science and Pollution Research* 27:14412–14431. https://doi.org/10.1007/s11356-020-07992-1.

Peng, B., Y. Tu, and G. Wei. 2018. Governance of electronic waste recycling based on social capital embeddedness theory. *Journal of Cleaner Production* 187:29–36.

Perkins, D. N., M. N. B. Drisse, T. Nxele, and P. D. Sly. 2014. E-waste: A global hazard. *Annals of Global Health* 80(4):286–295.

Petricca, C., Z. Moloo, and M. Stoisser. 2020. *Hazardous e-waste recycling in Agbogbloshie, Accra, Ghana*. Environmental Justice Atlas. Available online at: https://ejatlas.org/conflict/agbogbloshie-e-waste-landfill-ghana (Accessed on: 10 March 2022)

Pini, M., F. Lolli, E. Balugani, et al. 2019. Preparation for reuse activity of waste electrical and electronic equipment: Environmental performance, cost externality and job creation. *Journal of Cleaner Production* 222:77–89.

Pinto, V. N. 2008. E-waste hazard: The impending challenge. *Indian Journal of Occupational and Environmental Medicine* 12(2):65–70. https://doi.org/10.4103/0019-5278.43263

Popoola, O. E., A. O. Popoola, and D. Purchase. 2019. Levels of awareness and concentrations of heavy metals in the blood of electronic waste scavengers in Nigeria. *Journal of Health & Pollution* 9(21):1–10. https://doi.org/10.5696/2156-9614-9.21.190311

Pradhan, J. K., and S. Kumar. 2014. Informal e-waste recycling: Environmental risk assessment of heavy metal contamination in Mandoli industrial area, Delhi, India. *Environmental Science and Pollution Research* 21(13):7913–7928.

Purchase, D., G. Abbasi, L. Bisschop, et al. 2020. Global occurrence, chemical properties, and ecological impacts of e-wastes (IUPAC Technical Report). *Pure and Applied Chemistry* 92(11):1733–1767. https://doi.org/10.1515/pac-2019-0502

Rai, V., D. Liu, D. Xia, Y. Jayaraman, and J.-C. P. Gabriel. 2021. Electrochemical approaches for the recovery of metals from electronic waste: A critical review. *Recycling* 6:53. https://doi.org/10.3390/recycling6030053

Rene, E. R., M. Sethurajan, V. K. Ponnusamy, et al. 2021. Electronic waste generation, recycling and resource recovery: Technological perspectives and trends. *Journal of Hazardous Materials* 416:125664. https://doi.org/10.1016/j.jhazmat.2021.125664

Ribeiro, J. N., A. F. M. Barbosa, A. V. F. N. Ribeiro, et al. 2022. E-Waste and its consequence for environment and public health: Perspectives in Covid-19 pandemic times. *Global Journal of Health Science* 14(3):54–76. https://doi.org/10.5539/gjhs.v14n3p54

Robaina, M., J. Villar, and E. T. Pereira. 2020. The determinants for a circular economy in Europe. *Environmental Science and Pollution Research* 27(11):12566–12578.

Sahajwalla, V. and V. Gaikwad. 2018. The present and future of e-waste plastics recycling. *Current Opinion in Green and Sustainable Chemistry* 13:102–107.

Sengupta, D., I. M. S. K., Ilankoon, K. D. Kang, and M. N. Chong. 2022. Circular economy and household e-waste management in India: Integration of formal and informal sectors, *Minerals Engineering* 184:107661.

Sethurajan, M., E. D. van Hullebusch, D. Fontana, et al. 2019. Recent advances on hydrometallurgical recovery of critical and precious elements from end of life electronic wastes - a review, *Critical Reviews in Environmental Science and Technology* 49(3):212–275. https://doi.org/10.1080/10643389.2018.1540760

Siddiqi, M. A., R. H. Laessig, and K. D. Reed. 2003. Polybrominated diphenyl ethers (PBDEs): New pollutants- old diseases. *Clinical Medicine & Research* 1(4):281–290.

Singh, K., G. Arora, P. Singh, and A. Gupta. 2021. IoT-based collection vendor machine (CVM) for E-waste management. *Journal of Reliable Intelligent Environments* 7(1):35–47.

Singh, M., P. S. Thind, and S. John. 2018. Health risk assessment of the workers exposed to the heavy metals in e-waste recycling sites of Chandigarh and Ludhiana, Punjab, India. *Chemosphere* 203:426–433. https://doi.org/10.1016/j.chemosphere.2018.03.138

Song, Q. and J. Li. 2014. Environmental effects of heavy metals derived from the e-waste recycling activities in China: A systematic review. *Waste Management* 34(12):2587–2594.

Standaert, M. 2015. China's Notorious E-Waste Village Disappears Almost Overnight. Basel Action Network, Seattle, WA

Sthiannopkao, S. and M. H. Wong. 2013. Handling e-waste in developed and developing countries: Initiatives, practices, and consequences. *Science of the Total Environment* 463:1147–1153.

Sun, Y.-X., X.-J. Luo, L. Mo, et al. 2012. Hexabromocyclododecane in terrestrial passerine birds from e-waste, urban and rural locations in the Pearl River Delta, South China: Levels, biomagnification, diastereoisomer- and enantiomer-specific accumulation. *Environmental Pollution* 171:191–198. https://doi.org/10.1016/j.envpol.2012.07.026

Szűcs, T., and P. Szentannai. 2021. Developing an all-round combustion kinetics model for nonspherical waste-derived solid fuels. *Chemical Papers* 75(3):921–930.

Tansel, B. 2017. From electronic consumer products to e-wastes: Global outlook, waste quantities, recycling challenges. *Environment International* 98:35–45.

Thakur, A. 2015. Urban mining: Lift cover for your next ring. Available online at: https://www.fortuneindia.com/technology/urban-mining-lift-cover-for-your-next-ring/101212. (Accessed on: 15 October 2022).

Thakur, P. and S. Kumar. 2022. Evaluation of e-waste status, management strategies, and legislations. *International Journal of Environmental Science and Technology* 19(17):6957–6966.

The United Nations Global E-Waste Monitor. 2020. A Collaborative Product of Global E-Waste Statistics Partnership (GESP), the International Telecommunication Union (ITU), and the International Solid Waste Association (ISWA). Available online at: https://ewastemonitor.info/. (Accessed on: 26 March 2022).

Thorat, S. 2021. Review of life-cycle analysis of e-waste in India. *International Journal of Modern Agriculture* 10(2):838–857.

Turaga, R. M. R., K. Bhaskar, S. Sinha, et al. 2019. E-waste management in India: Issues and strategies. *Vikalpa* 44(3):127–162.

Vaccari, M., G. Vinti, A. Cesaro, et al. 2019. WEEE treatment in developing countries: Environmental pollution and health consequences-An overview. *International Journal of Environmental Research and Public Health* 16(9):1595.

Wakuma, K. A., and S. M. Woldemikael. 2021. Electronic waste management in Addis Ababa: The case of Bole and Nefas Silk Lafto sub-cities. *African Journal of Science, Technology, Innovation and Development* 13(2):235–246.

Wath, S. B., A. N. Vaidya, P. S. Dutt, and T. Chakrabarti. 2010. A roadmap for development of sustainable e-waste management system in India. *Science Total Environment* 409(1):19–32. https://doi.org/10.1016/j.scito tenv.2010.09.030

Website 1. 2022. Government notifies Battery Waste Management Rules, 2022. Available online at: https://pib.gov.in/PressReleasePage.aspx?PRID=1854433. (Accessed on: 20 September 2022).

Wen, Z., C. Zhang, X. Ji, and Y. Xue. 2015. Urban mining's potential to relieve China's coming resource crisis. *Journal of Industrial Ecology* 19(6):1091–1102.

Widmer, R., H. Oswald-Krapf, D. Sinha-Khetriwal, M. Schnellmann, and H. Boni. 2005. Global perspectives on e-waste. *Environmental Impact Assessment Review* 25(5):436–458.

Yang, S., S. Gu, M. He, X. Q. Tang, L. Ma, J. Xu, and X. Liu. 2020. Policy adjustment impacts Cd, Cu, Ni, Pb and Zn contamination in soils around e-waste area: Concentrations, sources, and health risks. *Science of the Total Environment* 741:140442–140450. https://doi.org/10.1016/j.scitotenv.2020.140442

Zlamparet, G. I., W. Ijomah, Y. Miao, A. K. Awasthi, X. Zeng, and J. Li. 2017. Remanufacturing strategies: A solution for WEEE problem. *Journal of Cleaner Production* 149:126–136.

Zeng, X., J. A. Mathews, and J. Li. 2018. Urban mining of e-waste is becoming more cost-effective than virgin mining. *Environment Science Technology* 52(8):4835–4841.

Zeng, X., T. Xiao, G. Xu, E. Albalghiti, G. Shan, and J. Li. 2021. Comparing the costs and benefits of virgin and urban mining. *Journal of Management Science and Engineering*. https://doi.org/10.1016/j.jmse.2021.05.002

Zhang, C. C., and F. S. Zhang. 2018. High copper recovery from scrap printed circuit boards using poly (ethylene glycol)/sodium hydroxide treatment. *Environmental Chemistry Letters* 16(1):311–317.

Zhou, J., N. Zhu, H. Liu, P. Wu, X. Zhang, and Z. Zhong. 2019. Recovery of gallium from waste light emitting diodes by oxalic acidic leaching. *Resources Conservation and Recycling* 146:366–372.

Zimmermann, F., M. T. Lecler, F. Clerc, A. Chollot, E. Silvente, and J. Grosjean. 2014. Occupational exposure in the fluorescent lamp recycling sector in France. *Waste Management* 34(7):1257–1263.

12 Role of Sulfate-Reducing Bacteria in Sustainable Acid Mine Bioremediation

Arunima Biswas and Shakuntala Ghorai
Raidighi College

12.1 INTRODUCTION

Mining is an important contributor to the economic growth of many nations, both developed and developing. It is an ancient practice. But the oldest explorations were not designed in a way to minimize adverse environmental impacts and reduce related hazards, which are inherent to mining operations. This was mostly due to a lack of scientific understanding and awareness. Sadly, that trend still continues in many countries. Across the globe, opinions are divided between those who firmly believe that mineral wealth is a blessing for a country and those who vehemently argue that its development is an unmitigated curse, resulting in depletion of natural resources, gross violation of safety regulations, and detrimental effects on workers, the local environment, and the ecosystem (Richards, 2009, 4; Alvarenga, 2021, 1).

Mining is always associated with the formation of huge amounts of waste, including toxic, corrosive, even flammable, and radioactive materials. Mining sites usually have piles of left-over and/or processing waste, tailings ponds, and hazardous chemicals. Mine wastes consist of ash, dust, gangue, overburden waste rock, industrial minerals, slag, tailings, sediment, mineral fuels, ore, chemicals, etc. It has been reported that one ton of mining waste is generated per ton of metal ore extracted (Ayangbenro et al., 2018, 1). If these are somehow released, then that can lead to perilous impacts on our water, air, and land resources.

The major complaints against the mining industry are its poor waste management and an utter indifference towards environmental safety that continued for decades. Waste materials that remain after the extraction of usable ores are often dumped on the surrounding land or allowed to run off from the mining site into adjoining ground and water. Acidic pH, high metal content, and toxic contamination can wipe out existing life forms and discourage future growth of plants, animals, and microbes (Western Mining Action Network, 2004).

Most worrisome are the mine waste tailings, which are the materials that remain after separating the valuable fraction of an ore from the uneconomic fraction (gangue). Mine tailing wastes are usually produced from the mill in a slurry or mixture of

DOI: 10.1201/9781003451457-12

fine mineral particles and water, and they are generally stored in 'tailings ponds' using an earth-fill embankment dam called a 'tailings dam' (Azam, 2010, 50). These ponds and dams can be dangerous sources of toxic chemicals, heavy metals, sulfides, and radioactive material. These constructions are especially risky as they are highly vulnerable to major breaches or leaks that cause environmental disasters (Anthony, 2019; Lyu, 2019, 1). Tailings dams are designed for permanent or long-term containment of these harmful wastes and thus continue to pose major challenges to the practice and concept of 'responsible mining'. An analysis of tailings dam failures by the UN indicates that such failure incidents are becoming more probable due to climate change effects. When they occur, they can destroy entire communities and livelihoods, thus prompting global concerns about the safety, management, and impacts of storing and managing large volumes of mine tailings.

One of the biggest problems associated with mining and mineral processing activities is the production of exceedingly acidic and toxic effluents with high concentrations of metals and metalloids, which originate during mining activities (specifically from metal mines or coal mines) and also from the oxidation of sulfides present in their waste deposits (Alvarenga, 2021, 2). Such acidic discharges are often generated from metal mines where the ore is a sulfide mineral or is associated with pyrite (FeS_2). In many such cases, the predominant metal ion may not be iron itself but some other metal like zinc, copper, or nickel. For instance, chalcopyrite, which has been the most important and most commonly mined ore of copper since the Bronze Age, is actually a combination of copper-iron sulfide ($CuFeS_2$) and occurs with a range of other sulfides. Thus, copper mines are often major culprits of acid discharge (Akcil, 2006, 1140).

In the unmined rock, which is located deep beneath the soil, the reactive minerals are sheltered and protected from chemical oxidation and are stable. However, upon excavation during mining activities, these become exposed to atmospheric oxygen and water. A series of biogeochemical processes ensue that can lead to huge acid production. Some of the major factors that determine the rate of acid generation include degree of saturation with water, availability of oxygen, pH conditions, temperature, chemical activation energy required to begin acid generation, sulfide minerals, surface area of exposed metal sulfide, and the activity of the microbial population present (Ayangbenro et al., 2018, 3). This acidic effluent, which has a high tendency to run off, is known as acid mine drainage (AMD) or Acid Rock Drainage (ARD). Such acidic discharge can easily spill, run away, and pollute nearby water and land resources, and it has become a serious global issue owing to its hazardous impact on the local environment and living organisms.

Actively operating mines are supposed to have a zero-discharge policy for AMD, but for abandoned and closed mines, such drainage continues to be a long-term menace. Also, when a mine is in active production, water tables are kept artificially low by pumping. However, when mines are closed and abandoned and the pumps are turned off, the consequent rebound of the water table can lead to polluted and toxic groundwater being discharged from the mining site, which is no longer in operation. Another added disadvantage is that often there is no legally responsible owner for such shut-down mines (Johnson, 2003, 47).

As the number of mining operations keeps on steadily increasing around the globe, it also magnifies the chances of related adverse environmental, social, and economic impacts. Consequently, it augments the cost of dealing with such a hazardous occurrence and the risks and expenses associated with its perpetual management and related restoration activities.

Mining companies, communities, and governments all recognize and admit that mine waste tailings, contaminated acidic effluents, and AMD-driven aquatic and terrestrial pollution can damage lives and livelihoods. On the other hand, it also leads to a negative opinion and mindset about the mining sector, which can damage its reputation and threaten its development.

12.2 FEATURES OF AMD

AMD is caused by the outflow of mine acidic water that can seep into nearby streams, rivers, soils, and groundwater, thus heavily polluting them. This usually occurs in rocks that contain an abundance of sulfide minerals. The most immediate and serious impacts of AMD are on natural waterbodies, but it has far-reaching and dire consequences as well. The acidic effluent often leaches out toxic metals from ore and waste rocks, which, combined with a reduced pH, can have a detrimental impact on the neighborhood environment, ecosystem, and all resident populations (ranging from humans to microbes) (Tomiyama, 2022).

Sulfides are mineral compounds composed of sulfur (S) in its lowest natural oxidation state (S^{2-}) and metals, e.g., iron (Fe), and/or semi-metals, e.g., arsenic (As). When oxidized, they generate a very acidic pH that often leads to the solubilization of the remaining constituents of these minerals (metals and/or semi-metals). This phenomenon can occur slowly and gradually under certain environmental conditions as part of the natural weathering process but is greatly aggravated due to man-made large-scale earth disturbances characteristic of mining processes, construction activities, etc. (Alvarenga, 2021, 2).

For example, AMD is formed when iron pyrite, the most abundant sulfide mineral ('fool's gold' or iron sulfide, FeS_2), is exposed to the atmosphere and reacts with air and water to form sulfuric acid and dissolved iron. Some or all of this iron can, under certain conditions, precipitate to form the characteristic red, orange, or yellow sediments in the bottom of nearby streams that contain this mine drainage. This so-called "yellow boy" feature is the most obvious visual indicator of such AMD (Alvarenga, 2021, 2). It usually means that iron and possibly other metals are precipitating, discoloring the water and coating the rocks and stream bed. What actually happens is that when the pH of AMD is raised beyond 3, either through contact with fresh water or by some neutralizing minerals, previously soluble ferric ions now precipitate as iron hydroxide, a yellow-orange solid that is colloquially known as "yellow boy". Other types of iron precipitates are also possible, like iron oxides, oxyhydroxides, and sulfates. Few organisms can survive in these streams and rocks. Thus, it leads to disruption of the local ecosystem.

Such AMD-affected water bodies often give off a bad odor (rotten egg smell), and the water itself may produce a stinging sensation on the skin and taste metallic and bitter. Some severely affected streams may look clear and sterile with almost no sign of life, and with iron and/or other metals completely dissolved in the acid water.

It should however be remembered that the natural oxidation process of pyrite is not entirely due to chemical reactions; it is rather biochemical in nature, with certain bacteria playing a crucial role, namely *Acidothiobacillus ferrooxidans*, *Acidothiobacillus thiooxidans*, and *Leptospirillum ferrooxidans*, some chemolithotrophic bacteria, etc. (Alvarenga, 2021, 3).

Although a number of contributing factors lead to AMD, pyrite oxidation is by far the most important and greatest contributor. Table 12.1 represents how the complex reactions proceed (equation 12.1). The oxidation of the sulfide to sulfate solubilizes the ferrous iron, which is subsequently oxidized to ferric iron (equation 12.2). These reactions can occur slowly, spontaneously, or be catalyzed by relevant acidophilic microorganisms. The ferric cations produced can also oxidize additional pyrite and reduce it to produce more ferrous ions (equation 12.3), thus augmenting the process. The net effect of these reactions is to release H^+, which makes the pH acidic and maintains the solubility of the ferric ion (Dold, 2014, 623; Alvarenga, 2021, 2). Thus, pyrite oxidation takes place in multiple steps, involving an oxygen-independent reaction (ferric iron attack on the mineral) and oxygen-dependent reactions. At pH values above 4, this process may be mediated either chemically or biologically (by some iron-oxidizing bacteria), while below pH 4, abiotic iron oxidation is generally negligible (Dold, 2014, 623; Evangelou, 1998, 197).

AMD can be many times more acidic than acid rain or battery acid and can burn human skin, kill diverse organisms, wreak havoc on our water and land resources, and even catch fire and cause explosions.

The most striking and scariest feature is the occurrence of a negative pH. Negative pH arises if water evaporates from a pool that is already acidic, thus increasing the concentration of hydrogen ions (Nordstrom, 2000, 254).

In addition to the acid contribution to surface waters, AMD can cause metals such as arsenic, cadmium, copper, silver, and zinc to leach from mine wastes and pollute nearby water sources, thus endangering the local community (Ayangbenro and Babalola, 2017, 1).

AMD may sometimes go unnoticed for years, and by the time it is detected, the damage is already done. Moreover, it can often occur for decades or even centuries after mining operations cease at a place. Acidic effluent is still known to be seeping from mines in Europe that were owned by the Romans prior to A.D. 476, thus emphasizing its serious, long-term adverse environmental implications (Alvarenga 2021, 3).

TABLE 12.1
Pyrite Oxidation, Most Important and Greatest Contributor to Acid Mine Drainage (Dold, 2014)

$$FeS_2 + 7/2O_2 + H_2O \rightarrow Fe^{2+} + 2SO_4^{-2} + 2H^+ \quad (12.1)$$

$$Fe^{2+} + 1/4O_2 + H^+ \rightarrow Fe^{3+} + 1/2H_2O \quad (12.2)$$

$$FeS_2 + 14Fe^{3+} + 8H_2O \rightarrow 15Fe^{2+} + 2SO_4^{2-} + 16H^+ \quad (12.3)$$

Tailings from such mines are also extremely hazardous. The toxicity of acid mine tailings waste (AMTW) is especially due to its low pH and high content of heavy metals. The combined effect is an alarming increase in the bioavailability of heavy metals, thus increasing the chances of biotoxicity and bioaccumulation, as discussed later (Ayangbenro and Babalola, 2017, 1).

Tailings ponds can also be a source of major acid drainage. For instance, in 1994, the operators of the Olympic Dam mine, a large polymetallic underground mine located in South Australia, admitted that their uranium tailings containment had released around 5 million m^3 of contaminated acid water into the subsoil, thus endangering the local habitat (Nuclear-risks.org, 2009).

On a different but related note, tailings dam failures can also lead to significant ecological damage and death, as in the case of the Brumadinho dam disaster in Brazil that took place on January 25th, 2019, where as many as 252 people are still unaccounted for and at least 134 are reported dead. The disaster released 12 million m^3 of iron waste, which drained into the Paraopeba River. It left 120 km of the river ecosystem incredibly acidic and toxic. The river water is no more portable and it is now called a "dead river". High levels of lead and chromium were also found near the collapse site (Anthony, 2019).

Another tragic example of when a tailings dam holding highly toxic, contaminated waters burst, leading to a catastrophe, is that of the 2000 Baia Mare cyanide spill in Romania. It involved the leakage of huge amounts of cyanide into the rivers by the gold mining company Aurul. The polluted waters spread far and wide, killing large numbers of fish in Hungary, Serbia, and Romania. The spill has been labeled by many as the worst and most tragic environmental disaster in Europe after the Chernobyl disaster (BBC, 2000).

The fiscal impacts of mitigating the damage caused by AMD can also be enormous. The Summitville mine was a gold mining site in the United States (Colorado) and is remembered for the worst cyanide spill in the history of the United States. The disaster badly affected all local waterways and the Alamosa River. The mining tactics involved the treatment of the pyritic ore with a sodium cyanide solution in order to leach the gold out of the ore. Soon, the acidic effluent and cyanide found their way into nearby water sources. The mining company declared bankruptcy and walked away, and taxpayers are paying \$120–150 million to clean and restore the site (USGS, 2016).

12.3 ENVIRONMENTAL IMPACT OF AMD

The formation of AMD and associated toxic contaminants is often described as the largest environmental challenge facing the mining industry on a global scale.

Water is essential for life on earth. Sustainable Development Goals adopted by the UN include the need for clean water (United Nations Resolution A/RES/71/313, 2017, 10). In order to ensure that, we need to keep our natural waterbodies like streams, rivers, lakes, and oceans uncontaminated. Water pollution from discharged mine effluent and seepage from mine waste tailings greatly threaten both the surface and ground water sources on which we all depend for our survival. Water has been rightly called "mining's most common casualty". AMD severely degrades water quality and makes it virtually unusable. Extended long-term impacts on aquatic life include significant losses of flora, fauna, and microbial species. This can lead to a local biodiversity crisis, thus threatening the structure and proper functioning of the surrounding ecosystem, putting it at risk of deterioration and, even, collapse. The acid can potentially harm our land resources as well. AMD thus endangers UN Sustainable Development Goal 15 "Life on Land" and Sustainable Development Goal 14 "Life Below Water" (United Nations Resolution A/RES/71/313, 2017, 18–19).

Due to the more disaggregated, disassembled, and concentrated nature of the acid-generating minerals in mine waste tailings, the AMD that flows from them may be more aggressive than that discharges from the mine itself.

Heavy metal contamination may occur when metals or metalloids such as arsenic, cobalt, cadmium, chromium, lead, zinc, copper, and silver (which may be either present in excavated rock or exposed in an underground mine) come into contact with water. Such metals can be leached out and then carried downstream as water flows over the rocks. The leaching process is greatly magnified in the low pH conditions of AMD. At low pH, heavy metals are stable in solution and mobile, but at increased pH, they remain adsorbed and thus remain immobile. So, an acidic pH greatly promotes the risk of heavy metals being transported to the surrounding environment. Even at very low concentrations, exposure to heavy metals in aquatic systems can stunt fish growth and development. Birds and/or other wild animals can be poisoned after drinking contaminated water and may fall sick or die. Heavy metal pollution is considered a major environmental threat because metal ions persist in the environment due to their nondegradable nature. Many heavy metals are toxic to living organisms at low concentrations and are capable of entering the food chain, where they accumulate and inflict damage at every trophic level. They can induce serious physical stress in plants and animals and contaminate our food sources. The increase in heavy metal concentration up the food chain, a phenomenon called 'biomagnification', is especially dangerous, causing severe public health concerns (Ayangbenro and Babalola, 2017, 3).

Exposure to heavy metals from mine waste has been implicated in many disorders and diseases that affect mine workers and the local community.

Chemicals used by mining companies for processing purposes (for example, to separate valuable metals and minerals from ore) may also spill or leak into streams, rivers, and groundwater along with the acidic discharge. Some of these chemicals, such as mercury, can persist in the environment for ages and continue to contaminate waterbodies. Cyanide, which is widely used in modern gold mining, is another potentially deadly chemical that can leach out and enter nearby water supplies along with AMD.

AMD is very acute, even at inactive and abandoned mining sites. A tragic example of the loss of biodiversity due to AMD is the case of the Tsolum River on Vancouver Island, British Columbia, Canada. For thousands of years, the river remained clean, unpolluted, full of various aquatic life forms, and rich in runs of different species of Pacific salmon. A copper mining company carried out active operations in the region for only 3 years and then abandoned the mining site. But they left behind a toxic legacy that resulted in drastic losses in the salmon population, which was vulnerable to AMD that continued from the closed mine. This, in turn, led to a sharp decline in fishery resources, resulting in huge financial losses as well.

Migratory creatures are also affected by AMD. The deaths of migratory birds have been documented at mine sites where contaminated water filled abandoned pits or accumulated in tailings ponds.

12.4 & 12.5 PREVENTION AND REMEDIATION STRATEGIES: PHYSICAL AND CHEMICAL APPROACHES AND THEIR DRAWBACKS

The mining industry has long since acknowledged that preventing catastrophic tailings dam failure incidents and minimizing AMD are fundamental to the practice of 'responsible mining', with a target of zero mortality and a warranty of maximum environmental protection. These are thought to be achievable through strong adherence to internationally established and accepted best practices. However, eliminating the possibility of all catastrophic incidents still remains a challenge with respect to existing engineering and technical knowhow and safety measures, which are clearly insufficient (Johnson, 2005, 3; Simate and Ndlovu, 2014, 1785).

These days, it is a leading practice to carry out a geochemical assessment of mine materials during the early stages of a mining operation to determine the potential for AMD, since many cases have been reported where AMD remained undetected for years and caused heavy damage. Sadly, in many countries, discharges from abandoned or closed mines are exempt from any regulatory control. This is a gross oversight, and the issue should be promptly addressed by environment experts and policymakers.

Since prevention is better than cure, many "source control" approaches have been evaluated and suggested to be effective in minimizing AMD.

Both water and oxygen play pivotal roles in augmenting AMD. So, exclusion of either or both can prevent such drainage from abandoned underground mines, which can be flooded and sealed. The dissolved oxygen present in flooding waters will be used up by existing microbes, and sealing will prevent further inflow of oxygen-laden water. For this strategy to be successful, the location of all shafts and adits must be known (Johnson, 2005, 5).

Also, the use of underwater storage for acid mine tailings can minimize the risk of dissemination. A shallow water cover can be used. Tailings can also be covered additionally with a layer of sediment, clay, or organic material, which will limit oxygen entry and offer protection against resuspension of the tailings due to the actions of wind and waves. Dry covers and clay sealing layers can be used for surface storage. However, the success of this approach will depend on the weather pattern of a particular region, as drying and cracking of the cover will reduce its efficacy (Johnson, 2005, 5).

Another approach involves blending acid-generating and acid-consuming materials together. For instance, solid-phase phosphates can be added to pyritic mine waste to precipitate ferric ions as ferric phosphate, thereby reducing its ability to oxidize sulfide minerals. A modified approach has led to the development of 'coating technology', where soluble phosphate is added together with hydrogen peroxide. The peroxide oxidizes pyrite, producing ferric iron, which, in turn, reacts with the phosphate to generate a protective coating of ferric phosphate on the surface. An alternate technique that involves the formation of an iron oxide or silica coating on pyrite surfaces has also been reported to be effective (Johnson, 2005, 5).

Many attempts are going on to standardize, both at laboratory scale and field scale, an effective inhibitory method that will help prevent the activities of iron-oxidizing and sulfur-oxidizing bacteria in mineral spoils and tailings by using biocides. This has generally involved the application of anionic surfactants such as sodium dodecyl sulfate. But, so far, this approach has not seen much success as a sustainable, long-term prevention process.

Due to the practical difficulties that are often faced during the implementation of "source control" approaches to prevent AMD at source, the other alternative left is to adopt biotic and abiotic approaches as remedial or counteractive measures.

Several conventional remedial technologies have been popular for decades.

An active chemical "migration control" strategy involves the continuous application of alkaline materials to the acidic mine waters. The addition of an alkaline material to AMD will neutralize its acidity and gradually raise its pH, accelerate the rate of chemical oxidation of ferrous iron, and cause many of the metals that are present in solution to precipitate as hydroxides and carbonates. To achieve this, either active aeration or the addition of a chemical oxidizing agent such as hydrogen peroxide is essential. The result is the production of an iron-rich sludge that may also contain various other metals (Johnson, 2005, 7).

The standard treatment strategy involves chemical precipitation using lime or other basic substances. The most commonly used commercial process for treating AMD is lime (CaO) precipitation or lime neutralization. Slaked lime, calcium carbonate, sodium carbonate, sodium hydroxide, and magnesium oxide and hydroxide can also be used with varying efficacy and expense. Major drawbacks include the high operation cost and disposal of the bulky sludge. The large volumes of wet sludge often require drying facilities to concentrate the metal hydroxide sludge. "High-density sludge process" can reduce the cost of disposal or storage. Other approaches include the use of calcium silicate neutralization methods, carbonate neutralization methods, ion exchange processes, metal sulfide precipitation techniques, the use of water treatment plants or reagents, etc.

The use of anoxic limestone drains is also becoming a popular practice. In this approach, mine water is forced to flow through a bed of limestone gravel within a drain that is impermeable to both air and water. The aim is to add alkali to AMD while maintaining the iron in its reduced form to avoid the oxidation of ferrous iron and precipitation of ferric hydroxide on the limestone, which otherwise severely reduces the effectiveness of the neutralizing agent. Inside the drain, the partial pressure of carbon dioxide is increased, enhancing the rate of limestone dissolution and consequently generating a high degree of alkalinity at a much lower cost. However, these drains are effective on a short-term basis and cannot treat all types of acid effluents. These are more useful when used in combination with other remedial methods (Johnson, 2005, 7; Kleinmann, 1998).

Many remedial strategies effectively use biological organisms, mostly microbes, as discussed separately in the following sections.

12.6 BIOREMEDIATION VIA SULFATE-REDUCING BACTERIA

Bioremediation of AMD using various treatment methods has been quite effective for some time due to their low cost, reuse of waste materials, eco-friendly nature, lower maintenance, etc. Sulfate-reducing bacteria utilize sulfate in the absence of oxygen while obtaining energy from the oxidation of organic compounds and producing H_2S via the reduction of sulfate (Deng et al., 2018, 6369). These bacteria are heterotrophs and facultative anaerobes that catabolize organic materials by utilizing sulfates, sulfites, thiosulfates, sulfur, or other oxidized sulfides as electron acceptors. Up until now, 18 genera and more than 40 species have been studied (Deng et al., 2018, 6371). The most common genera include *Desulfovibrio* (no spores, curve rods) and *Desulfotomaculum* (straight or curved rods). Others are *Desulfomicrobium* (rod-shaped, no spores), *Desulfobollus* (curved), *Desulfobacter* (round, rod-shaped, without spores), *Desulfococcus* (spheroidal, no spores), *Desulfosarcina* (stacking, without spores), *Desulfoarculus* (curved), *Desulfomonile* (rod-shaped, spheroidal), *Desulfobulbus* (oval, lemon-shaped, no spores), *Desulfacinam* (spheroidal or oval type), *Desulfobacula* (oval), *Desulfonema* (screw shape, no spores), *Desulfobacterium* (round, rod-shaped), and *Thermodesulfobacterium* (small shape, arc, rod), etc. Sulfate reducers grow between −5°C and 75°C (Jamil and Clarke, 2013, 572), but some can survive below and above the range as well. They are mostly mesophiles, having a temperature preference between 25°C and 40°C for optimum growth, except for *Thermodesulfobacterium*, which requires 65°C–70°C. Sulfate reducers are dependent on methanogens and acetogens. Methanogens and some sulfate reducers are mesophiles, so H_2S production will vary with the variation in temperature. Most of the sulfate-reducing bacteria are netrophiles, with the most appropriate pH for growth found to be 6.5–7.8. For optimal growth, sulfate reducers require a pH range of 6.8–7.2, and pH below 5 is found to be inhibitory. Maximum H_2S is ensured at a pH range of 5–6. They can tolerate salinity up to 240 g/L, with an optimum level of 100 g/L for best growth. Most of the halophilic sulfate reducers are mild halophiles that can tolerate 1%–4% of salinity. Ferrous ion is an essential component of various enzymes in these sulfate reducers, so the concentration of Fe^{2+} is critical for their growth. Carbon sources utilized by the sulfate-reducing bacteria vary, and they act as electron donors for the reduction and dissimilation of sulfates. C3 and C4 fatty acids are mostly used, along with some volatile fatty acids and easily fermented substrates like hydrogens.

12.6.1 Mechanisms of H_2S Production

Production of hydrogen sulfide by sulfur-reducing bacteria happens in two ways. They can assimilate sulfate or degrade sulfur-containing organic compounds (mostly cysteine). Sulfur assimilation from sulfate is lower in amount and is readily assimilated into organic sulfur compounds, whereas the degradation of organic sulfur yields a large amount of H_2S. Sulfate metabolism yielding H_2S is completed in three consecutive steps: (1) decomposition, (2) electron transfer, and (3) Oxidation, respectively (Deng et al., 2018, 6372).

12.6.1.1 Decomposition

Sulfate ions accumulated outside of the cell enters the sulfate-reducing bacterial cell under anaerobic conditions. Meanwhile, organic compounds are decomposed to carbon dioxide, water, and acetate through substrate-level phosphorylation, producing adenosine triphosphate. This triphosphate is consumed during sulfate activation by ATP sulfurylase to produce pyrophosphoric acid (PPi) and adenosine-5′-phosphosulfate (APS). PPi is unstable and quickly hydrolyzes to inorganic phosphate (Pi). Reactions involved in sulfate activation are shown in Table 12.2a (equations 12.4–12.6).

12.6.1.2 Electron Transfer Stages

During the decomposition of organic compounds to acetate, carbon dioxide, and water in sulfate-reducing bacteria, electrons are transferred step by step through a transport chain composed of flavoprotein, cytochrome C3, and others. APS produced in the decomposition step is continuously converted to adenosine monophosphate by APS reductase, while sulfur is converted to sulfite as shown in Table 12.2b (equations 12.7–12.9). Sulfite is further converted to pyrosulfite ($S_2O_5^{2-}$), which is extremely unstable and produces one intermediate dithionite ($S_2O_4^{2-}$). This intermediate is quickly reduced to trithionate ($S_3O_6^{2-}$), which further produces thiosulfate ($S_2O_3^{2-}$) and sulfite (SO_3^{2-}).

12.6.1.3 Oxidation

Conversion of sulfate to sulfide requires the transfer of eight electrons. The electrons are transferred to the oxidized sulfur entities $S_2O_3^{2-}$, SO_4^{2-} and SO_3^{2-}, etc. at the expense of ATP. The first sulfate is adsorbed to ATP phosphatase by ATP thiokinase, resulting formation of APS. The sulfated part of APS is reduced to sulfite along with the production of AMP. Sulfite is the initial reduction product of sulfate, which finally produces H_2S by sulfite reductase. The overall reaction is given in Table 12.2c (equation 12.10).

Production of H_2S from SO_3^{2-} may proceed via the following pathways (Deng et al., 2018, 6373).

a. Trithionate and thiosulfate formation involving three succeeding two-electron transfers, as mentioned in Table 12.2d.
b. It (SO_3^{2-}) may lose six electrons directly, and the intermediates mentioned above are not formed, which is defined as a coordinated six-electron reaction as depicted in Table 12.2e (equation 12.11).
c. The reverse reaction may also happen during the continuous two-electron transfer process.

In bacterial sulfate reduction, lactate is utilized as an electron donor and SO_4^{2-} behaves as an electron acceptor. Three enzymes were engaged in the reduction of SO_3^{2-} to S^{2-} -(1) $S_3O_6^{2-}$ forming enzyme. (2) $S_3O_6^{2-}$ reductase or thiosulfate forming enzyme, and (3) thiosulfate reductase. Accumulation of a large amount of S^{2-} or HS^- in the bacterial cell results in the release of the entity outside of the cell through

TABLE 12.2a
Reactions Involved in Sulfate Activation during Decomposition of Organic Compounds (Deng et al., 2018)

$$SO_4^{2-} + ATP + 2H^+ \xrightarrow{\text{ATP sulfurylase}} APS + PPi \quad (12.4)$$

$$PPi + H_2O \xrightarrow{\text{Pyrophosphatase}} 2Pi \quad (12.5)$$

$$SO_4^{2-} + ATP + 2H^+ + H_2O \longrightarrow APS + 2Pi \quad (12.6)$$

TABLE 12.2b
Stages of Electron Transfer during the Decomposition Process (Deng et al., 2018)

$$\text{E-FAD} + \text{Electron Carrier (Reduced)} \leftrightarrow \text{E-FADH}_2 + \text{Electron Carrier (Oxidized)} \quad (12.7)$$

$$\text{E-FADH}_2 + APS \leftrightarrow \text{E-FADH}_2\left(SO_3^{2-}\right) + AMP \quad (12.8)$$

$$\text{E-FADH}_2\left(SO_3^{2-}\right) \leftrightarrow \text{E-FAD} + SO_3^{2-} \quad (12.9)$$

TABLE 12.2c
Overall Reaction of Conversion of Sulfate to Sulfide (Deng et al., 2018)

$$SO_4^{2-} + 8e^- + 10H^+ \rightarrow H_2S \uparrow + 4H_2O \quad (12.10)$$

TABLE 12.2d
Formation of Trithionate and Thiosulfate from Sulfite (Deng et al., 2018)

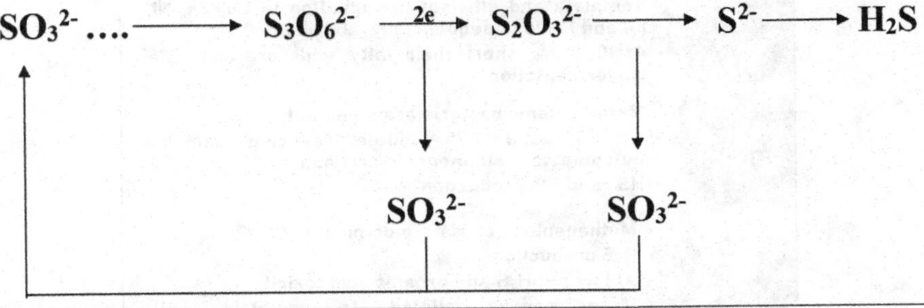

TABLE 12.2E
Coordinated Six Electron Reaction during Conversion of Sulfite to Sulfide (Deng et al., 2018)

$$SO_3^{2-} + 6e^- + 8H^+ \rightarrow H_2S \uparrow + 3H_2O \quad (12.11)$$

the bacterial membrane. Once the concentration of S^{2-} increases in the environment, it initiates changes in the redox potential of the surrounding environment and the concentration of related ions.

12.7 REMEDIATION POTENTIAL OF SULFATE-REDUCING BACTERIA

AMD remediation by sulfate-reducing bacteria is associated with other microbial communities like methanogens, acetogens, etc. Microbial efficiency is dependent on various factors such as growth stage, nutrient availability, physiological state of bacterial cells, presence of ions, concentration of biomass, and environmental conditions like pH, ionic strength, temperature, etc. Sulfate reduction by sulfate-reducing bacteria in strong AMD gradually generates an alkaline condition, which can create a successive alkalinity-producing system for better growth of the microbial population. Sulfate-reducing bacteria may have an autotrophic, litho-autotrophic, or hetertrophic respiration type of life under anaerobic conditions (Hussain et al., 2016, 4). Autotrophs use CO_2 as a carbon source and obtain electrons from the oxidation of H_2, whereas heterotrophs utilize various organic substrates as their carbon source. These bacteria can utilize different organic and inorganic compounds as terminal electron acceptors in addition to sulfite, sulfate, thiosulfate, and tetrathionate. Sugars, amino acids, alcohols, monocarboxylic acids, dicaroboxylic acids, and aromatic compounds can serve as potential electron donors for sulfate-reducing bacteria, but they prefer low-molecular-weight organic compounds as carbon and energy sources.

Sulfate-reducing bacteria preferentially grow at a pH 6–8, whereas AMD or mine waste-laden water generally shows a pH less than 3 (Nancucheo et al., 2017, 3). As Figure 12.1 describes, various bacterial population and chemical species dominate at

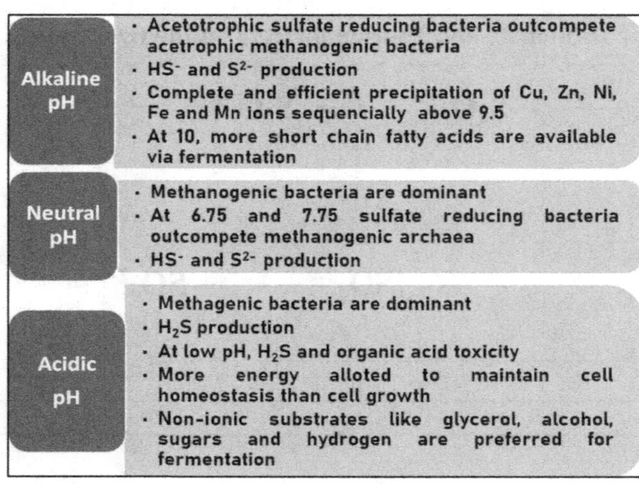

FIGURE 12.1 Dominant bacterial species at different pH ranges. (Dominant bacterial and chemical species are available at different pH ranges during the degradation of acid mine drainage).

different pH ranges in the mine drainage. This necessitates an intermediate system where the acidity of the wastewater should be somewhat neutralized so that it does not pose a threat to the viability of the microbial cells. So, acidophilic or acid-tolerant sulfidogenic bacteria may play a critical role in AMD treatment and the subsequent metal recovery process from waste.

Acid-tolerant sulfate-reducing bacteria of genus the *Desulfovibrio*, strain S4, were isolated from an AMD-containing pond in Vietnam. It was effective at pH 2 in the presence of high concentrations of various heavy metals, including Zn^{2+} and Cu^{2+}. Metal tolerance of these deltaproteobacterial species was also established at a concentration of 100 mg/L. These acid-tolerant bacteria were populated at low pH and reduced sulfate to sulfide, which eventually raised the pH of the medium. The precipitated metal sulfides favored the growth of other acid- and high-metal-intolerant sulfate-reducing bacterial population like S10, as experimentally proven by co-inoculation of the acid- and metal-tolerant strain S4 with the neutrophilic, metal-sensitive strain in the study (Nguyen et al., 2020, 1011). This metal- and acid-tolerant strain posed an effective alternative as seed culture during the treatment process of AMD in bioreactors.

Sulfate-reducing bacteria are in use to treat heavy metal-laden wastewater like Cr (III), Cr (VI), $Zn^{2+,}$ U (VI), Cu^{2+}, Mn (IV), Fe (III), etc. effectively (Wei-ting et al., 2014, 113).

In a pilot and field-scale study conducted by EPA and USDoE under the Mine Waste Technology Program, AMD was treated by sulfate-reducing bacteria in packed-bed bioreactors (Canty, 1998, 61). In continuous upflow operations, AMD retention time was 5 days for a period of 60 days at 8°C. Metal removal efficiency on a pilot scale was achieved at 99% for zinc, 99% for aluminum, 96% for manganese, 98% for cadmium, and 96% for copper. Poor removal was observed in the cases of iron and arsenic. Metal adsorption and sulfate reduction were detected in reactors. The in-situ effectiveness of this innovative technology was evaluated via field-scale testing. Flooded subsurface mine workings of the Lilly/Orphan Boy Mine near Ellison, Montana, were selected for that purpose. Nearly 70%–100% of the removal of aluminum, cadmium, copper, and zinc was estimated during one to one and a half years of in-situ treatment. Iron and arsenic removal efficiency, sulfate reduction, sulfide generation, etc. were similar to those of pilot-scale testing.

Synthetic wastewater containing 1,000 and 1,500 mg/L of sulfate, 100 mg/L of Fe^{2+}, 20 mg/L of Zn^{2+}, and 5 mg/L of Cu^{2+} was treated in an anaerobic stirred batch reactor for 218 days using sulfate-reducing bacteria (Castro Neto et al., 2018, 543). Bacterial inoculum was granular sludge biomass generated during treatment of poultry slaughterhouse wastewater in an upflow anaerobic sludge blanket reactor. Both the organic carbon source and electron donor were ethanol. During the treatment, the sulfate-to-chemical oxygen demand (COD) ratio was kept at 1. More than 99% of the metal was removed with a simultaneous increase in pH to 7.4 from the initial value of 4. Sulfate removal was 43%–65%, with sulfide concentration reaching 56.6 mg/L.

Sulfate-reducing bacteria can be utilized in the treatment of antimony (V)-containing wastewater (Chen et al., 2020, 213). The process is highly pH-dependent. When the pH reaches 6.5, most aqueous Sb {Sb (III)-Sb (II) complex} gets precipitated as Sb_2S_3. Compared to lactate, ethanol acted as a better carbon source considering the cost factor and favorable Sb removal.

HS^- produced during sulfate reduction results in a lowering of the pH of the solution. This acidity is neutralized by bicarbonates released during lactic acid degradation under anaerobic conditions. This automated pH control renders most reactor operations efficient with regard to pH correction. Biogenic H_2S reacts with heavy metals present in the mixture, and as their solubility is low, they ultimately precipitate. Cd^{2+}, Cu^{2+}, Ni^{2+}, Pb^{2+}, U^{6+}, Sb^{5+}, etc. form insoluble sulfides and they can be efficiently removed from the waste-laden water because of their high densities, low solubility, favorable dewatering and settling properties, etc. Metal sulfides are precipitated at different pHs. These sulfate-reducing bacteria are capable of removing heavy metals via the precipitation of carbonates and hydroxides. Extracellular polymeric substances (EPS) secreted from the cell contribute to the removal of heavy metals (Xu and Chen, 2020, 1800). These polymeric substances contain nonpolar amino acids (mass fractions of 12% Ala, 11% Gly, and 7% Phe) and acidic amino acids (mass fractions of 14% Glu and 13% Asp), which facilitate precipitation of metal sulfides because of the large negative charges in Glu and Asp. The main entities in EPS that react with heavy metals are Trp-like substances. Different parts of EPS are capable of binding with different heavy metals, like protein, and polysaccharide-binding sites can absorb Zn^{2+}, whereas polysaccharides can absorb Cu^{2+}. As EPS strongly absorbs metal sulfides on their surface, they can reduce the toxicity of metal sulfides toward sulfate-reducing bacteria. For this reason, EPS-secreting sulfate-reducing bacteria can be favorably applied to the treatment of heavy metal-rich wastewater. In addition to that, active heavy metals can react with some organic ligands and alter their forms. Moreover, microbial enzymes can convert active metals to their inactive state, resulting in a reduction in heavy metal toxicity. As presented in Figure 12.2, removal of sulfate may be possible in one or more ways by sulfate-reducing bacteria.

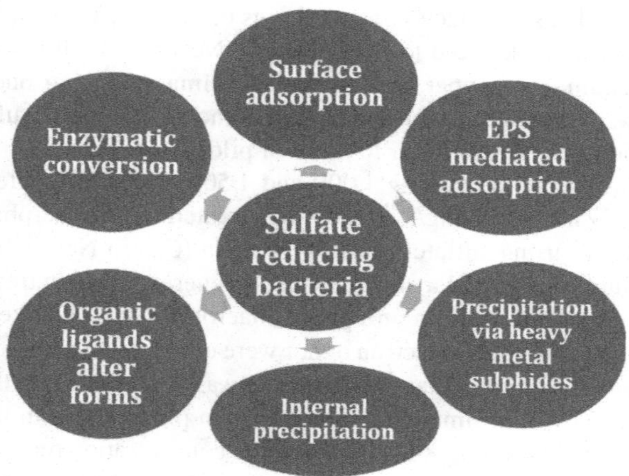

FIGURE 12.2 Removal of sulfate by sulfate-reducing bacteria. (Removal of sulfate by sulfate-reducing bacteria may be possible in various ways, like precipitation via metal sulfides, adsorption via extracellular polymeric substances, internal precipitation, enzymatic conversion, surface adsorption, alteration of forms by organic ligands, etc.).

12.8 FACTORS AFFECTING REMEDIATION EFFICIENCY

The metal removal efficiency of sulfate-reducing bacteria is controlled by several parameters like acidity, oxidation-reduction potential, sulfide concentration, temperature, ration of chemical oxygen demand to sulfate, solid retention time, hydraulic retention time (HRT), heavy metal toxicity, growth condition, abundance of microbial species, diversity of microorganisms, interaction (especially synergistic) between them, etc.

12.8.1 pH

The pH of the reaction vessel determines several factors during the treatment process using sulfate-reducing bacteria. It affects microbial metabolic mechanism via the dissociation and homeostasis of electron donors (Janyasuthiwong et al., 2016). Glycerol, hydrogen, carbohydrates, alcohols, etc. nonionic substrates that are more favorable for fermentation at low pH values than others. At alkaline pH, acetotrophs outcompete acetotrophic methanogens, whereas at acidic or neutral pH, methanogenic bacteria are dominant. Again, at pH 6.75 and 7.75, sulfate-reducing bacteria outcompete methane-producing archaea (Dai et al., 2017). As protons facilitate sulfate reduction step, this process is energetically favorable at a lower pH. Acidity enhances the sulfate reduction rate and increases the cell numbers of sulfate reducers. A major part of cellular energy is utilized to sustain pH homeostasis strategies to balance large differences between extracellular and intracellular pH values (Willis et al., 2019). In addition to this, other events like alterations in lipid constituents, expression of positively charged surface proteins, proton pumps, and amino acid decarboxylases require a major share of cellular energy, leaving less energy for cellular growth. In bioreactor operations, precipitates are separated to avoid mixing sulfate reducers with acidic water. Offline bioreactor units are used to recycle water, and measures are taken to isolate and enrich acidophilic sulfate reducers to perform at a lower pH. pH plays a critical role in the size, quality, precipitation, and separation of metal sulfides, influencing the charge and binding properties of the metal. In the case of immobilized sulfate-reducing bacteria, alteration of the optimum pH results in decreased biosorption and bioprecipitation. Neutrophilic sulfate reducers perform best at a pH range of 7.0–7.8, and the maximum sulfate reduction rate is obtained in the range 7.0–7.5. Sulfate reduction is hampered by these neutrophiles when pH is either acidic at 5 or alkaline at 9, and at pH below 2, no activity is observed. This barrier can be overcome by acidophiles, which can perform at pH values 2–4 (Zhao et al., 2017). Acidity in the reaction vessel is introduced by the produced H_2S and acetate, but it HCO_3^- increases alkalinity. pH greatly influences the sulfate reduction rate and its conversion to acetate and H_2S. Metal sulfides are separated via precipitation by altering the pH of the reaction vessel. Although treatment of acid mine water in an anaerobic reactor is continued at a pH of 7, heavy metals like Cu, Ni, Fe, Zn, Mn, etc. cannot be precipitated at this pH, but complete precipitation of these metals via the formation of metal sulfides is achieved at an alkaline pH greater than 9.5. First, Cu is removed, followed by Zn, Ni, Fe, and Mn, depending on their solubilities (Dev et al., 2017). It has been observed that Mn precipitation is somehow affected by the concentration of Fe in the influent, as Mn can be converted to a less soluble form through reduction by Fe. Figure 12.1 shows the dominant bacteria species available at different pH levels in the fermentation medium. Sulfate-reducing bacterial isolates *Desulfotomaculum orientis* from estuary dam ecosystem sediments in Denpasar was able to survive at pH 3 and showed abundant growth at pH 4–7 (Suyasa et al., 2019, 395). Those motile, spore-forming, straight rods precipitated heavy metal ions from acid mine waste. Organic matter was used as compost during treatment, and sulfate reducers were able to reduce sulfate by 92.77%–94.52% while the pH of the system was increased from 3 to 6.7. Soluble heavy metals such as Cd (II), Pb (II), Mn (II), Zn (II), Cu (II), and Fe (III) were reduced to much less soluble or insoluble forms by 85.45%–90%, 95.10%–97.95%, 99.79%–99.89%, 99.87%–99.90%, 99.23%–99.43%, and 99.93%–99.96%, respectively. Considering capital cost, removal efficiency, chemical oxygen demand, coulombic efficiency, etc., optimum sulfate removal efficiency by sulfate reducing bacteria was achieved at a pH of 4.5 (Liang et al., 2013, 147). *Paludibacter* sp. was found to play the most important role in sulfate removal at this pH, using ethanol as the electron donor for the reduction of sulfate in acclimatized sludge. *Desulfuromonadaceae* sp., *Desulfobulbaceae* sp., and *Desulfovibrio* sp. were involved in power generation.

12.8.2 Temperature

Majority of sulfate-reducing bacteria are mesophiles, but some are psychrophiles and therophiles in nature (Kim et al., 2015). So, the optimum temperature for sulfate reduction and metal separation is 37°C, and temperatures higher than this will lead to protein denaturation and inactivation of cellular machinery (Krukenberg et al., 2016). Psychrophilic sulfate-reducing bacteria best grow at 7°C–18°C, but the maximum sulfate reduction rate is achieved at a temperature 2°C–9°C higher than that. Thermophilic sulfate-reducing bacteria prefer to grow at 65°C–70°C, and moderately thermophiles grow best around 50°C. Low temperatures reduce rate of metal and sulfate removal, along with alkalinity and microbial activity. Nielsen et al. (2018) reported that the rate of heavy metal and sulfate removal decreased from 79% to 90% in the summer to 0%–39% in the winter. This trouble can be overcome by adopting different operating processes, like suspended and immobilized biomass systems, at different temperatures. At lower temperatures (<15°C) immobilized systems showed better heavy metal removal capacity compared to the suspended system (Hao et al., 2016). At elevated temperatures, rates of protonation and deprotonation of functional groups and microbial activity increase, but the solubility of H_2S decreases in the wastewater. At high temperatures sulfate-reducing bacteria and methanogenic bacteria compete with each other, and this condition favors the growth of sulfate-reducing bacteria. Microorganisms adapt to this elevated temperature by altering their membrane lipid ordering, composition of fatty acids, nature of the head group of the phospholipids, chain length, degree of unsaturation of the acyl chains, etc.

Unsuitable temperatures for mesophiles and thermophiles result in alteration of gene expression, enzyme activity, and the production and formation of various lipid/protein assemblies (Xu and Chen, 2020). Mesophilic sulfate-reducing bacteria metabolize slowly at cold temperatures (Virpiranta et al., 2019). Virpiranta et al. characterized one sulfate-reducing bacterial consortium from an Artic sediment sample from northern Finland. Sulfate-reducing bacteria isolated from the region belong to the δ-proteobacteria. The most common sulfate reducer was *Desulfobulbus*. Other members of the family of *Desulfobulbaceae* were also detected. *Desulfovibrio* and *Desulfotignum* were among the other genera. The majority of the cultivated consortia reduced sulfate at a temperature of 6°C using succinic acid as a carbon source. Rate of sulfate reduction at this temperature was noted to vary from 13 to 42 mg/L/d. Growth of the microbial corsortia was fluent in the presence of lactate and succinic acid but not with ethanol. This type of contortia may be applied for the biological treatment of sulfate-rich wastewater at low temperatures.

12.8.3 Hydraulic Retention Time and Solids Retention Time (HRT and SRT)

This factor affects microbial activity and reaction mixture composition, and it varies with different treatment processes and environments. A low value of HRT is not supportive of microbial activity and growth as acidic wastewater can wash out biomass. In fluidized-bed bioreactors, acidity increases because of incomplete oxidation of carbon sources and the accumulation of acetate. A higher amount of dissolved oxygen induces a higher oxidation-reduction potential (>−100 mv), which results in decreased sulfate reduction and availability of sulfate-reducing bacteria, the elimination of an anaerobic environment, and reduced heavy metal removal efficiency. Aoyogi et al. (2017) demonstrated in a packed-bed bioreactor that optimum HRT was 6 and 20 hours under neutral and acidic conditions, respectively. Long hours of HRT are beneficial for microbial growth as they increase the degradation of organic nitrogen and carbon and make low-molecular-weight organic substrates available to the growing bacteria. HRT favors sequential batch anaerobic digestion and improves the activity of sulfate-reducing bacteria and biomass retention rate. Sulfate-rich wastewater treated in an anaerobic sequential batch reactor showed outcompetence of sulfate reducing bacteria over methanogenic sulfate-reducing bacteria, including a 75% sulfate reduction rate with an HRT of 12 hours (Moon et al., 2015). Treatment of source-diverted black water with HRT of 2.2 hours in an Up-flow anaerobic sludge bed reactor yielded more than 80% of COD removal, and sulfate-reducing bacteria outcompeted methanogenic sulfate reducers. Whereas, AMD treated in the same bioreactor favored a greater sulfate removal rate (92%) with 16 hours of HRT compared to 24 hours (<80%) (Cunha et al., 2019). SRT is also critically important for the sulfate reduction process. A long duration of SRT is beneficial for an improved rate of sulfate reduction (Huang et al., 2020).

12.8.4 Carbon Sources

As the AMD is scarce in organic carbon content, the sulfate reduction rate is critically controlled by the availability of organic carbon sources. External or added carbon sources have enhanced the treatment rate of AMD (Castro Neto et al., 2018, 544). Reduction of sulfate requires a substantial amount of energy-rich reducing agents, so the composition of organic sources is an important determinant of the bioremediation process. Multiple organic carbon sources that are readily degradable biologically and create low redox environments will serve better in sustaining the microbial milieu of the AMD ecosystem. Less complex organic carbon sources such as alcohols, acids, and sugars are readily consumed by sulfate reducers and require a shorter HRT of 3–5 days for the precipitation of metal sulfides. Whereas complex organic sources like organic compost, wood or paper waste, and by-products from food production are further degraded before being utilized by sulfate reducers and show a double HRT of 7–10 days for metal sulfide precipitation. Utilizing dairy and chicken manure and sawdust as organic carbon sources, sulfate removal was achieved by 79%, 64%, and 50%, respectively, in the presence of sulfate-reducing bacteria (Ayangbenro et al., 2018). Heavy metals like Cd, Cu, Fe, Mn, Ni, and Zn were removed during 35 days of treatment. In another experiment, As, Cu, Fe, Ni, and Zn-containing AMD was treated in an up-flow sludge blanket reactor under anaerobic conditions for 500 days using ethanol as a carbon and electron source. *Desulfovibrio desulfuricans* and *Desulfomicrobium baculatum* were able to remove 99% of sulfate and 98%–100% of Cu, Fe, Ni, and Zn.

Estuary sediment was utilized as a sulfate-reducing bacterial source for the treatment of acid mine water. Experimental sets were in different combinations, such as T1 (sediment and compost), T2 (sediment), T3 (compost), and T4 (control-only acid mine water), etc. (Fahruddin et al., 2020, 191). In each set, the treatment process was associated with an increase in pH and bacterial growth. Maximum bacterial growth and an increase in pH were observed in the case of set T1. Sulfate concentrations were reduced by 78%, 56%, 21%, and 5% in T1, T2, T3, and T4, respectively. So, sulfate reduction in acid mine wastewater can be achieved using estuary sediment as an inoculum of sulfate-reducing bacteria and compost.

AMD with high content of Cr^{6+}, Cr^{3+}, SO_4^{2-}, and acidity is treated in the presence of sulfate-reducing bacteria belonging to *Desulfotomaculum* and iron scrap through three dynamic columns prepared using sugar cane slag, corn cob, and sunflower straw as carbon sources, respectively (Wang et al., 2021, 163). Sulfate reducers were able to remove 96.9%, 67.1%, and 54.3%, respectively, of Cr^{6+}, Cr^{3+}, and SO_4^{2-} using sugar cane slag. The average release of TFe and COD was 4.4 and 287.3 mg/L, respectively, in column 1 at an average pH of 6.98. This column was very effective in Cr^{6+} removal, and the pH value was adjusted suitably by this column. The corn cob-made dynamic column was running at 6.23 and effectively removed Cr^{6+}, Cr^{3+}, and SO_4^{2-} by 96.1%, 85.6%, and 74.9%, respectively, releasing TFe and COD of 15.8 and 636.5 mg/L on average. Although this column was not good enough at adjusting pH, it efficiently removed Cr^{3+} and SO_4^{2-}. Another dynamic column made of sunflower straw was not efficient at releasing TFe and COD but very efficiently adjusted pH (Wang et al., 2021, 174). The average release of TFe and COD via this column was 1.1 and 158.1 mg/L, respectively. It was running at pH 7.96 and removed Cr^{6+}, Cr^{3+}, and SO_4^{2-} at a rate of 95.2%, 57.8%, and 32.5%, respectively. Column packed with sugar cane slag was better in comparison to others with respect to removal of Cr^{6+}, Cr^{3+}, and SO_4^{2-} ions and release of TFe, COD, and adjustment of acidity.

12.8.5 Ratio of Chemical Oxygen Demand to Sulfate (COD/SO_4^{2-})

The heavy metal removal efficiency of sulfate-reducing bacteria greatly depends on the ratio of chemical oxygen demand to sulfate concentration. The nature of the organic substrate is therefore very important during the treatment process, especially when bioremediation is running with a mixed culture of sulfate-reducing bacteria. Under standard conditions, 1 g of sulfate reduction requires 0.67 g of COD, and if this ratio is less than 0.67, all the electrons are taken up by sulfate (Mulopo and Schaefer, 2013). When the ratio exceeds 0.67, sulfate-reducing bacteria compete with other microorganisms for common electron donors. Sulfate-reducing bacteria and methanogenic bacteria actively compete when the ratio is between 1.7 and 2.7. Below 1.7, sulfate reducers outcompete methanogens, but the reverse happens when the ratio is greater than 2.7. A higher ratio favors biomass production and acetate accumulation. As a result, it facilitates the growth of acetogenic bacteria, which are resistant to high concentrations of H_2S. It can be concluded that, up to a certain level, a higher COD-to-sulfate ratio leads to highly efficient sulfate reduction and heavy metal removal by the sulfate-reducing bacteria. The ratio of COD to SO_4^{2-} determined growth of sulfate-reducing bacteria along with the process of sulfate removal as investigated in an anaerobic batch reactor (Hardyanti et al., 2018). Artificial coal acid mine wastewater containing 2,000 mg SO_4^{2-}/L was treated with laundry septic tank sediments. The best sulfate removal with 46.58% efficiency and a reduction rate of 29.128 mg/L/day was achieved when the COD to SO_4^{2-} ratio was 2. Reduced efficiency was noted at a higher ratio.

12.8.6 Metal Ions

Very high concentrations of metallic ions such as iron, copper, manganese, zinc, etc. are able to inhibit the growth of sulfate reducers, exerting their toxicity (Jamil and Clarke 2013, 572). The nature of the sulfate reducer also determines the extent of inhibition by toxic metal ions. The inhibitory effect is found to be increasing in the following order:

$$\text{Sulfate} < \text{Thiosulfate} < \text{Sulfite} < \text{Total sulfide} < H_2S$$

Cell vitality can be enhanced by adding inorganic cations like calcium, magnesium, iron, etc. into the reaction mixture, which increase salinity and combine with anionic sites on the cell surface. So the cell surface is not available for interaction with heavy metals, and their toxicity is also reduced. Mg^{2+} is found to have an impact on the activity and growth of sulfate-reducing bacteria. At high concentrations of this ion, microbial activity was enhanced via improved diffusion of SO_4^{2-} into the bacteria and mass transfer. In addition to reducing the toxicity of sulfides via precipitation, Fe^{2+} acts as a trace element for the sulfate reducers and aids in the synthesis of some enzymes involved in the removal of contaminants. Insoluble FeS has a greater ability to adsorb and remove heavy metals. FengJuan et al. (2018) demonstrated that 200 mg/L Fe^{2+} improves the removal efficiency of As from 78% to 98% and slightly increases the removal rate of Sb from 98.8% to 99.4%. Moreover, they have noted that NaCl performed better compared to $MgCl_2.6H_2O$ with regard to enhancement of growth rate and activity of sulfate-reducing bacteria. A better affinity between sulfate reducers and NaCl was observed at pH 7.6 than with $MgCl_2.6H_2O$ at pH 7. Zero-valent ions are found to be highly efficient in improving microbial activity and the removal of heavy metals during the treatment process (Zhang et al., 2016). They act as electron donors, and microbes adhered to them gain electrons from H_2 generated from zero-valent ion corrosion, which reduces COD for reduction of SO_4^{2-}. These ions create a more reducing environment by decreasing oxidation-reduction potential. Similar trends were reported earlier in the case of *Desulfobacterium* sp. and *Desulfovibrio* strains whose reducing capacities were driven by H_2, although H_2 and organic compounds were both electron donors.

12.8.7 Sulfide Concentration

Sulfides negatively impact the process of heavy metal removal at an excess and scarce concentration. All forms of sulfide, such as H_2S, HS^-, and S^{2-}, exhibit toxicity to the sulfate-reducing bacteria at different levels, and undissociated H_2S is the most toxic form. Its toxicity arises from its neutrality, which permits it to enter the cell without the help of any specific receptor. Inside the cell, this undissociated form acts as a decoupler perturbing sulfur assimilation, intracellular pH, damaging DNA, and denaturing proteins involved in cellular metabolism. It also inhibits enzymes like cytochrome oxidase, which ultimately blocks the transfer of electrons to oxygen. H_2S enhances acetic acid accumulation, which results in effluents with high COD content (Kushkevych et al., 2019). Sulfides produced by sulfate-reducing bacteria have reducing abilities and they can act on some oxidized metals, ameliorating their toxicity. Sulphide to heavy metal ratio is indicative of the rate of metal removal from the waste-laden water and the extent of toxicity exerted by the heavy metals on sulfate-reducing bacteria. When M/S^{2-} is low (<1), sulfate reduction produces large amounts of sulfides that precipitate heavy metals efficiently but create a toxic environment for microbial growth. On the other hand, when this ratio is >1, metal removal and sulfate reduction rates are reduced by metal toxicity. Bacterial growth is inhibited initially but later increases as heavy metals and sulphates are removed from the fermentation media via precipitation. Hydrogen-utilizing sulfate-reducing bacteria can tolerate higher H_2S concentrations than autoclasic sulfate-reducing bacteria. A high COD-to-SO_4^{2-} ratio is favorable for acetogenic bacterial growth but not for acetoclastic bacteria. In the absence of acetoclastic sulfate-reducing bacteria, the effluent shows high acidity and COD due to the accumulation of acetic acid. Some essential metals, like Fe, required for the growth of sulfate-reducing bacteria are also precipitated by H_2S. Sulfide toxicity of the bacteria can be eliminated by the use of more complex electron donor molecules.

12.8.8 Oxidation Reduction Potential

Microbial activity of sulfate-reducing bacteria depends on their oxidation reduction potential. Sulfate-reducing bacteria are microaerophilic, so a low concentration of oxygen promotes greater microbial activity. Reducing environments with oxidation reduction potencial (ORP) ≤ −100 mV are favorable for the microbial activity, but oxidizing environments with ORP > −100 mV inhibit the activity of sulfate-reducing bacteria (Zhang et al., 2016). Redox substances like nitrate, nitrite, and zero-valent iron affect oxidation–reduction potential. Zero-valent iron favors sulfate-reducing bacterial activity and reducing sulfate because of the decrease in oxidation–reduction potential in solution. On the other hand, the presence of nitrate enhances ORP, creating an oxidizing environment that is unfavorable for the reduction of sulfate by sulfate-reducing bacteria.

12.8.9 Types of Bioreactors

Continuous stirred tank reactors are reliable, consistent, and rapid in reaching equilibrium conditions but are not so efficient in biomass retention (Jamil and Clarke, 2013, 569). The anaerobic contact process demonstrates better retention of biomass compared to continuous stirred tank reactors, but pumping biomass in this reactor breaks down flocks and sludge. Anaerobic filter reactor configuration is unsuitable sometimes as it aids in building pressure gradients, but it offers certain advantages like low shear forces, longer sludge retention time, down-flow gravitational mode, etc. Up-flow anaerobic sludge blanket reactors have many advantages except biomass flushing out. It is possible to attain high treatment rates via this reactor without channeling of flow, compacting of sludge, or clogging. Fluidized-bed reactors are advantageous as they consist of a large surface area for the growth of sulfate-reducing bacteria, maintain small pressure gradients, and efficiently retain biomass. Moreover, influent concentrations are diluted due to recycle flow in the reactor. But this bioreactor provides less volume for biomass compared to an up-flow anaerobic sludge blanket reactor, and biomass is detached due to sheer force. Carrier fluidization in the reactor requires energy. At the industrial scale, two patented bioreactors have been used for bioremediation of acid mine waste utilizing sulfate-reducing bacteria such as BioSulphide R (BioteQ Environmental Technologies Inc., Canada) and Thiopaq R (Paques, The Netherlands) (Hussain et al., 2016, 5). Among the bioreactors, up-flow anaerobic sludge blanket and fluidized bed reactor systems are most commonly applied (Ahmat and Mamindy-Pajany, 2021).

12.8.10 Biotic Components

Microbial milieu of the ecosystem fostering AMD bioremediation is composed of sulfate-reducing bacteria, methanogens, acetogens, etc. where sulfate-reducing bacteria need help from other microbial communities to oxidize complex organic compounds and hydrolyze complex carbohydrates to simpler forms for consumption. If the sulfate concentration is higher, then sulfate-reducing bacteria win over other competitors for carbon sources. Microbial consortia are advantageous over pure cultures as they have a lower risk of contamination by other microorganisms and more adaptability to changes in environmental conditions (Virpiranta et al., 2019, 282). Moreover, in microbial milieus substrates are extensively utilized by the components of consortia compared to a single culture. The partial oxidation products of complex organic compounds can be taken up by other members of the consortium as carbon and energy sources. Diverse groups of microbes are able to degrade a wide range of organic compounds.

One arsenic-rich AMD in France was investigated for three long years to determine the effect of environmental conditions on the structure of the microbial community (Giloteaux et al., 2013, 724). At the most acidic pH of 1.4 and high iron and arsenic concentrations of 6,379 and 524 mg/L, respectively, *Desulfohalobiaceae*-related sequences were detected. At the source and downstream stations, the predominant strains were mostly similar to the family *Desulfobulbaceae*. Moreover, new dsrAB sequences with no isolated representatives were detected at pH near 3.4 with iron, sulfate, and arsenic concentrations of 854, 3,134, and 110 mg/L, respectively, at the downstream station COWG.

One abandoned Cu-Zn mine sediment treated with anaerobic bacteria at a pH range of 4–7.5 demonstrated that sulfate-reducing bacteria were active in a moderately acidic environment in the underground mine workings. At pH 4, preferential precipitation of Cu and Cd was observed over Zn. Phospholipid fatty acid and denaturing gradient gel electrophoresis analysis of the mine sediments coupled with sequence and phylogenetic analysis of 16S rDNA gene segments showed a moderately diverse microbial community structure composing iron and sulfate-reducing bacteria *Desulfosporosinus* and *Desulfitobacterium* (Church et al., 2007, 1).

In the bioreactor, sulfate-reducing bacteria also face competition and synergy with other bacterial species for electron donors. The usual competitors are acetobacteria for butyrate and propionate, whereas sulfate reducers compete with methanonic archaea for acetate and hydrogen. A high COD-to-SO_4^{2-} ratio favors the growth of methagens and acetogens, so in that environment, sulfate reducers are dominated by the said microbes, and sulfate-reducing bacteria have fewer electrons available to reduce sulfur. Although waste materials are advantageous to utilize in bioreactors, they are not fully hydrolyzed by monocultures. Mixed carbon and nitrogen sources are better exploited by rich microbial population. When substrates are complex in nature, better sulfate reduction efficiency is expected from sulfate reducers co-existing with other synergistic bacterial population.

12.9 POSSIBILITIES AHEAD

12.9.1 Mitigating Challenges Regarding Selection of Appropriate Carbon Sources

Undoubtedly, this may be attributed to being the hardest of all difficulties faced while applying the sulfate-reducing bacteria in the bioremediation of AMD and related processes. Any treatment process is burdened with the thought of associated costs, and this remediation technique is no exception. Moreover, the availability of the substrates to the consumer sulfate-reducing bacteria and the overall efficiency of the process are also taken into account while selecting the appropriate carbon sources. Wastewaters with a high concentration of organic compounds, or COD, such as sulfate-rich sewage, usually serve better as electron donors to sulfate reducers. However, AMD, which contains a low amount of dissolved organic carbon, cannot successfully support the growth of the said microbes. To ascertain complete reduction of sulfate using those sulfate-reducing bacteria by keeping the ratio of COD/SO_4^{2-} at 0.67 (basic value), it requires the addition of some essential electron donors during the treatment process. Incomplete degradation of the organic carbon sources leads to higher COD values in the effluent, which ultimately affects the performance of the bioremediation process and its application at the industrial level.

Low-molecular-weight organic compounds like lactate and ethanol are always the best choices as direct carbon sources, but their frequent use is not encouraged due to the cost factor. Low cost nonedible plant parts and agricultural residues such as leaf mulch, rice/maize straw, organic-rich soil, compost, oak leaves, oak chips, animal manure, peanut shells, silage, reed canary grass and its hydrolysate, algae and wood waste etc. have been exploited as carbon sources for sulfate-reducing bacteria in both laboratory scale and field-scale treatment of AMD.

12.9.2 Alleviating Heavy Metal Tolerance of Sulfate Reducers via Immobilization/Adsorption/Chelation

Toxicity of heavy metal to the sulfate-reducing bacteria poses an immense burden on the overall sulfate reduction and bioremediation efficiency of the sulfate-reducing bacteria. Several strategies have been adopted to alleviate the challenge, like immobilization of sulfate reducing bacteria and masking metal toxicity via additives, etc. (Zhang et al., 2018). Factors facilitating efficient sulfate removal may include providing a large surface area for bacterial growth, the presence of carriers for adsorbing sulfate and heavy metals, eliminating the toxicity of heavy metals toward sulfate reducers, maintaining homeostasis in the cell in a heavy metal environment to avoid cell death, etc. Suspended sulfate-reducing bacterial systems in continuous bioreactors face frequent environmental fluctuations and less cell retention. In addition to these, rich carbon sources generate high COD in the effluent. Immobilization of sulfate-reducing bacteria can overcome all these hurdles to a great extent, as gathered from different studies. Sulfate reduction efficiency was enhanced by 33%–40% after immobilization. Sulfate-reducing bacteria are immobilized in two ways: (1) without any internal carbon source in the beads, only sulfate-reducing bacteria are fixed via biofilm formation and (2) sulfate-reducing bacteria and the carbon source are fixed at the same time, mainly via gel entrapment. Immobilization without carbon sources generates high COD in the effluent, which can be eliminated by using carbon sources.

Good carrier molecules possess a hydrophobic surface, an iron reduction state, a suitable porosity, etc. The total surface area provided by the biofilm rather than the thickness and biomass is the deciding factor for bioreactor performance (Ting et al., 2019). Sand, biomass support particles, glass beads, etc. are useful because of their hydrophobicity and higher contact angles. Particles with enhanced adhesive properties such as macroporous supports and nano-scale materials also get adhered to the sulfate reducing bacteria due to their highly rough surface. Copper-iron bimetallic particles are more efficient than single iron particles (Dong et al., 2020). Zero-valent copper acts as an electronic shuttle to mitigate the barrier to electronic transfer exerted by insoluble film or iron sulfide deposits on the iron surface. Negative charges on the surface of recycled aggregate bio-carriers are also capable of adsorbing heavy metals and are quite suitable as carrier molecules. Polyvinyl alcohol, sodium alginate and microalgae are excellent in adsorbing sulfate and heavy metals. Sodium alginate is easily available, cost-effective, and strongly binds with metal ions. Microalgae with many functional groups on its surface like amino, hydroxyl, carboxyl and sulfate groups of the polysaccharides and proteins polymers represent appropriate binding substrate for heavy metals. Immobilization also facilitates longer microbial retention times and the recycling of beads.

Immobilized particles prepared via polyvinyl alcohol (PVA) and boric acid embedded crosslinking methods were utilized to treat AMD from coal mines containing concentrations of SO_4^{2-}, Cr^{6+}, Cr^{3+}, Mn^{2+}, Cu^{2+}, and Zn^{2+} at 816, 9.9, 20.1, 1.6, 0.8, and 1.2 mg/L, respectively (Wang et al., 2021, 1). Nano-zero-valent iron was used to aid the removal action of sulfate-reducing bacteria. A series of single factor tests and orthogonal tests were carried out to determine the optimal ratio of each matrix component, like sulfate-reducing bacteria dosage, nano-Fe0 dosage, corn cob particle size (immobilized particles), COD release, TFe release, pH effect, cob dosage, etc. Experimental results indicated that the best treatment was achieved with removal rates of SO_4^{2-}, Cr^{6+}, and Cr^{3+} at 82.99%, 99.78%, and 38.78%, respectively, in the presence of 4% nano-Fe0 particles, 30% sulfate-reducing bacteria, and 60 mesh corn cob particles with a dose of 3%. At the same time, TFe and COD release rates were 4.26 and 1,033.4 mg/L, respectively. The overall pH of the medium reached 8.04 from the initial value of 4.

Naturally occurring carboxylic acids like citric acid form various complexes with metals because of their multidentate chelating properties depending on the metal-to-acid ratio, pH of the medium, and type of metal present. In the presence of citric acid, bivalent heavy metal ions like Cu^{2+}, Ni^{2+}, Cd^{2+}, etc. form complexes with citrate 3^- ions, and uranium forms tridentate and binuclear complexes. When uranium and iron are both present, a ternary iron–uranium–citric acid complex is formed. Moreover, Ni^{2+}–citrate complex formation is preferred rather than Ni^{2+}–lactate complex formation and citrate eventually removes free Ni^{2+} via complex formation, alleviating Ni toxicity to the sulfate-reducing bacteria and promoting their growth (Xu and Chen, 2020). Citrate is utilized as a chelator to take up heavy metals from soil, sewage, etc. It is readily fermented by sulfate-reducing bacteria into actetate, CO_2, and hydrogen.

In another experiment, coal gangu was combined with sulfate-reducing bacteria to enhance the effectiveness of the bacteria for sulfate reduction (Dong et al., 2020, 1). Coal gangu contains various amounts of several oxides, such as SiO_2, Al_2O_3, Fe_2O_3, MgO, CaO, Na_2O, K_2O, P_2O_5, SO_3, CO_2, etc. Four dynamic columns were prepared using four different sources of coal gangu like (1) spontaneous combustion coal gangu from Gaode open-pit mine, (2) spontaneous combustion coal gangu from Haizhou open-pit mine, (3) original coal gangu from Haizhou open-pit mine, and (4) original coal gangu from a coal mine in Jincheng City. Results demonstrated that Column 3 performed best compared to others. It removed 84.41% COD along with 72.73% SO_4^{2-}, 98.70% Fe^{2+}, and 79.97% Mn^{2+}. The effluent oxidation reduction potential was −262.83 mV, whereas the pH value increased to an average of 9.09. AMD was fixed in the four columns prepared with coal gangu, as proved by injecting deionized water, which showed no secondary release at the end of the experiment. Coal gangu, being a convenient and cheap material with strong impact resistance and high treatment efficiency, can be utilized as an excellent alternative to a low-cost adsorbent for removing metals from mine wastewater.

12.9.3 Integrated Approach

AMD is an excellent source for valued resources like ferric hydroxide, barium sulfate, gypsum, sulfur or sulfuric acid, and some rare earth metals. Moreover, some nanoscale biogenic materials and minerals are available for their optical and electronic functionalities. Precipitation of heavy metals via biogenic sulfide production using sulfate-reducing bacteria is a highly profitable and attractive option, especially for the treatment of sulphate-rich wastewater, considering environmental sustainability. Sulfate-reducing bacteria have been utilized in the treatment of heavy metals and sulfate-laden wastewaters coming from the textile, pulp, paper, tanning, and mining industries for some time.

One integrated anerobic/aerobic system was designed to treat sulfate-rich wastewater from AMD via reduction of sulfate and oxidation of sulfide (Loreto-Muñoz et al., 2021, 1005). Anaerobic treatment of sludge was performed separately under sulfate-reducing conditions and aerobic sludge was activated in a glass column using sucrose as a carbon source. Integration of an anaerobic sulfate-reducing reactor and an aerobic sulfide-oxidizing reactor was completed after sludge activation. The integrated system was operated with a sulfate-rich synthetic solution with an increasing sulfate concentration SO_4^{2-} of 0.4–5.97 g/L.d, COD of 2–4 g/L.d, and a decreasing COD to SO_4^{2-} ratio of 5–0.67. During adaptation of anaerobic sludge under sulfate-reducing conditions, sulfate-reducing activity was completely replaced by methanogenic activity after 43 days at a COD-to-SO_4^{2-} ratio of 1 g/g. The COD, sulfate, and produced sulfide removal efficiency of this integrated system was calculated to be 90%, 60%, and 99%, respectively. At the end of the operation of this integrated system, 91% of the electron flow was directed toward sulfidogenic activity. This novel system indicated that a two-step anaerobic/aerobic biological system can be effective against sulfate-rich wastewater with a high organic load.

A coupled membrane-free microbial fuel cell with a permeable reactive barrier system was developed to utilize sulfate-reducing bacteria as catalysts to generate electricity directly from organic compounds in wastewater, sludge, etc. (Hai et al., 2016, 110). Inoculum was prepared using sewage sludge obtained from an anaerobic sludge digester at a municipal wastewater treatment plant. The system operated for five periods at a HRT of 48 hours and removed 99.5% of the initial concentrations of Cu^{2+}, Zn^{2+}, and Pb^{2+}. Average sulfate removal percentages of 51.2%, 39.8%, and 33.1% were achieved in effluents of 1,000, 2,000, and 3,000 mg/L, respectively. The coupled system was able to generate electricity from AMD continuously.

12.10 CONCLUDING REMARKS

AMD from mine tailings poses a severe environmental threat throughout the world and requires painstakingly rigorous and lengthy treatment processes to eliminate it. Biologically, they can be tackled in both aerobic and anaerobic ways using iron-oxidizing and sulfate-reducing bacterial communities. Methods in each case may vary depending on climate conditions, the nature and characteristics of the mine tailings, and the characteristics of microbial communities. Anaerobic bioremediation processes utilizing sulfate-reducing bacterial consortia or pure culture can be achieved by constructing wetlands or through bioreactors. Constructed wetlands are eco-friendly and cost-effective, with less use of hazardous chemicals, but they require large spaces and lack uniformity in performance and recovery of valuable elements. On the other hand, anaerobic bioreactors are more compact, easier to control, and advantageous in selectively recovering valuable metals and metalloids, but they are costly and require a steady supply of organic matter. Both methods need systems to increase pH upstream and suffer failure in the presence of mercury and a rapid influx of acidic, toxic, and aerobic waters.

The high acidity of AMD is detrimental to the growth of most of the sulfate-reducing bacteria surviving at pH 6–8, which can be overcome by seeding the inoculums of the bioreactor with some acidophilic or acid-tolerant sulfate-reducing bacteria at the early stage of remediation. So, careful design of the inoculums and seed cultures is essential for the bioreactor's performance. Development and management of microbial consortia is laborious as bioreactor conditions differ greatly depending on season, location, AMD load, etc. Sulfate-reducing bacteria, in combination with sulfidogenic bacteria, result in the recovery of valuable compounds from the bioreactors at the end. This method is cost-effective in the long run.

In some cases, remediation methods (Bruneel et al., 2019) take the help of plant species where plant-harboring microorganisms play dominant roles in scavenging and extracting heavy metals from the rhizospheric soil and above-ground plant biomass. Plant growth-promoting rhizobacteria and endophytic bacteria deserve particular attention in this regard. Inorganic or organic compounds secreted by the microbial community may acidify, chelate, or reduce the available metals or metalloids and metabolize them consequently. The majority of plant species grown around mining sites (80% of them) are colonized by arbuscular mycorrhizal fungi that are metal-tolerant and influence the availability and uptake of metal and metalloids by host plants through the production of siderophores, organic acids, and polymeric substances such as glycoprotein, glomaline, etc.

Current remediation strategies adopt additional modifications of the previous techniques to make the whole process more economical and eco-friendly, like incorporation of sulfide oxidizing bacteria to convert the polluting hydrogen sulfide gas into elemental sulfur for industrial application, coupling of microbial fuel cells during sulfate reduction via sulfate reducing bacteria for production of electricity from organic wastes, addition of substrates to enhance cell vitality and microbial activity and to alleviate metal tolerance of sulfate reducers, use of low-cost substrates like coal gangu as metal adsorbent during reduction of sulfate by bacterial corsortia, etc.

No uniform, standardized protocol is available for biological treatment of AMD. Every mine tailing has its own distinct composition and characteristics, like temperature, pH, metal and metalloid concentrations, etc. Microbial population of sulfate-reducing bacteria varies depending on pH, temperature, and other environmental factors, and their efficiency of sulfate reduction also differs. In majority of the cases, chemical, physical, and biological treatment methods are combined to achieve maximum remediation of chemical hazards from mine tailings and recovery of valuable chemicals from the waste.

12.11 ACKNOWLEDGEMENT

The authors are grateful to Dr. Sasabindu Jana, Principal, Raidighi College, for his constant encouragement and support.

REFERENCES

Ahmat, Adoum Mahamat and Yannick Mamindy-Pajany. 2021. "Over-sulfated soils and sediments treatment: A brief discussion on performance disparities of biological and non-biological methods throughout the literature." *Waste Management and Research*. 39(4):528–545. https://doi.org/10.1177/0734242X20982053

Akcil, Ata and Soner Koldas. 2006. "Acid Mine Drainage (AMD): Causes, treatment and case studies." *Journal of Cleaner Production*. 14(12-13): 1139–1145. https://doi.org/10.1016/j.jclepro.2004.09.006

Alvarenga, Paula, Guerreiro Nadia, Simões Isabel, Imaginario Maria Jose and Patricia Palma. 2021. "Assessment of the environmental impact of acid mine drainage on surface water, stream sediments, and macrophytes using a battery of chemical and ecotoxicological indicators." *Water*. 13(10): 1436. https://doi.org/10.3390/w13101436

Aoyogi, Tomo, Takaya Hamai, Tomoyuki Hori, Yuki Sato, Mikio Kobayashi, Yuya Sato, Tomohiro Inaba, Atsushi Ogata, Hiroshi Habe and Takeshi Sakata. 2017. "Hydraulic retention time andpH affect the performance and microbial communities of passive bioreactors for treatment of acid mine drainage". *AMB Express*. 7(142):2017. https://doi.org/10.1186/s13568-017-0440-z

Ayangbenro, Ayansina S and Olubukola O Babalola. 2017. "A new strategy for heavy metal polluted environments: A review of microbial biosorbents." *International Journal of Environmental Research and Public Health*. 14(1): 94.https://doi.org/10.3390/ijerph14010094

Ayangbenro, Ayansina S, Oluwaseyi S Olanrewaju and Olubukola O Babalola. 2018. "Sulfate-reducing bacteria as an effective tool for sustainable acid mine bioremediation." *Frontiers of Microbiology*. 9:1986. https://doi.org/10.3389/fmicb.2018.01986

Azam, Shahid and Qiren Li. 2010. "Tailings dam failures: A review of the last one hundred years." *Geotechnical News*. 50–53.

BBC. 2000. "Death of a river". https://news.bbc.co.uk/2/hi/europe/642880.stm

Broadle, Anthony. 2019. "Vale confirms tailings dam break at Feijao mine, echoing 2015 Samarco disaster". *The Australian Financial Review*. January 26.

Bruneel, Odile, N Mghazli, L Sbabou, M Héry, C Casiot and A Filali-Maltouf. 2019. "Role of microorganisms in rehabilitation of mining sites, focus on Sub Saharan African countries". *Journal of Geochemical Exploration*. 205:106327. https://doi.org/10.1016/j.gexplo.2019.06.009.hal-02615510

Canty, Marietta. 1998. "Overview of the sulfate-reducing bacteria - Demonstration project under the mine waste technology program". *Mineral Processing and Extractive Metallurgy Review*. 19(1):61–80. https://doi.org/10.1080/08827509608962429

Castro Neto, E S, A B S Aguiar, R P Rodriguez and G P Sancinetti. 2018. "Acid mine drainage treatment and metal removal based on a biological sulfate-reducing process." *Brazilian Journal of Chemical Engineering*. 35(02):543–552. https://doi.org/10.1590/0104-6632.20180352s20160615

Chen, Jingjing, Guoping Zhang, Chao Ma and Dongli Li. 2020. "Antimony removal from wastewater by sulfate-reducing bacteria in a bench-scale upflow anaerobic packed-bed reactor." *Acta Geochim*. 39(2):203–215. https://doi.org/10.1007/s11631-019-00382-6

Church, Clinton D, Richard T Wilkin, Charles N Alpers, Robert O Rye and R Blaine McCleskey. 2007. "Microbial sulfate reduction and metal attenuation in pH4 acid mine water". *Geochemical Transactions*. 8(10):1–14. https://doi.org/10.1186/1467-4866-8-10

Cunha, Mirabelle Perossi, Rafael Marcel Ferraz, Giselle Patricia Sancinetti and Renata Piacentini Rodriguez. 2019. "Long-term performance of a UASB reactor treating acid mine drainage: Effects of sulfate loading rate, hydraulic retention time and COD/SO_4^{2-} ratio." *Biodegradation*. 30(1):47–58. https://doi.org/10.1007/s10532-018-9863-8

Dai, Xiaohu, Chongliang Hu, Dong Zhang, Lingling Dai and Nina Duan. 2017. "Impact of a high ammonia-ammonium-pH system on methane-producing archaea and sulfate-reducing bacteria in mesophilic anaerobic digestion". *Bioresource Technology*. 245:598–605. https://doi.org/10.1016/j.biortech.2017.08.208

Deng, Q, X Wu, Y Wang and M Liu. 2018. "Activity characteristics of sulfate reducing bacteria and formation mechanism of sulphide." *Applied Ecology and Environmental Research*. 16(5):6369–6383. https://doi.org/10.15666/aeer/1605_63696383

Dev, Subhabrata, Shantonu Roy and Jayanta Bhattacharya. 2017. "Optimization of the operation of packed bed bioreactor to improve the sulfate and metal removal from acid mine drainage". *Journal of Environmental Management*. 200:135–144. https://doi.org/10.1016/j.jenvman.2017.04.102

Dold, Bernhard. 2014. "Evolution of acid mine drainage formation in sulphidic mine tailings." *Minerals*. 4(3): 621–641. https://doi.org/10.3390/min4030621

Dong, Yanrong, Junzhen Di, Xianjun Wang, Lindan Xue, Zhenhua Yang, Xuying Guo, and Mingwei Li. 2020. "Dynamic experimental study on treatment of acid mine drainage by bacteria supported in natural minerals." *Energies*. 13(2): 439. https://doi.org/10.3390/en13020439

Dong, Yanrong, Junzhen Di, Zhenhua Yang, Yuanling Zhang, Xianjun Wang, Xuying Guo, Zhennan Li, and Guoliang Jiang. 2020. "Study on the effectiveness of sulfate-reducing bacteria combined with coal gangue in repairing acid mine drainage containing Fe and Mn." *Energies* 13(4): 995. https://doi.org/10.3390/en13040995

Evangelou, V P. 1998. "Pyrite chemistry: The key for abatement of acid mine drainage." *Acidic Mining Lakes*. Environmental Science Series. Berlin, Heidelberg: Springer/Geller W, Klapper H, Salomons W. 197–222. https://doi.org/10.1007/978-3-642-71954-7_10

Fahruddin, Fahruddin, Asadi Abdullah, Nurhaedar and Nursiah La Nafie. 2020. "Estuary sediment treatment for reducing sulfate in acid mine water." *Environment and Natural Resources Journal*. 18(2):191–199. https://doi.org/10.32526/ennrj.18.2.2020.18

FengJuan, Liu, Zhang GuoPing, Liu ShiRong, Fu ZhiPing, Chen JingJing and Ma Chao. 2018. "Bioremoval of arsenic and antimony from wastewater by a mixed culture of sulfate-reducing bacteria using lactate and ethanol as carbon sources." *International Biodeterioration and Biodegradation*. 126:152–159. https://doi.org/10.1016/j.ibiod.2017.10.011

Giloteaux, Ludovic, Robert Dran, Corinne Casiot, Odile Bruneel, Francoise Elbaz-Poulichet and Marisol Goni-Urriza. 2013. "Three-year survey of sulfate-reducing bacteria community structure in Carnoulès acid mine drainage (France), highly contaminated by arsenic". *FEMS Microbiology and Ecology*. 83(2013):724–737. https://doi.org/10.1111/1574-6941.12028

Hai, Tang, Pu Wen-Cheng, Cai Chang-Feng, Xu Jian-Ping, and He Wen-Jun. 2016. "Remediation of acid mine drainage based on a novel coupled membrane-free microbial fuel cell with permeable reactive barrier system." *Polish Journal of Environmental Studies*. 25(1): 107–112. https://doi.org/10.15244/pjoes/60891.

Hao, Tianwei, Hamish R Mackey, Gang Guo, Rulong Liu and Guanghao Chen. 2016. "Resilience of sulfate-reducing granular sludge against temperature, pH, oxygen, nitrite and free nitrous acid." *Applied Microbiology and Biotechnology*. 100(19). 8563–8572. https://doi.org/10.1007/s00253-016-7652-z

Hardyanti, Nurandani, Sudarno Utomo, Angelica Oktaviana, Katrin Serafina and Junaidi Junaidi. 2018. "Effect of COD/SO_4^{2-} supply ratio variations of sulfate-reducing bacteria of sulphood raise in acid mine drainage." *E3S Web of Conferences*. 73:05009. ICENIS 2018. https://doi.org/10.1051/e3sconf/20187305009

Huang, Hao, Basanta Kumar Biswal, Guang-Hao Chen and Di Wu. 2020. "Sulfidogenic anaerobic digestion of sulfate-laden waste activated sludge: Evaluation on reactor performance and dynamics of microbial community." *Bioresource Technology.* 297(2020):122396. https://doi.org/10.1016/j.biortech.2019.122396

Hussain, Ali, Ali Hassan, Arshad Zavid and Zaved Iqbal Qazi. 2016. "Exploited application of sulfate-reducing bacteria for concominant treatment of metallic and non-metallic wastes: A mini review." *3 Biotech.* 6(119): 2016. https://doi.org/10.1007/s13205-016-0437-3.

Jamil, Illi, and William P. Clarke. 2013. "Bioremediation for acid mine drainage: Organic solid waste as carbon sources for sulfate reducing bacteria: A review." *Journal of Mechanical Engineering and Sciences.* 5:569–581. https://dx.doi.org/10.15282/jmes.5.2013.3.0054

Janyasuthiwong, Suthee, Eldon R Rene, Giovanni Esposito and Piet N L Lens. 2016. "Effect of pH on the performance of sulfate and thiosulfate-fed sulfate reducing inverse fluidized bed reactors". *Journal of Environmental Engineering.* 142 (9): https://doi.org/10.1061/(ASCE)EE.1943-7870.0001004

Johnson, D Barrie, and Kevin B. Hallberg. 2005. "Acid mine drainage remediation options: A review." *Science of the Total Environment.* 338(1–2): 3–14. https://doi.org/10.1016/j.scitotenv.2004.09.002

Johnson, D Barrie. 2003. "Chemical and microbiological characteristics of mineral spoils and drainage waters at abandoned coal and metal mines." *Water, Air, & Soil Pollution: Focus.* 3: 47–66. https://doi.org/10.1023/A:1022107520836

Kim, In Hwa, Jin-Ha Choi, Jeong Ock Joo, Young-Kee Kim, Jeong-Woo Choi, Byung-Keun Oh. 2015. "Development of recycled aggregate bio-carrier with sulfate reducing bacteria for the elimination of heavy metals from seawater." *Biotechnology and Bioprocess Engineering*. 25(9):1542–1546. https://dx.doi.org/10.4014/jmb.1504.04067

Kleinmann, R L P, R S Hedin, and R W Nairn. 1998. "Treatment of mine drainage by anoxic limestone drains and constructed wetlands." *Acidic Mining Lakes: Acid Mine Drainage, Limnology and Reclamation*. Berlin: Springer/Geller A, Klapper H, Salomons W. 303–19.

Krukenberg, Viola, Katie Harding, Michael Richter, Frank Oliver Glockner, Harald Gruber-Vodicka, Birgit Adam, Jasmine S Berg, Katrin Knittel, Halina Tegetmeyer, Antje Boetius and Gunter Wegener. 2016. "Candidatus *Desulfofervidus auxilii*, a hydrogenotrophic sulfate-reducing bacterium involved in the thermiphilic anaerobic oxidation of methane." *Environmental Microbiology*. 18(9):3073–3091: https://doi.org/10.1111/1462-2920.13283

Kushkevych, Ivan, Dani Dordevic and Monica Vitezova. 2019. "Toxicity of hydrogen sulphide toward sulfate-reducing bacteria *Desulfovibrio piger* Vib-7". *Archives of Microbiology*. 201(3):339–397. https://doi.org/10.1007/s00203-019-01625-z

Liang, Fangyuan, Yong Xiao and Feng Zhao. 2013. "Effect of pH on sulfate removal from wastewater using a bioelectrical system." *Chemical Engineering Journal*. 218(2013):147–153. https://doi.org/10.1016/j.cej.2012.12.021

Loreto-Muñoz, C D, F J Almendariz-Tapia, A R Martin-Garcia, R Sierra-Alvarez, V Ochoa-Herrera, and O Monge-Amaya. 2021. "Sulfate-rich wastewater treatment using an integrated anaerobic/aerobic biological system." *Revista Mexicana De Ingeniería Química.* 20(2), 1005–17. https://doi.org/10.24275/rmiq/IA2332

Lyu, Zongjie, Chai Junrui, Xu Zengguang, Yuan Qin, and Jing Cao. 2019. "A comprehensive review on reasons for tailings dam failures based on case history." *Advances in Civil Engineering.* 2019(9):1–18. https://doi.org/10.1155/2019/4159306

Moon, Chungman, Rajesh Singh, Sathyanarayan S. Veeravalli, Saravanan R. Shanmugam, Subba R. Chaganti, Jerald A. Lalman, and Daniel D. Heath. 2015. "Effect of COD:SO_4^{2-} ratio, hrt and linoleic acid concentration on mesophilic sulfate reduction: reactor performance and microbial population dynamics." *Water*. 7(5): 2275–2292. https://doi.org/10.3390/w7052275

Mulopo, Jean and L Schaefer. 2013. "Effect of the addition of zero valent iron (Fe0) on the batch biological sulfate reduction using grass cellulose as carbon source". *Applied Biochemistry and Biotechnology.* 171(8): 2020–2029. https://doi.org/10.1007/s12010-013-0500-z

Nancucheo, Ivan, and D Barrie Johnson. 2014. "Removal of sulfate from extremely acidic mine waters using low pH sulfidogenic bioreactors". *Hydrometallurgy.* 150:222–226. https://doi.org/10.1016/j.hydromet.2014.04.025

Nancucheo, Ivan, Jose A P Bitencourt, Prafulla K Sahoo, Joner Oliveira Alves, Jose O Siqueira and Guilherme Oliveira. 2017. "Recent developments for remediating acidic mine waters using sulfidogenic bacteria." *BioMed Research International.* 2017:7256582. https://doi.org/10.1155/2017/7256582

Nguyen, Hai Thi, Huong Lan Nguyen, Minh Hong Nguyen, Thao Kim Nu Nguyen and Hang Thuy Dinh. 2020. "Sulfate reduction for bioremediation of AMD facilitated by an indigenous acid and metal tolerant sulfate-reducer." *Journal of Microbiology and Biotechnology.* 30(7):1005–1012. https://doi.org/10.4014/jmb.2001.01012.

Nielsen, Guillaume, Ido Hatam, Karl A Abuan, Amelie Janin, Lucie Coudert, Jean Francois Blais, Guy Mercier and Susan A Baldwin. 2018. "Semi-passive in-situ pilot scale bioreactor successfully removed sulfate and metals from mine impacted water under subartic climate conditions." *Water Research.* 140:268–279. https://doi.org/10.1016/j.watres.2018.04.035

Nordstrom, Darrell Kirk, Charles N Alpers, Carol J Ptacek and David Blowes. 2000. "Negative pH and extremely acidic mine waters from Iron Mountain, California." *Environmental Science & Technology.* 34(2):254–258. https://doi.org/10.1021/es990646v

Nuclear-risks.org. 2009. "Olympic Dam, Australia." https://www.nuclear-risks.org/en/hibaku-sha-worldwide/olympic-dam.html

Richards, Jeremy P. 2009. *Mining, Society, and a Sustainable World.* Heidelberg: Springer.

Simate, Geoffrey S, and Sehliselo Ndlovu. 2014. "Acid mine drainage: Challenges and opportunities." *Journal of Environmental Chemical Engineering.* 2(3):1785–1803. https://doi.org/10.1016/j.jece.2014.07.021

Suyasa, W Budiarsa, Iryanti E Supprihatin, G A Dwi Adi Suastuti and G A Sri Kunti. 2019. "Deposition of heavy metals on sulfate reducing bacteria bioreactor treatment." *Nature Environment and Pollution Technology.* 18(2):395–402.

Ting, Zhang, Li Yun, Feng Xinhui and Zhou Xiaolong. 2019. "Preparation of macroporous magnetic polymer carrier and study on immobilized sulfate reducing bacteria". *Journal of Functional Materials.* 50(3): 3090–3095. https://doi.org/10.3969/j.issn.1001-9731.2019.03.015

Tomiyama, Shingo, and Toshifumi Igarashi. 2022. "The potential threat of mine drainage to groundwater resources." *Current Opinion in Environmental Science & Health.* 27:100347. https://doi.org/10.1016/j.coesh.2022.100347

United Nations Resolution A/RES/71/313, 2017. Resolution adopted by the General Assembly on 6 July 2017, Work of the Statistical Commission pertaining to the 2030 Agenda for Sustainable Development, 1–25

USGS. 2016. "The summit ville mine and its downstream effects." Last modified on Dec 2016. https://pubs.usgs.gov/of/1995/ofr-95-0023/summit.htm

Virpiranta, Hanna, Sanna Taskila, Tiina Leiviska, Jaakko Ramo and Juha Tanskanen. 2019. "Development of a process for microbial sulfate reduction in cold mining waters-cold acclimation of bacterial consortia from an Artic mining district." *Environmental Pollution.* 252(2019):281–288. https://doi.org/10.1016/j.envpol.2019.05.087

Wang, Xianjun, Junzhen Di, Bing Liang, Yu Yang, Yanrong Dong and Mingxin Wang. 2021. "Study on treatment of acid mine drainage by nano zero-valent iron synergistic with SRB immobilized particles." *Environmental Engineering Research.* 26(5):200333. https://doi.org/10.4491/eer.2020.333

Wang, Xianjun, Junzhen Di, Yanrong Dong, Yu Yang, Bing Liang, Fankang Meng, Tingting Wang, Wenbo An, Zengxin Li, and Jianzhi Guo. 2021. "The dynamic experiment on treating acid mine drainage with iron scrap and sulfate reducing bacteria using biomass materials as carbon as carbon source." *Journal of Renewable Materials*. 9(1):163–177. https://doi.org/10.32604/jrm.2021.011678

Western Mining Action Network. 2004. *Environmental Impacts of Mining*. WMAN. 2004. https://wman-info.org/resource/environmental-impacts-of-mining/#_edn4Fact Sheet. Updated December 28, 2004.

Wet-ting, Chen, Zhang Hong-guo, Luo Ding-gui and Chen Yong-heng. 2014. "Research on treating acid wastewater containing heavy metals by sulfate-reducing bacteria." *International Conference on Material and Environmental Engineering (ICMAEE 2014)*. Atlantis Press. 112–115.

Willis, Graciana, Ivan Nancucheo, Sabrina Hedrich, Alejandra Giaveno, Edgardo Donati and David Barrie Johnson. 2019. "Enrichment and isolation of acidtolerant sulfate-reducing microorganisms in the anoxic, acidic hot spring sediments from Copahue volcano, Argentina". *FEMS Microbiology Ecology*. 95:12. https://doi.org/10.1093/femsec/fiz175

Xu, Ya-Nan, and Yinguang Chen. 2020. "Advances in heavy metal removal by sulfate-reducing bacteria." *Water Science and Technology*. 81(9):1797–1827. https://doi.org/10.2166/wst.2020.227

Zhang, Daoyong, Jianlong Wang, Jiayue Zhao, Yayun Cai and Qinghua Lin. 2016. "Comparative study of nickel removal from synthetic wastewater by a sulfate reducing bacteria filter and a zero valent iron-sulfate reducing bacterial filter." *Geomicrobiology Journal*. 33(3–4):318–324. https://doi.org/10.1080/01490451.2015.1052116

Zhang, Hongguo, Huosheng Li, Meng Li, Dinggui Luo, Yong-Heng Chen, Diyun Chen, Hailing Luo, Zhenxin Chen and Keke Li. 2018."Immobilizing metal-resistant sulfatereducing bacteria for cadmium removal from aqueous solutions." *Polish Journal of Environmental Studies*. 27(6):2851–2859. https://doi.org/10.15244/pjoes/83666

Zhao, Jing, Di Fang, Penfei Zhang and Lixiang Zhou. 2017. "Long-term effects of increasing acidity on low-pH sulfate-reducing bioprocess and bacterial community." *Environmental Science and Pollution Research International*. 24(4):4067–4076. https://doi.org/10.1007/s11356-016-8147-2

13 Algae as a Bioremediation of Heavy Metal-Associated Water Pollution

Subhankar Dey
New Alipore College

Biplob Kumar Modak
Sidho Kanho Birsha University

13.1 INTRODUCTION

Water is the primary and renewable resource supporting the largest ecosystem on earth. Almost all types of living forms need water for survival, and nothing can substitute it. The surface, as well as groundwater resources, play a pivotal role in agriculture, livestock production, forestry, fisheries, industrial activities, hydropower generation, navigation, recreational activities, etc. However, in the last few decades, the pollution of this natural resource has drawn our attention. There are several natural and anthropogenic reasons known to us that negatively influence the quality of this life-saving natural resource. One of the major threats to water pollution is heavy metals (HMs) (Gupta, 2016). HMs are naturally occurring elements having metallic properties with an atomic number >20 or an atomic density of 3.5 g/cm^3 (Sharma et al., 2015). There are around 53 known HMs that are common constituents of soil and the earth's crust. In general, HMs are considered toxic materials because of their high reactivity, which interferes with several metabolic reactions in many biological systems. Some of the HMs react with biomolecules and develop ligands, whereas others interrupt the essential metabolic pathways by mimicking essential metals (Babula et al., 2008). HMs like mercury (Hg), arsenic (As), cadmium (Cd), and lead (Pb) are transported by runoff from industries and agricultural fields into various water sources that turn out to be very toxic and impose serious health problems on all living organisms. This HM contamination has been known to be a prime environmental concern for over 5,000 years due to its ill-effects on the environment. In the middle of the 19th century, the use of HMs was found to be at a peak level, thus accelerating the rate of adverse effects on living organisms (Gupta, 2016; Sharma et al., 2015).

Thus, under this global threat, the treatment of HMs contaminated water becomes very essential, and to overcome the problem, various scientific methods have been

DOI: 10.1201/9781003451457-13

introduced. Earlier, several chemical remediations such as precipitation, flotation, ion exchange, reverse osmosis, ultrafiltration, nanofiltration, adsorption, evaporation, cementation, coagulation, and flocculation were adopted. On the contrary, bioremediation is found to be one of the best techniques for the removal of toxic contaminants from wastewater. Bioremediation using living or dead algal biomass is known as phycoremediation (Suresh Kumar et al., 2015). Micro and macroalgae are a diverse group of aquatic photosynthetic organisms. They require very few and simple nutrients for their growth; thus, they are abundantly found around a wide range of natural water bodies. Along with the many other utilities (such as biofuel and biofertilizer), they also hold the great power of HM removal from contaminated water. With the increasing demand for sustainable development and the advancement of technologies, algae have gained more priority. Several artificial algae cultivation setups, such as a high-rated algal pond and photo-bioreactor, have been established to mitigate large quantities of wastewater (Lim et al., 2010). The recent development of genetically engineered microalgae is now becoming a special and advanced tool in the bioremediation of HM-associated water pollution (Rajamani et al., 2007).

13.2 HAZARDOUS EFFECTS OF HEAVY METALS

13.2.1 Effects of HMs on Human Health

Some HMs and metalloids such as Pb, Hg, Cd, As, selenium (Se), barium (Ba), copper (Cu), zinc (Zn), nickel (Ni), chromium (Cr), and silver (Ag) are well known for their catastrophic impact on human health (AzehEngwa et al., 2019). The uptake of HMs in the human body occurs via several routes. Pb is initially carried by red blood cells (RBC) to the kidney and liver, further redistributing the hairs, bones, and teeth as a phosphate salt (AzehEngwa et al., 2019). Mercury causes Minamata disease, which can be taken as the best example of HM-associated problems in human health. Similarly, Cd is also highly toxic and causes severe renal problems called "itai-itai disease" (Suresh Kumar et al., 2015). The regulatory limit for some selected HMs, their source, and their impact on human health are shown in Table 13.1. It has been well documented that HMs develop oxidative stress and cause genetic or cellular damage by generating free radicles, and the mechanism of free radicle formation is metal species-specific such as iron (Fe), Cu, and Cr, which generate hydroxyl radicals (OH), whereas Co and As generate superoxide (O2$^{\bullet-}$). The HMs also frequently interrupt the normal functioning of several cell signaling, cell cycle regulation, DNA methylation, and DNA repair proteins, thus causing carcinogenicity (Figure 13.1) (Kim et al., 2015).

13.2.2 Effects of HMs on Aquatic Life

HM pollution in the aquatic environment is primarily a consequence of anthropogenic activities and remains a major concern worldwide. Almost all the natural water bodies are contaminated with certain levels of HMs. Some of the HMs, such as copper (Cu), molybdenum (Mo), manganese (Mn), Fe, Ni, and Zn, are essential up to a limited concentration for aquatic organisms; however, other HMs and metalloids,

TABLE 13.1
The Permissible Range, Source, and Toxic Effects of HMs on Human Health

HMs	Allowable Values (mg/L) WHO	Allowable Values (mg/L) US EPA	Source	Hazardous Effects to Human Health
Pb	0.05	-	Battery, pipe, ceramic, glass production	• Excessive exposure in children causes impaired development and reduced intelligence • Short-term memory loss • Learning disabilities and coordination problems • Risk of cardiovascular disease
Se	0.04	0.05	Natural deposits, releases from copper smelting	• Dietary exposure of ~300 μg/day affects endocrine function • Impairment of natural killer cells activity • Hepatotoxicity and gastrointestinal disturbances • Hair and fingernail changes • Damage to the peripheral nervous system • Fatigue and irritability
Hg	0.01	0.05	Pharmaceutical, paper, pulp, ore, battery	• Loss of memory • Hair issues • Visual disturbance • Lung and kidney failure • Autoimmune diseases
Ba	0.7	2.0	Mineral deposits, disposal of drilling wastes, smelting of copper, motor vehicle parts manufacturing	• Cardiac arrhythmias • Respiratory failure • Gastrointestinal dysfunction • Elevated blood pressure
Cu	1.0	0.25	Fertilizer, paints, pigments, tannery	• Brain and kidney damage • Elevated levels result in liver cirrhosis • Chronic anemia • Stomach and intestinal irritation
Zn	3.0	1.0	Mining, steel fabrication, galvanization, stabilizers, coal combustion	• Gastrointestinal disorder • Neurological damage • Loss of appetite • Dizziness, nausea, and fatigue
Ni	0.015	0.2	Battery, mining, coinage, electroplating, glass, paints	• Allergic skin diseases • Cancer of the lungs, nose, sinuses, or throat through continuous inhalation • Chronic asthma • Immunotoxic, neurotoxic, genotoxic
Cd	0.003	0.005	Fertilizer, battery, power plants, mining, smelting, fuel combustion	• Carcinogenic impact • Disruption in endocrine regulation • Lung damage • Fagile bones • Affects calcium regulation in biological systems

(Continued)

TABLE 13.1 (*Continued*)
The Permissible Range, Source, and Toxic Effects of HMs on Human Health

HMs	Allowable Values (mg/L) WHO	US EPA	Source	Hazardous Effects to Human Health
Cr	0.05	0.05 Cr (VI) 0.01 Cr (III)	Synthetic dyes, steel production, textile, ceramics	• Lung cancer • Hemorrhage • Vomiting • Severe diarrhea • Hair loss
Ag	0.1	0.1	In soil as insoluble silver chloride (AgCl) and silver sulfide (Ag_2S) Majority from use as a bacteriostat in water treatment devices	• Exposure may cause skin and other body tissues to turn gray or blue-gray • Breathing problems • Lung and throat irritation and stomach pain
As	0.01	0.05	Glass, mining, textile, paper, insecticides, phosphate fertilizers, mining, coal combustion	• Affects essential cellular processes such as oxidative phosphorylation and ATP synthesis • Lung and kidney cancer • Liver tumor • Nausea

Source: Modified from Znad et al. (2022); Salama et al. (2019).
US EPA, United States Environmental Protection Agency; WHO, World Health Organization.

such as As, Pb, Cd, Hg, stannum (Sn), strontium (Sr), and titanium (Ti), are non-essential and toxic even in low concentrations (David and Cosio, 2021; Jais et al., 2017). Studies have revealed that the accumulation of HMs is organ-specific and varies among species; thus, some HMs show more bioaccumulation and toxicity in some species than others, such as muscles, which show bioaccumulation of almost all types of HMs, but show more sensitivity towards Cu than to Cd (David and Cosio, 2021; Al-Ghanim et al., 2015). Hyperplasia of the respiratory epithelium, fusion of primary and secondary gill lamella, cloudy swelling in hepatocytes, formation of melanomacrophage centers, cirrhosis, and necrosis of the liver (Poleksic et al., 2010), and inflammatory cell aggregation, atrophy, and coagulative necrosis in renal tubules (Alesci et al., 2022) are common changes that are observed in fish due to HM toxicity.

13.3 CONVENTIONAL VS. NOVEL APPROACHES IN HEAVY METAL REMEDIATION

The conventional approaches to HM removal from wastewater include various chemical procedures like precipitation (such as hydroxide, sulfide, and chelating precipitation), flotation, ion exchange, membrane filtration (such as reverse osmosis,

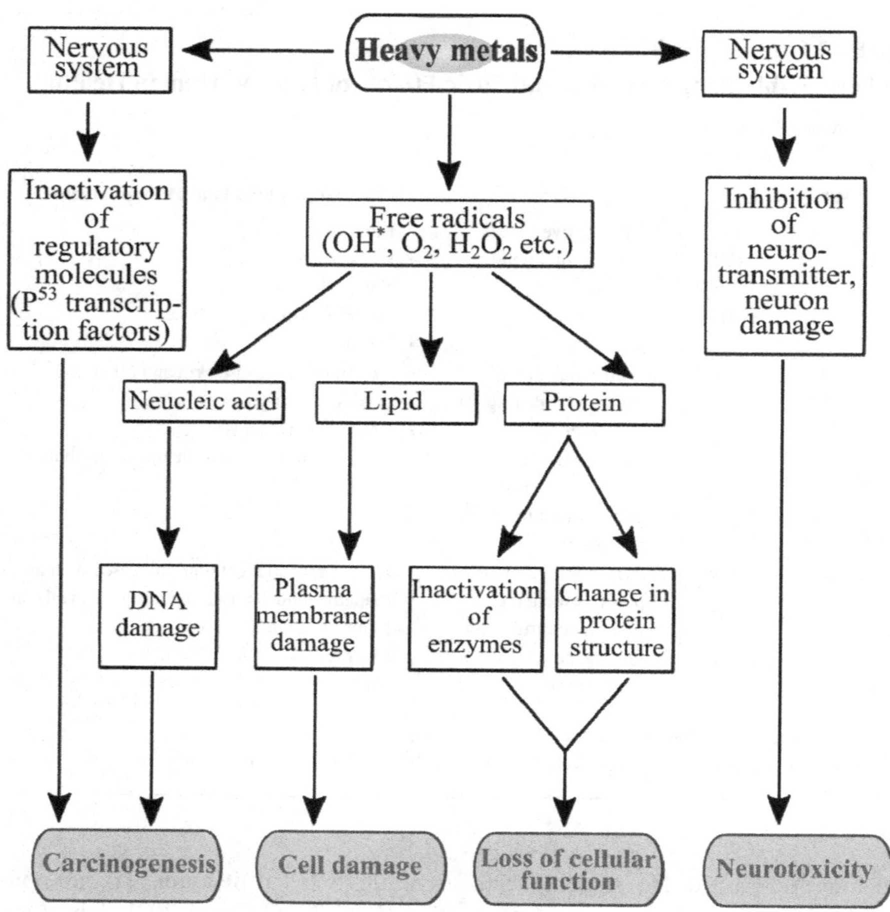

FIGURE 13.1 Impact of HMs on human.

ultrafiltration, and nanofiltration), adsorption, evaporation, cementation, coagulation, and flocculation. Various commonly used chemical remediation of HM are presented in Figure 13.2. Although these conventional procedures have been around for a long time, these technologies have lots of drawbacks (Table 13.2). Chelation is a popular method of remediation of HMs; chelating agents such as ethylenediaminetetraacetic acid are used for the precipitation of HMs in wastewater. However, the excess use of such chelating agents may have an adverse impact, and the removal of some metals (such as Ni) becomes difficult by this process (Singh, Renu, and Agarwal, 2017). Chemical adsorption of HMs by inorganic (such as bauxite red muds, ash, slag, water treatment sludge, clay, ores, and natural minerals) or organic (such as organic waste of plants or animals) adsorbents is another widely used and effective remediation, but this process requires high labor input and becomes expensive too (Suresh Kumar et al., 2015). Coagulation process involves the removal of HMs with the use of metal coagulating agents (such as magnesium chloride, polyaluminum chloride, aluminum

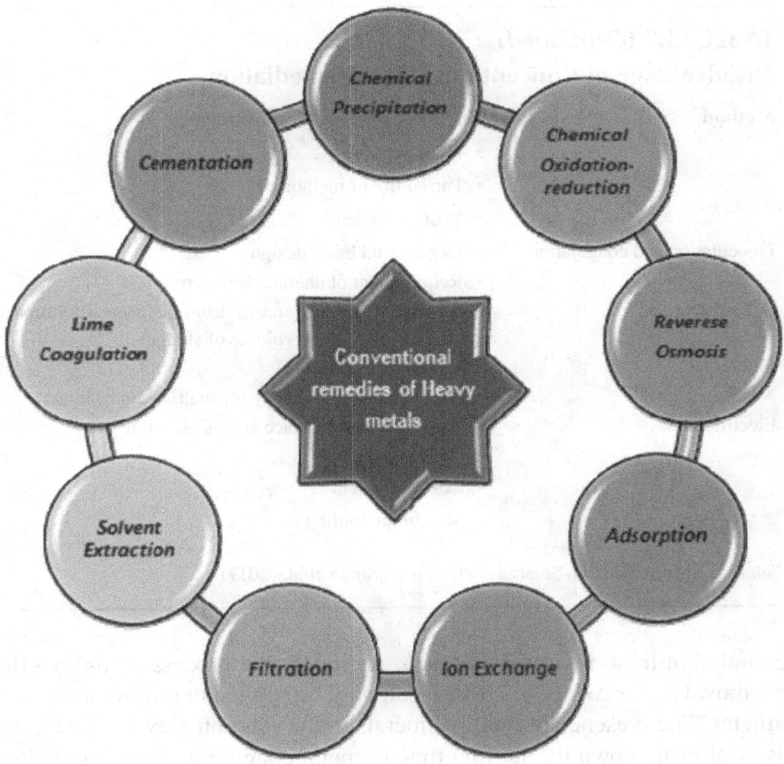

FIGURE 13.2 Conventional remedies of HM.

TABLE 13.2
Disadvantage of Conventional HMs Remediation

Method	Disadvantages
Chemical precipitation (e.g., Hydroxide precipitation, Chelating precipitation, etc.)	• Difficulty in separation • pH dependent process • Production of huge secondary waste • Formation of amphoteric metal hydroxides • Adverse influence by completing agent • High requirement of chemical reagents
Ion exchange	• High operational cost • Sensitive to particles • Prone to fouling of resin by precipitates and organics • Oxidation of resin by chemicals
Membrane	• Metallic fouling • No selectivity to alkaline metals

(*Continued*)

TABLE 13.2 (*Continued*)
Disadvantage of Conventional HMs Remediation

Method	Disadvantages
	• High pressure • Partial life of membrane • Cost-inefficient
Flocculation and coagulation	• Depends on basin design • Requirement of chemicals (electrolytes) • Inefficient for HM removal from multimetal solution • Production of large volume of sludge
Flotation	• Less selective for HMs
Nanofiltration	• High rejection efficiency for multivalent cations
Electrodialysis	• Large electrode surface area is essential • Time-consuming • Cost-inefficient • Membrane fouling

Source: Modified from Salama et al. (2019); Singh et al. (2017).

sulfate, and aluminum hydroxide oxides). Although this process is highly effective for the removal and recovery of HMs, it can only be applied to remove a single metal contaminant. The presence of multiple metals in the solution may hinder the overall process by slowing down the settling time of metal coagulum. The coagulation process is also highly pH-sensitive and produces a large volume of sludge. It has been observed that almost all of these conventional approaches are often ineffective or expensive, especially when the solution contains multiple metals in a range of 1–100 mg/L (Nourbakhsh et al., 1994; Singh et al., 2017).

In the recent era, the development of sustainable, cost-effective, eco-friendly, and recycling approaches has gained more attention from researchers to avoid the disadvantages of the conventional chemical methods of HM removal from wastewater or any other contaminated natural water bodies. The non-conventional methods are introduced not only for the removal of HMs but also to prevent HM pollution and their recovery for other essential purposes. Bioremediation widely includes two processes, *viz.*, biosorption and bioaccumulation. Biosorption is a metabolically independent process that passively aggregates HMs ions over the cell surface, whereas bioaccumulation is a metabolic pathway-dependent active transportation and aggregation of HMs in the cytosol. Bioremediation, which includes the use of several biological compounds that are derived from bacteria, plants, animals, fungi, or algae (such as crustacean-derived chitosan, plant-derived tannin, etc.), has gained huge popularity. However, in viewing the availability, cost of maintenance, recycling mode of use, resilience nature of growth in different types of wastewater, and promotion of decarbonization, algae have indeed gained priority from the scientific community, and this has become a subject of interest in the field of research and development (R&D) or research and technology development (R&TD) (Suresh Kumar et al., 2015; Pacheco et al., 2020). Like plants, algae highly utilize nitrate and phosphate for their normal cell division and growth, and certain HMs are also important for their normal functioning (such as iron for photosynthesis and chromium for metabolism). As a marine ecosystem is bestowed on these HMs, some marine microalgae have developed an efficient mechanism to aggregate HMs for their use. This unique property is like a natural gift for developing countries where the amount of HM effluent is increasing day by day due to the setting up of more industries. The bioremediation of HMs involving algae has several advantages over others. Some algae can metabolize and remove certain HMs even in the absence of oxygen. Some algae utilize oxygen; they can oxidize HMs and remove carbon dioxide in the aquatic system, which in turn helps in the sequestration of carbon along with phycoremediation. Microalgae help in the removal of obnoxious odors present in the HM effluent and convert them into a rich algal smell. Thus, algal remediation specifically has been considered a sustainable and eco-friendly way for the detoxification of wastewater and has turned out to be less costly and abundantly available than any other technology due to its virtually low investment in 'capital equipment' (Dwivedi, 2012).

13.4 ROLE OF MICROALGAE IN HEAVY METAL REMOVAL

The term phycoremediation can be defined as an application of algae in the treatment process of wastewater pollution (Jais et al., 2017), and now it has drawn the major attention of many researchers. Algae are microscopic, submicroscopic, or macroscopic photosynthetic organisms found in both fresh and marine water. Microalgae can molecularly discriminate between essential and nonessential elements; thus, they selectively absorb essential ones for their normal growth and physiology (Perales-Vela et al., 2006). The use of algae in the preparation of pharmaceutical, food, and cosmetic products has been a popular approach for a long time. They are also found to be good bioadsorbents due to their high biochemical contents (Altenor et al., 2012; Vijayaraghavan and Yun, 2008). Due to their high demand, they are now also cultivated artificially (Nazal, 2019). Marine algae are commonly called seaweeds, and they are classified mainly into three groups: Chlorophyta (green algae), Rhodophyta (red algae), and Phaeophyta (brown algae). Among them, Phaeophyta show the highest bioadsorption properties. The phaeophyta contains many bioactive compounds of different functional groups, such as amine, imidazole, carboxylic acid, hydroxyl, thioether, sulfhydryl phosphate, and phenolic, in their cell wall, which allow interaction and selective binding of different metals in the bioadsorption process (Nazal, 2019; Podgorskii et al., 2004). Microalgae have an edge over others because of their high specificity, highly efficient metal binding ability (they can bind up to 10% of biomass), and rapid growth rate. Other advantages include ease of handling, ease of year-round occurrence, no toxic waste generation, eco-friendliness, applicability to high- or low-contaminated water, and recycling properties (Monteiro et al., 2012). The living microalgae or the dead microalgal biomass are both extremely useful for the removal of HM contaminants from water. Microalgae require minimum attention or effort; they can grow at a low level of oxygen and nutrients, and some can also live in anaerobic conditions. Dead biomass of algae does not require any nutrients or oxygen, so it needs almost zero effort in maintenance (Monteiro et al., 2012; Rajamani et al., 2007;

Figueira et al., 1999). Various major modes of phycoremediation of HMs are presented in Figure 13.3. The biotechnological manipulation of microalgae not only provided more power for HM binding but also helped in the development of fluorescent HM biosensors. The tendency of metal binding to the algae depends on two factors: properties of the metal (e.g., atomic weight, reduction potential, or ionic size) and properties of the algae (e.g., surface area, functional groups present in the cell wall, production of metal-binding biomolecules) (Çetinkaya Dönmez et al., 1999; Suresh Kumar et al., 2015).

A compendium table of different algae along with their ability for HM removal has been presented in Table 13.3. Among many microalgae, *Chlorella vulgaris, C. miniata, C. salina, Chlamydomonas reinhardtii, Scenedesmus* spp., *Cyclotellacryptica, Chlorococcum* spp., *Phaeodactylum tricornutum, Porphyridium purpureum, Spirogyra* spp., *Stichococcus bacillaris*, and *Stigeoclonium tenue* are found to be most potent. However, some cyanobacteria (e.g., *Aphanothece halophytica, C. paris, Lyngbyaspiralis, Phormidiummolle, Spirulina, Stigonema* spp., and *Tolypothirxtenuis*) are also documented for their high metal adsorption characteristics, so they could be used for HMs removal in large-scale setups (Tüzün et al. 2005; Brinza, Dring, and Gavrilescu 2007). It has been observed that mostly the photosynthetic algae produce some active peptides that bind with HMs and form an organometallic complex, which is further ingested and internalized into the vacuoles to neutralize the deleterious impact of HMs (Perales-Vela et al., 2006; Cobbett and Goldsbrough, 2002).

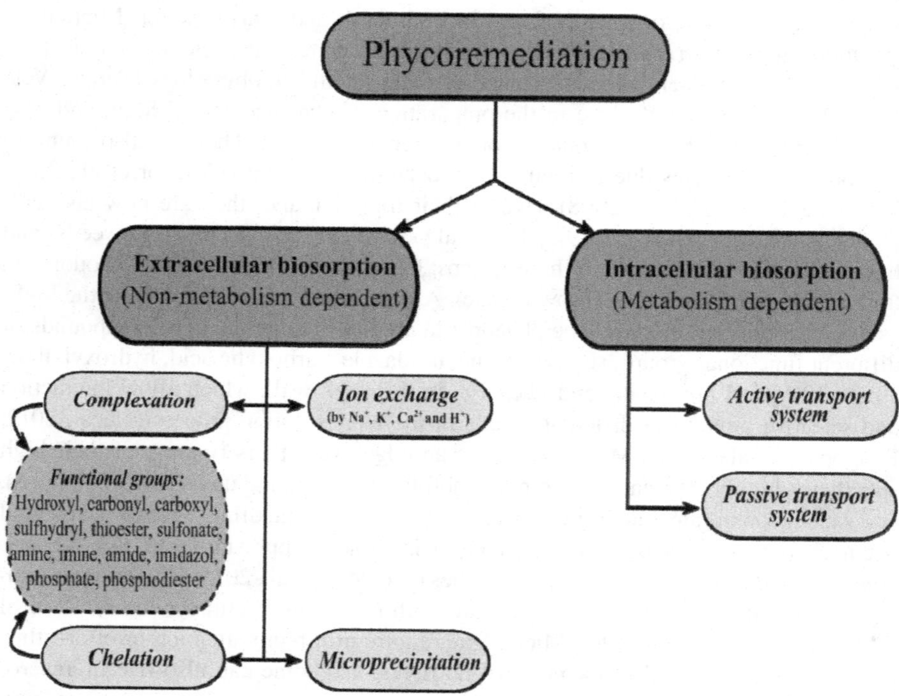

FIGURE 13.3 Strategies of phycoremediation in HM removal.

TABLE 13.3
List of Algae with HM Removal Efficiency

Metal	Speciation	Algae Species	pH	Type of Algae Biomass	Metal Uptake (mg/g)
Cd	Cd^{2+}	AER *Chlorella*	3–7	Nonliving	7.74
		AER *Porphyridium*	3–7	Nonliving	7.55
		AER *Spirulina*	3–7	Nonliving	7.28
		Aulosira fertilissima	5	Nonliving	14.57
		Calothrix parietina TISTR 8093	7	Nonliving	79
		Chaetoceros calcitrans	8	Live	1055.27
		Chlamydomonas reinhardtii	5.5	Cell wall	5.75
		Chlamydomonas reinhardtii	5.5	Without cell wall	3
		Chlamydomonas reinhardtii	6	Ca-alginate	28.9
		Chlamydomonas reinhardtii	6	Immobilized	79.7
		Chlamydomonas reinhardtii	6	Nonliving	42.6
		Chlamydomonas reinhardtii			2.3
		Chlorella homosphaera			8.4
		Chlorella pyrenoidosa	6.8–7.0		2.8
		Chlorella sorokiniana	5	Nonliving	33.5
		Chlorella sorokiniana	5	Immobilized	192
		Chlorella sp. HA-1			21.6
		Chlorella vulgaris	4		86.6
		Chlorella vulgaris	4	Nonliving	62.3
		Chlorella vulgaris	4	Nonliving	33.72
		Chlorella vulgaris	6	Nonliving	12.45
		Chlorella vulgaris	3–7	Nonliving	8.41
		Chlorella vulgaris	6.8–7.0		2.6
		Chlorella vulgaris BCC 15	7	Nonliving	76
		Chlorella vulgaris CCAP211/11B	7	Nonliving	62
		Coelastrum sp.	7.5	Nonliving	32.8
		Cyclotella cryptica	6	Nonliving	22.24
		Desmodesmus pleiomorphus		Nonliving	58.6
		Desmodesmus pleiomorphus (ACOI 561)	4	Live	85.3
		Desmodesmus pleiomorphus (L)	4	Live	61.2
		Hydrodictyon reticulatum	5	Nonliving	7.2
		Isochrysis galbana		Live	0.02
		Parachlorella sp.	7	Live	96.2
		Phaeodactylum tricornutum	6	Nonliving	1.24

(*Continued*)

TABLE 13.3 (*Continued*)
List of Algae with HM Removal Efficiency

Metal	Speciation	Algae Species	pH	Type of Algae Biomass	Metal Uptake (mg/g)
		Phormidium spp.	5	Nonliving	9.6
		Pithophora odeogonia	5	Nonliving	13.07
		Planothidium lanceolatum	7	Live	275.51
		Porphyridium cruentum	3–7	Nonliving	8.84
		Porphyridium purpureum	6	Nonliving	0.42
		Pseudochlorococcum typicum	7	Live	5.48
		Scenedesmus abundans	7.8–8	Live	574
		Scenedesmus abundans			0.64
		Scenedesmus acutus IFRPD 1020	7	Nonliving	110
		Scenedesmus obliquus			11.4
		Scenedesmus obliquus		Nonliving	60.8
		Scenedesmus obliquus CNW-N	6	Nonliving	24.4–108.5
		Scenedesmus subspicatus	6	Nonliving	7.29
		Spirogyra hyalina		Nonliving	18.18
		Spirogyra insignis	6	Nonliving	22.9
		Spirogyra neglecta	5	Nonliving	27.95
		Spirulina platensis	6	Nonliving	12.08
		Spirulina platensis	3–7	Nonliving	8.06
		Spirulina platensis	7–8	Nonliving	357
		Spirulina platensis		Live	44.56
		Spirulina platensis		Immobilized	47.89
		Spirulina platensis TISTR 8217	6	Alginate immobilized	70.92
		Spirulina platensis TISTR 8217	7	Nonliving	98.04
		Spirulina platensis TISTR 8217	4–7	Silica-immobilized	36.63
		Spirulina spp.	7.5	Nonliving	0.46
		Spirulina spp.			0.463
		Synechocystis sp.		Live	199.83
		Tetraselmis chuii	8	Live	13.46
		Tetraselmis chuii		Live	292.6
		Tetraselmis chuii		Nonliving	210.54
		Tolypothrix tenuis TISRT 8063	7	Nonliving	90

(*Continued*)

TABLE 13.3 (*Continued*)
List of Algae with HM Removal Efficiency

Metal	Speciation	Algae Species	pH	Type of Algae Biomass	Metal Uptake (mg/g)
Co	Co^{2+}	*Chlamydomonas reinhardtii*	5.5	Cell wall	0.89
		Chlamydomonas reinhardtii	5.5	Without cell wall	1.3
		Oscillatoria angustissima	4	Nonliving	15.32
		Spirogyra hyalina		Nonliving	12.82
		Spirulina spp.	7.5	Nonliving	0.01
		Chlorella miniata	3	Nonliving	14.17
		Chlorella miniata	4	Nonliving	28.72
Cr	Cr^{3+}	*Chlorella miniata*	4.5	Nonliving	41.12
		Chlorella spp.		Nonliving	98
		Chlorella spp.			9.62
		Spirulina sp.		Live	304
		Spirulina spp.		Nonliving	167
		Spirulina sp.(HD-104)		Live	306
	Cr^{6+}	*Chlamydomonas angulosa*	8.2	Nonliving	5.32
		Chlorella minutissima	2	Immobilized	57.33
		Chlamydomonas reinhardtii	2	Nonliving native	18.2
		Chlamydomonas reinhardtii	2	Heat-treated	25.6
		Chlamydomonas reinhardtii	2	Acid-treated	21.2
		Chlorella vulgaris	2	Nonliving	23
		Chlorella spp.			9.62
		Chlorella vulgaris	4	Nonliving	23.6
		Dunaliella sp. 1	2	Nonliving	58.3
		Dunaliella sp. 2	2	Nonliving	45.5
		Nostoc muscorum	3	Nonliving	22.92
		Oscillatoria nigra	8.2	Nonliving	1.86
		Oscillatoria tenuis	8.2	Nonliving	7.35
		Phormidium bohneri	8.2	Nonliving	8.55
		Scenedesmus obliquus	2	Nonliving	15.6
		Spirulina sp.		Live	333
		Spirulina spp.		Nonliving	143
		Spirulina platensis	2	Immobilized	49
		Synechocystis spp.	2	Nonliving	19.2
		Ulothrix tenuissima	8.2	Nonliving	4.56
	$Cr_2O_7^{-2}$	*Chlorella* spp.		Nonliving	104
		Spirulina sp.(HD-104)		Live	226
Cu	Cu^{2+}	*Anabaena cylindrica*	4.0–5.0	Live	12.62
		Anabaena spiroides	4.0–5.0	Live	8.73
		Asterionella formosa	4.0–5.0	Live	1.1

(*Continued*)

TABLE 13.3 (*Continued*)
List of Algae with HM Removal Efficiency

Metal	Speciation	Algae Species	pH	Type of Algae Biomass	Metal Uptake (mg/g)
		Asterionella formosa	4.0–5.0	Nonliving	0.53
		Aulacoseira varians	4.0–5.0	Live	2.29
		Aulacoseira varians	4.0–5.0	Nonliving	3.03
		Aulosira fertilissima	5	Nonliving	21.77
		Ceratium hirundinella	4.0–5.0	Live	2.3
		Ceratium hirundinella	4.0–5.0	Nonliving	5.75
		Chlamydomonas reinhardtii	5.5	Cells with Cell wall	6.42
		Chlamydomonas reinhardtii	5.5	Cells without cell wall	7.54
		Chlorella fusca	6	Live	3.2
		Chlorella miniata	6	Nonliving	23.26
		Chlorella pyrenoidosa	7	Live	2.4
		Chlorella spp.	7	Immobilized	33.4
		Chlorella spp.		Live	220
		Chlorella spp.		Nonliving	108
		Chlorella vulgaris	2	Nonliving	16.14
		Chlorella vulgaris	4	Nonliving	37.6
		Chlorella vulgaris	4	Nonliving	34.89
		Chlorella vulgaris	4.5	Nonliving	40
		Chlorella vulgaris	4.5	Nonliving	48.17
		Chlorella vulgaris	4.5	Free	76.71
		Chlorella vulgaris	4.5	Immobilized	63.08
		Chlorella vulgaris	5	Nonliving	1.8
		Chlorella vulgaris	6	Nonliving	10.9
		Chlorella vulgaris	6		7.5
		Chlorella vulgaris	6	Nonliving	18.72
		Chlorella vulgaris	4.0–5.0	Live	3.63
		Chlorella vulgaris	4.0–5.0	Nonliving	4.26
		Closterium lunula	7	Live	0.5
		Cyclotella cryptica	6	Nonliving	26.28
		Eudorina elegans	4.0–5.0	Live	3.96
		Eudorina elegans	4.0–5.0	Nonliving	2.13
		Hydrodictyon reticulatum	5	Nonliving	8.72
		Isochrysis galbana		Live	0.11
		Microcystis aeruginosa	4.0–5.0	Live	8.21
		Microcystis aeruginosa	4.0–5.0	Nonliving	2.47

(*Continued*)

TABLE 13.3 (*Continued*)
List of Algae with HM Removal Efficiency

Metal	Speciation	Algae Species	pH	Type of Algae Biomass	Metal Uptake (mg/g)
		Microcystis spp.	9.2	Nonliving	0.003
		Phaeodactylum tricornutum	6	Nonliving	1.67
		Phormidium spp.	5	Nonliving	10.1
		Pithophora odeogonia	5	Nonliving	23.08
		Planothidium lanceolatum	7	Live	134.32
		Porphyridium purpureum	6	Nonliving	0.27
		Scenedesmus obliquus	4.5	Nonliving	20
		Scenedesmus obliquus	7	Live	1.8
		Scenedesmus subspicatus	6	Nonliving	13.28
		Scenedesmus quadricauda	4	Nonliving	2.8
		Spirogyra insignis	4	Nonliving	19.3
		Spirogyra neglecta	5	Nonliving	40.83
		Spirulina sp.		Live	389
		Spirulina platensis	6	Nonliving	10.33
		Spirulina platensis	6		10
		Spirulina platensis	9	Live	0.85
		Spirulina spp.		Nonliving	100
		Spirulina spp.			0.271
		Spirulina sp.(HD-104)		Live	576
		Synechocystis spp.	4.5	Nonliving	23.4
Fe	Fe^{3+}	*Chlorella vulgaris*	2	Nonliving	24.52
		Microcystis sp.	9.2	Nonliving	0.03
Hg	Hg^{2+}	*Calothrix parietina* TISTR 8093	7	Nonliving	19
		Chlamydomonas reinhardtii	6	Ca-alginate immobilized	35.9
		Chlamydomonas reinhardtii	6	Ca-alginate immobilized	106.6
		Chlamydomonas reinhardtii	6	Nonliving	72.2
		Chlorella vulgaris BCC 15	7	Nonliving	18
		Chlorella vulgaris CCAP211/11B	7	Nonliving	16
		Cyclotella cryptica	4	Nonliving	11.92
		Phaeodactylum tricornutum	4	Nonliving	0.51
		Porphyridium purpureum	4	Nonliving	0.51

(*Continued*)

TABLE 13.3 (*Continued*)
List of Algae with HM Removal Efficiency

Metal	Speciation	Algae Species	pH	Type of Algae Biomass	Metal Uptake (mg/g)
		Pseudochlorococcum typicum	7	Live	15.13
		Scenedesmusacutus IFRPD 1020	7	Nonliving	20
		Scenedesmus subspicatus	4	Nonliving	9.2
		Spirogyra hyalina		Nonliving	35.71
		Spirulina spp.	7.5	Nonliving	1.34
		Tolypothrix tenuis TISRT 8063	7	Nonliving	27
Ni	Ni^{2+}	*Arthrospira (Spirulina) platensis*	5.0–5.5	Nonliving	20.78
		Aulosira fertilissima	5	Nonliving	4.16
		Chlamydomonas reinhardtii	5.5	Cell wall	0.4
		Chlamydomonas reinhardtii	5.5	Without cell wall	0.63
		Chlorella miniata	6	Nonliving	20.37
		Chlorella miniata	7.4	Live	1.37
		Chlorella spp.		Live	122
		Chlorella spp.		Nonliving	183
		Chlorella vulgaris	4.5		58.4
		Chlorella vulgaris	4.5	Free	59.29
		Chlorella vulgaris	4.5	Immobilized	111.41
		Chlorella vulgaris	4.7	Nonliving	24.06
		Chlorella vulgaris	5	Live	15.4
		Chlorella vulgaris	5	Nonliving	15.6
		Chlorella vulgaris	5	Immobilized	28.6
		Chlorella vulgaris	5	Nonliving	42.3
		Chlorella vulgaris	6	Nonliving	12.06
		Chlorella vulgaris	7.4	Live	0.64
		Chlorella vulgaris	5.0–5.5	Nonliving	29.29
		Hydrodictyon reticulatum	5	Nonliving	13.86
		Phormidium spp.	5	Nonliving	5.7
		Pithophora odeogonia	5	Nonliving	11.81
		Scenedesmus obliquus	5	Nonliving	18.7
		Spirogyra insignis	6	Nonliving	17.5
		Spirogyra neglecta	5	Nonliving	26.3
		Spirulina sp.		Live	1378
		Spirulina spp.	7.5	Nonliving	0.19
		Spirulina spp.		Nonliving	515

(*Continued*)

TABLE 13.3 (*Continued*)
List of Algae with HM Removal Efficiency

Metal	Speciation	Algae Species	pH	Type of Algae Biomass	Metal Uptake (mg/g)
		Spirulina sp. (HD-104)		Live	1108
		Synechocystis spp.	5	Nonliving	15.8
Pb	Pb^{2+}	*Anabaena flosaquae*			70
		Arthrospira (Spirulina) platensis	5–5.5	Nonliving	102.56
		Aulosira fertilissima	5	Nonliving	31.12
		Calothrix parietina TISTR 8093	7	Nonliving	45
		Chlamydomonas reinhardtii	5	Nonliving	96.3
		Chlamydomonas reinhardtii	6	Ca-alginate immobilized	230.5
		Chlamydomonas reinhardtii	6	Ca-alginate immobilized	380.7
		Chlorella vulgaris	4	Nonliving	97.38
		Chlorella vulgaris	4	Nonliving	90
		Chlorella vulgaris	6	Nonliving	17.13
		Chlorella vulgaris	5–5.5	Nonliving	131.36
		Chlorella vulgaris BCC 15	7	Nonliving	127
		Chlorella vulgaris CCAP211/11B	7	Nonliving	39
		Cyclotellacryptica	6	Nonliving	36.68
		Hydrodictyon reticulatum	5	Nonliving	24
		Microcystis novacekii	5	Nonliving	80
		Oscillatoria laetevirens	5	Live	21.6
		Phaeodactylum tricornutum	6	Nonliving	1.49
		Phormidium spp.	5	Nonliving	13.6
		Pithophora odeogonia	5	Nonliving	71.13
		Porphyridium purpureum	6	Nonliving	0.32
		Pseudochlorococcum typicum	7	Live	4.49
		Scenedesmus acutus IFRPD 1020	7	Nonliving	90
		Scenedesmus subspicatus	6	Nonliving	38.71
		Spirogyra hyalina		Nonliving	31.25
		Spirogyra insignis	5	Nonliving	51.5
		Spirogyra neglecta	5	Nonliving	90.19
		Spirogyra spp.	5	Nonliving	140.84
		Spirulina (Arthrospira) platensis	7	Live	188

(*Continued*)

TABLE 13.3 (*Continued*)
List of Algae with HM Removal Efficiency

Metal	Speciation	Algae Species	pH	Type of Algae Biomass	Metal Uptake (mg/g)
		Spirulina maxima	5.5	Intact biomass	32
		Spirulina maxima	5.5	Pretreated biomass	42
		Spirulina platensis	6	Nonliving	16.97
		Stigeoclonium tenue	6.8	Nonliving	0.86
		Stigeoclonium tenue	8.2	Nonliving	0.38
		Synechocystis spp.		Live	155.63
		Tolypothrix tenuis TISRT 8063	7	Nonliving	31
Zn	Zn^{2+}	*Arthrospira (Spirulina) platensis*	5.0–5.5	Nonliving	33.21
		Aulosira fertilissima	5	Nonliving	19.15
		Chlorella homosphaera			15.6
		Chlorella spp.	7	Immobilized	28.5
		Chlorella vulgaris	4	Nonliving	24.19
		Chlorella vulgaris	4	Nonliving	24.5
		Chlorella vulgaris	6	Nonliving	6.42
		Chlorella vulgaris	5.0–5.5	Nonliving	43.41
		Cyclotella cryptica	6	Nonliving	242.9
		Desmodesmus pleiomorphus	5	Nonliving	360.2
		Euglena gracilis			7.5
		Hydrodictyon reticulatum	5	Nonliving	3.7
		Isochrysis galbana		Live	0.3
		Phaeodactylum tricornutum	6	Nonliving	14.52
		Phormidium spp.	5	Nonliving	9.4
		Pithophora odeogonia	5	Nonliving	8.98
		Planothidium lanceolatum	7	Live	118.66
		Porphyridium purpureum	6	Nonliving	2.01
		Scenedesmus obliquus		Nonliving	22.3
		Scenedesmus obliquus		Nonliving	6.67
		Scenedesmus quadricauda		Nonliving	5.03
		Scenedesmus subspicatus	6	Live	72.06
		Spirogyra insignis	6	Nonliving	21.1
		Spirogyra neglecta	5	Nonliving	31.51
		Spirulina platensis	6	Nonliving	7.36
		Spirulina spp.	7.5	Nonliving	0.17

(*Continued*)

TABLE 13.3 (*Continued*)
List of Algae with HM Removal Efficiency

Metal	Speciation	Algae Species	pH	Type of Algae Biomass	Metal Uptake (mg/g)
		Stigeoclonium tenue	6.8	Nonliving	0.88
		Stigeoclonium tenue	8.2	Nonliving	0.77
As	As	*Chlamydomonas reinhardtii*	9.5	Live	4.63
		Chlorella vulgaris	5.5	Live	3.89
		Maugeotia genuflexa	6	Nonliving	2.4
		Scenedesmus almeriensis	9.5	Live	5.0
		Ulothrix cylindricum	6	Live	2.45

Source: Modified from Suresh Kumar et al. (2015); Leong and Chang (2020).

Algae need some HMs (such as Mn^{2+}, Ni^{2+}, Cu^{2+}, Mo^{2+}, Fe^{2+}, and Zn^{2+}) in trace quantities for their normal growth and development, and some HMs are not required at all as these are highly toxic to them. However, it has been observed that some algae have an exceptional power of HM tolerance; these algae can not only survive in HM-polluted water, but they can also entrap and neutralize HMs within their bodies (Kotrba 2011). The metal removal capacity and mechanism vary amongst different species of algae; they may be metabolic cycle-dependent or independent. There is controversy regarding the efficiency of HM removal by living algae and dead algal biomass. Many have suggested the higher efficiency of living algal cells than that of dead biomass during wastewater treatment. Experiment-based studies have demonstrated the removal of 91% and 98% of Cu^{2+} and Ca^{2+}, respectively from municipality wastewater by *Spirulina* sp. (Al-Homaidan et al. 2015; Anastopoulos and Kyzas 2015); *Chlorella minutissima* showed 62%, 84%, 74%, and 84% removal efficiency against Zn^{2+}, Mn^{2+}, Cd^{2+}, and Cu^{2+}, respectively (Yang et al., 2015); 99% and 85% removal of Cu^{2+} and Zn^{2+}, respectively, by *Cladophorafracta* (Mahdavi, Ulrich, and Liu 2012); 46%, 34%, 48%, and 50% removal of Cu^{2+}, Ni^{2+}, Zn^{2+}, and Co^{2+}, respectively by *Oedogonium* sp. (Bakatula et al., 2014); 100% removal of arsenic (As) by filamentous chlorophyta *Cladophora* sp. (Jasrotia et al., 2014). However, several contrasting studies have pointed out the better HM removal efficiency of dried biomass algae. Matsunaga et al. (1999) evaluated the role of 191 marine microalgae in the removal of HMs from contaminated marine water, and they found *Chlorella* spp. NKG16014 strain to be the most potent agent for Cd removal (48.7%). Further analysis has revealed that this species accumulated 67% Cd intracellularly and 25% Cd was adsorbed on the cell surface, suggesting that HM removal magnitude is higher in the case of dried algal cells than living cells. Thus, algal biomass could be a good substitution for conventional adsorbent materials. A comparative discussion of biosorption and bioaccumulation has been presented in Table 13.4.

TABLE 13.4
Biosorption vs. Bioaccumulation of Metal by Algae

Characteristics	Biosorption	Bioaccumulation
Algal involvement	Living or nonliving algal biomass	Living algae
Mode of action	Passive and metabolism independent	Mostly active and metabolism dependent
Removal rate	Most mechanisms take place at a fast rate	Slower rate than biosorption
Selectivity	Poor, can be increased by modification/biomass transformation	Better than biosorption
Versatility	Reasonably good. The binding sites can accommodate a variety of ions	Not very flexible. Prone to be affected by high metal/salt condition
Metal recovery	HM recovery is possible with adequate eluent	Even if possible, biomass cannot be used for other purposes
pH	Strongly affects the sorption capacity of HMs; however, the process can occur within a wide pH range	Significant pH change can strongly affect living cells
Regeneration and reuse	Biosorbents can be regenerated and reused in many cycles	Partial reuse because of intercellular accumulation
Energy required	Usually low	Energy needed for cell growth
Cost	Usually low; biomass can be obtained from industrial waste, and cost is mostly associated with transportation and production of biosorbent	Process occurs in the presence of living cells that have to be sustained

Source: Modified from Salama et al. (2019); Diaconu et al. (2013).

13.4.1 Role of Living Algae

Living algae are highly potent and can remove toxic HMs from the aquatic environment. They can ingest HMs in higher concentrations, even greater than the concentration present in the surrounding water (Suresh Kumar et al., 2015). Algae use different metabolic processes to ingest HMs, and among these biosorption is one of the complex metabolic activities that helps to remove a significant amount of HMs (Ajayan et al., 2011). However, many authors have emphasized that bioaccumulation and this process comprise two sequential steps: an initial rapid binding of metals to the cell wall (it is a passive, non-metabolic, and essentially reversible process) and a gradual uptake of metals inside the cell (it is an active, metabolic process). Different algae species show high specificity in the binding of HMs due to the presence of various functional groups in different algal cell walls. This passive process includes ion exchange, chemisorption, adsorption, complexation, chelation, coordination, microprecipitation, entrapment (in the cell wall polysaccharide network), and diffusion. The metabolically dependent second step is an active transport mechanism that involves the transportation of HM ions across the plasma membrane and the subsequent aggregation of metals within the cytoplasm. This transport is facilitated by several mechanisms, such as surface precipitation, the oxidation–reduction reaction, cell surface crystallization, or diffusion (due to a high concentration of metal ions outside the cell). After entering the cell, HM ions are eventually trapped and compartmentalized into the different subcellular organelles. According to many authors, the transport of metal ions occurs because of two possible mechanisms: i) metal ions compete for multivalent ion carriers or are actively transported by binding with low-molecular-weight thiols (e.g., cysteine) and ii) binding of metal ions to metallothioneins (chelating proteins) followed by endocytosis (Monteiro et al., 2012). Algae not only can ingest the HMs, but rather they also play some vital metabolic reactions to neutralize or minimize the toxic impact of HMs intracellularly or extracellularly, and detailed investigations have revealed that molecular mechanism. It has been observed that ion exchange is the most dominant process of metal neutralization; however, complexation and microprecipitation are the most efficient ones. After binding of metal ions with cell surface precipitation of insoluble metal, complexes take place thereon; extracellular masking of HM ions by binding with the algal cells excreted metabolites is a next-level process of neutralization that occurs via the complexation. The hyper-efficient enzymatic response of algal cells converts the toxic metal ions into less toxic ones within the cell; further, these metals can be eliminated outside the cell by an energy-mediated efflux pump. The volatile metal species ions can also be eliminated by the process of vaporization. However, it has been suggested that the algae-mediated remediation of HMs is a pH-sensitive process that mostly occurs between pH 4 and 9 (Mehta and Gaur, 2005; Monteiro et al., 2012).

13.4.2 Details Mechanism of the Living Algae-Mediated HMs Phycoremediation

13.4.2.1 Extracellular Bioremoval of HMs

Under the stress response, the algal cells develop some extracellular polymeric substances (EPS), and the living microalgae can ingest HMs over the cell wall or in the expressed EPS by the process of biosorption. The biosorption of HMs on the cell surface is a metabolic cycle-independent process that relies on the physicochemical properties of the algal cell wall; however, the biosorption via EPS is a metabolically dependent process. To avoid metal toxicity, algae can not only synthesize various EPS but also maintain the level of expression of the biopolymers (Naveed et al., 2019; Ubando et al., 2021).

13.4.2.1.1 Role of the Cell Wall in Biosorption

The cell wall is one of the physical barriers for algal cells; it not only protects algae from HM toxicity but also helps in the removal of HMs from the aquatic environment. A large proportion of HM biosorption depends on the intrinsic chemical composition of the cell wall. Algal cells commonly contain diverse negatively charged functional groups that help in the binding of positively charged metallic ions. The macro-biomolecules (protein, carbohydrate, and lipid) provide different functional groups such as amino ($-NH_2$), carboxyl (–COOH), hydroxyl (–OH), sulfhydryl (–SH), and phosphate ($-PO_4^{3-}$) that develop an overall a negative charge on the cell wall and thus produce a strong binding affinity for metal cations via counter-interaction (Suresh Kumar et al., 2015). The binding of metal ions could be pH-dependent, which allows displacement of protons for bond formation; however, metal ions could also bind electrostatically to unprotonated carboxyl oxygen and sulfate. Many authors have confirmed the presence of different active polyelectrolyte compounds (such as peptidoglycan, teichuronic acid, teichoic acid, polysaccharides, and proteins) on the algal cell wall that facilitate HM uptake as well as removal by the algae (Chojnacka et al., 2005). Table 13.5 represents the different major classes of chemical binding groups present in the algal cell surface biomolecules and their binding abilities with different metallic elements.

13.4.2.1.2 Physicochemical Interactions of HMs on the Cell Surface

The basic understating of the physicochemical interaction of HMs with cell surfaces is challenging due to the complex nature of the algal cell surface. However, many researchers have pointed out chelation and complexation as the most potent mechanisms of HM interaction on the cell wall. Cell walls may contain several ions such as Na^+, K^+, Ca^{2+}, and Mg^{2+}, which allow reversible substitution with HM ions through a process known as ion exchange. The binding of HMs on cell walls may be facilitated by van der Waals and electrostatic interactions. Microprecipitation is another way that also involves the uptake of HMs by both active and passive pathways (Mantzorou et al., 2018).

13.4.2.1.3 Extracellular Polymeric Substances (EPS)-Mediated Interactions with HMs

EPS are extracellular biopolymers that have a high molecular weight and can be obtained from various microorganisms, including microalgae species. EPS includes proteins, sugars, lipids, humic substances, and other extracellular inorganic components that can be broadly classified into three major types *viz.*, soluble EPS in growth media (SL-EPS), tightly bound EPS (TB-EPS), and loosely bound EPS (LB-EPS). Many studies have documented that the EPS expression on algal cell walls increases under stress conditions, including metal toxicity. Yu et al. (2019) and Li et al. (2021) have reported a significant rise in the expression of LB-EPS in *Chlamydomonas reinhardtii* after exposure to Cd and Pb. Similar findings by Zhang et al. (2015) have shown a hike in EPS in Cu-enriched *Chorella* sp. Thus, EPS is found to be associated with extracellular protection from HM pollutants in algae by enhancing the extracellular adsorption of HMs. In addition, that EPS has abundant charged hydrophobic groups that also help in extracellular biosorption (Danouche et al., 2021).

13.4.2.2 Intracellular Bioaccumulation of HMs

In contrast to biosorption, the bioaccumulation of HMs is an intracellular process. This process is metabolic cycle-dependent and widely involves the cell membrane and subcellular cell organelles. The HM ion may accumulate inside the cell in an active or passive way; the active transport requires adequate energy expenditure.

TABLE 13.5
The Major Classes of Chemical-Binding Moiety in Algal Cell Surface Biomolecules and Their Ability to Bind with Different Metallic Elements

Binding Groups	Metallic Components
O^{2-}	Li, Be, Na, Mg, K, Ca, Sc, Rb, Sr, Y, Cs, Ba, La, Fr, Ra, Ac,
OH^-	Al, Lanthanides, Actinides
H_2O	
CO_3^{2-}	
SO_4^-	
$ROSO_3^-$	
NO_3^-	
HPO_4^{2-}	
PO_4^{3-}	
ROH	
$RCOO^-$	
C=O	
ROR	
RNH_2	Ti, V, Cr, Mn, Fe, Co, Ni, Cu, Zn, Ga, Cd, Sn, Sb, As
R_2NH	
R_3N	
$=N^-$	
CO-N-R	
O_2	
O_2^-	
O_2^{2-}	
H^-	Rh, Pd, Ag, Lr, Pt, Au, Hg, Tl, Pb, Bi
I^-	
R^-	
CN^-	
CO	
S^{2-}	
RS^-	
R_2S	
R_3As	

Source: Modified from Bilal et al. (2018).

13.4.2.2.1 Role of the Cell Membrane in HM Flux

In algae, the cell membrane plays a pivotal role in protection from HM toxicity by neutralizing the HMs inside or outside the cell. Cell membrane transporters are responsible for the influx or efflux of HMs in the algal cell. Researchers have categorized the cell membrane transporters into two broad categories, *viz.*, Group A and Group B. Group A transporters move HM ions into the cytosol, thus increasing the intracellular HM concentration. The Group A transporters include several cell surface protein members of different families, such as FTR (Fe TRansporter), CTR

(Cu TRansporter), NRAMP (Natural Resistance Associated Macrophage Proteins), and ZIP (Zrt- and Irt-like Proteins) families. This type of receptor can also be observed in subcellular organelles (such as the vacuole) and plays a similar role in metal uptake. In comparison to Group A, the Group B transporters play the opposite role. They decrease the load of HMs in the cytosol by efflux or secretory pathway-associated exocytosis of HMs. These transporters belong to several protein families, such as P1B-type ATPases, CDF (Cation Diffusion Facilitator), FPN (FerroPortiN), and Ccc1 (Ca^{2+}-sensitive Cross-Complementer 1)/VIT1 (Vacuolar Iron Transporter 1) (Blaby-Haas and Merchant, 2012).

13.4.2.2.2 Metallothioneins (MTs)-Mediated Chelation of HMs

MTs are small metal-binding polypeptides with a molecular weight of 6–7 kDa, and they can be majorly categorized into two groups: gene-encoded proteins (further classified into MT classes I and II) and enzymatically synthesized short-chain polypeptides (MT class III) (Cobbett and Goldsbrough, 2002). Class III MTs are also known as phytochelatin. These MTs are primarily involved in the regulation of intracellular concentrations of metals at regular levels. Many researchers have identified the presence and active synthesis of class III MT in algae. *Nannochloropsis*, *Chlorella*, *Aureococcus*, *Thalassiosira*, *Symbiodinium*, and *Ostreococcus* are found to be the most potential genera of microalgae that actively synthesize MTs (Balzano et al., 2020). MTs are cysteine-rich polypeptides that complex "soft" metal ions in thiol clusters. The class III MT is found to be associated with the formation of stable metal complexes in the cytosol that reduce the toxicity of free HM ions. Both in vivo and in vitro studies have shown a significant level of MT synthesis in several algae under stimulation by HMs (such as As, Ni, Cd, Ag, Bi, Pb, Zn, Cu, Hg, and Au). MTs are found to maintain zinc and copper homeostasis in several microalgae (Suresh Kumar et al., 2015).

13.4.2.2.3 HM Sequestering by Polyphosphate Bodies (PPBs)

PPBs are electron-dense acidic vacuoles, also known as acidocalcisomes. These structures are predominantly found in algae (and many other eukaryotes) and are commonly associated with the intracellular storage of nutrients. However, the binding of HM ions to PPBs and subsequent trafficking in the cytosol is a well-documented phenomenon in algae. PPBs help in sequestering several HMs such as Cu, Co, Cd, Hg, Mg, Ni, Pb, Sr, Ti, and Zn. Their role in HM remediation mostly depends on the storage and detoxification of HMs (Dwivedi, 2012).

13.4.2.2.4 HM Sequestration and Compartmentalization in Other Subcellular Organelles

Chloroplasts, mitochondria, and vacuoles are proven to be major sites for HM detoxification in algae. In a Pb, Hg, and Cd-induced experimental setup, the accumulation of HM-binding biomolecules around the pyrenoids of chloroplasts in algae reveals the vitality of chloroplasts in HM removal (Shanab et al., 2012). Isolation of high concentrations of Cd and Cu from the chloroplasts of *Chlamydomonas reinhardtii*, *Euglena gracilis*, and *Oocystis nephrocytioides* supports earlier evidence of the involvement of the chloroplast in HM accumulation (Suresh Kumar et al., 2015).

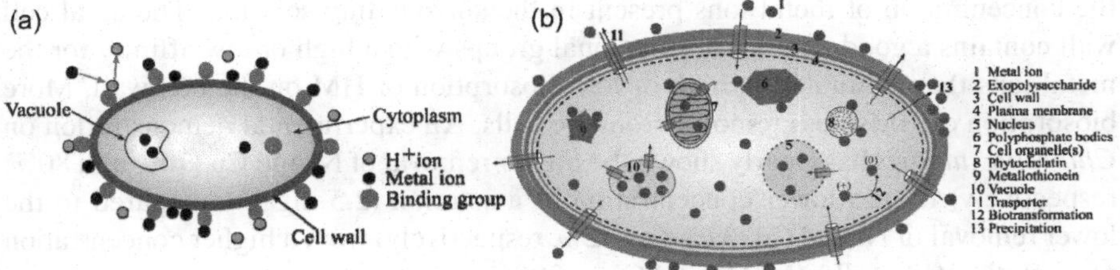

FIGURE 13.4 A schematic representation of various processes of uptake of HMs by the algal cells.

Similar findings have also been recorded for mitochondrial vacuoles. The presence of class III MT and HMs in such cell organelles suggests one of the possible pathways of HM entrapment in algae: first, MT are synthesized within the cytosol under metal stress, which is followed by binding of HMs with MTs, which are then transported into the mitochondria (or chloroplasts or vacuoles) (Perales-Vela et al., 2006).

13.5 ROLE OF NONLIVING ALGAE

Dead or nonviable algal biomass is equally popular as live algae for HM remediation. They can remove a bulk amount of HM in a very short time. Their mechanism of action mostly relies on simple, quick, and metal species-specific biosorption. Thus, researchers have shifted their attention towards the dead algae to make suitable tools for wastewater purification. Many studies have highlighted the chemical composition and metal binding properties of different algae species, and it has been observed that dead algal biomass (Pb) at maximum magnitude is followed by Ni, Cd, and Zn. *Chlamydomonas reinhardtii* is one of the potential green microalgae that has been recorded for maximum uptake of Hg, Cd, and Pb at around 72.2, 42.6, and 96.3 mg/g dry biomass, respectively (Tüzün et al. 2005). Studies on *Vaucheria* spp. biomass have revealed that the biomass releases Ca^{2+}, K^{+}, Na^{+}, and Mg^{2+} ions at the time of biosorption of HMs (such as Zn, Mn, Cu, and Co). This result clearly indicates that ion exchange is the most dominant mechanism of metal binding for biosorption in algal biomass (Michalak and Chojnacka 2010). Different mechanisms of metal uptake by living and non-living microalgae have been illustrated in Figure 13.4.

13.6 FACTORS AFFECTING PHYCOREMEDIATION OF HEAVY METALS

The bioaccumulation of HMs by living cells or biosorption by living or non-living algal biomass may be influenced by several factors, which are discussed here.

13.6.1 Abiotic Factors

13.6.1.1 Initial Metal Ion Concentration

The initial concentration of HMs is one of the major regulatory factors in the biosorption of HMs. The rate of biosorption into biomass is directly proportional to the concentration of metal ions present in the surrounding solution. The algal cell wall contains a good number of functional groups with a high or low affinity for the metal ions that facilitate the high or low biosorption of HM on the cell wall. More biosorption of HMs, less removal from the cells. An experimental demonstration on *Chlorella vulgaris* has clearly shown the higher removal of Ni and Cu (70% and 80%, respectively) from a lower concentration of a solution (2.5 mg/L) compared to the lower removal of Ni and Cu (37% and 42%, respectively) from a higher concentration of a solution (10 mg/L) (Mehta and Gaur, 2005).

13.6.1.2 pH

The solubility of HMs in the aqueous medium and their toxicity depend on the pH of the medium, which influences the adsorption of metal by algal biomass. pH maintains the chemistry of metals, thus regulating their binding affinity to different functional groups present on cell surfaces. Various scientific studies have revealed the variable optimum pH of different HMs for selective binding to algal species. A compendium table with optimal pH for the uptake of seven common toxic metals has been presented in Table 13.3. Precipitation of metals occurs at higher pH levels, resulting in a decrease in their removal. Thus, maintenance of pH becomes essential for algae-metal interaction (Suresh Kumar et al., 2015).

13.6.1.3 Metal Speciation

HMs exist in wastewater in different forms, like complexes (with inorganic and organic ligands), free aquo ions, and adsorbed on particulate phases. However, the toxicity and biosorption tendency of HMs mostly depend on the concentration of free aquo ions. The availability and binding of metal ions are determined by the metal species and pH of the solution. Normally, the concentration of free metal ions remains low at the high pH range (7–9), and ion concentration increases at low pH. An experimental study on *Spirulina* spp. showed better accumulation of Cr^{6+} (333 mg/g) than Cr^{3+} (304 mg/g). Thus, it indicates a direct relationship between metal species and their biosorption (Suresh Kumar et al., 2015; Mehta and Gaur, 2005).

13.6.1.4 Temperature

The role of temperature in metal biosorption in algal biomass is a matter of debate. Several studies have clearly stated that increasing temperature enhances the rate of biosorption. For example, the biomass of *Chlorella vulgaris* has shown an increase in the biosorption of metal ions from 15°C (48.1 mg/g) to 45°C (60.2 mg/g). It suggests the endothermic nature of biosorption. It is believed that high temperatures break the bonds in adsorption biomolecules that induce metal ion binding. However, several studies have reported the opposite phenomenon. For example, biosorption of Cd^{2+} by *Sargassum* sp. biomass has been shown to decrease with an increase in temperature (Mehta and Gaur, 2005).

13.6.1.5 Presence of Anions and Cations

Contrasting statements exist regarding the role of anions and cations in the biosorption of metals. The lowering in the accumulation of HMs in algal biomass has been observed in the presence of sulfate, nitrate, carbonate, orthophosphate, ethylenediaminetetraacetic acid, and chloride ions. However, an increase in the accumulation of cations (such as nitrate and ammonium) has been observed in the presence of metal ions in the medium. The presence of other cations, including metal ions, also significantly affects metal sorption by algae. Even mutual interference may occur between two or more HM ions present in the same solution. Thus, industrial effluents that contain a large volume of light-weight metal ion mixtures may affect the biosorption potential of algae. However, some studies have pointed out that the metal ion mixture has no significant impact on biosorption. For example, *Macrospora* can selectively accumulate Pb and Ni from a multi-metal mixture (Mehta and Gaur, 2005).

13.6.2 Biotic Factors

13.6.2.1 Biomass concentration

The increase in biomass directly increases the binding site for metal ions; thus, biomass concentration is directly proportional to metal biosorption. However, excess biomass may react inversely because of the aggregation of biomass that blocks the active binding site for metals (Suresh Kumar et al., 2015).

13.6.2.2 Species

The metal uptake capacity varies among the different genera or species of algae. Even species from the same genus may show diversity in metal biosorption. For example, *Chlorella miniata*, *C. vulgaris*, and *C. reinhardtii* are reported to remove divalent ions of Hg, Pb, Ni, Zn, Cu, and Cd, whereas *C. vulgaris* and *Spirulina platensis* are reported to remove trivalent ions of Cr and Fe (Suresh Kumar et al., 2015).

13.6.2.3 Size and Volume of Microalgae

Size and volume are crucial factors for any type of cell. Cellular growth, metabolism, and other biochemistry directly or indirectly depend on these two factors. Small algal cells have more surface area, so they show a higher rate of photosynthesis and growth. Within the small cells, transportation of inorganic or organic molecules occurs more frequently. It has been observed that small-cell algae are more sensitive to copper in comparison to larger algae (Suresh Kumar et al., 2015).

13.7 TRANSGENIC MICROALGAE IN BIOREMEDIATION OF HEAVY METALS

The use of genetically modified algae is a next-generation approach to the bioremediation of HM. The transgenic algae are modulated in such a way that they can express more cell surface or cytosolic metal-binding proteins (Rajamani et al., 2007). For example, the development of transgenic microalgae *Chlorella* spp. DT with the *merA* gene (from *Bacillus megaterium* strain MB1) has proven to express a protein called mercuric reductase (MerA) that converts toxic Hg^{2+} to volatile Hg^{o} (Huang et al., 2006). However, it has been observed that such transgenic algae could accelerate the biogeochemical cycling of HMs and may result in the accumulation of HMs in the food chain. Thus, many assessments and safety measures are required before using such transgenic algae in the practical field. In that prospect, the development of nonviable transgenic algae could be a good option (Suresh Kumar et al., 2015).

13.8 APPLICATION OF BATCH AND CONTINUOUS SYSTEMS IN HEAVY METAL REMOVAL

The removal of harmful HMs from wastewater is a basic need to have a healthy environment for sustainable development. Along with the removal of HMs from wastewater, the check-in source is equally important. In that regard, the application of free and immobilized algae in batch or continuous systems is very useful. Until now, many experiments have been performed in the 'batch system'. A batch reactor system contains algal biomass, and alongside it, it keeps HM solution for a certain amount of time until the complete biomass gets saturated by binding with HMs. The mixing of metal solutions in this system is done either by air or a magnetic stirrer. The metal-bound biomass is later separated by centrifugation or filtration, and the biomass is recovered by a series of sorption–desorption reactions (Pradhan and Rai, 2001). However, this process is not suitable for large industries where a huge amount of effluent is to be treated. The separation of liquid biomass is another challenge for maintaining a batch reactor. In comparison to a batch system, a 'Continuous system' is more feasible for treating a large volume of solution. A continuous system preferably includes immobilized algal biomass as an ion-exchange medium. A continuous stirred tank, moving bed, fluidized bed, and packed bed columns are the different setups of the continuous system (Figure 13.5). For large-scale wastewater management, a packed column is the best option. In an up-flow packed column, the fluid is passed from bottom to top, and in a down-flow column, the solution flows down into the column. The packed column contains either live or dead algal biomass that continuously accumulates metal ions from the solution; thus, the effluent or outflow contains minimum or no metal at all. After a long ring, metal concentration gradually increases in the effluent; this is called the 'breakthrough' point. When the system reaches a breakthrough, the flow is stopped. Several trials have verified the 99% efficiency of this continuous system until it reaches a breakthrough (Mehta and Gaur, 2005).

13.9 CONCLUSION

In the 21st century, the availability of pure and consumable water is a global crisis. According to WHO, around 2.2 billion people experiencing high water stress are deprived of accessing safe and managed drinking water (WHO/UNICEF, 2019). Agricultural practices consume about 70% of global water needs; thus, this field is also under tremendous water threat. The constraint on water resources is largely due to the pollution of water by various toxicants, including HMs. Water pollution is shown as a more predominant trend in the last few decades. Thus, to protect the earth from such a calamity, we should have to adopt sustainable development and management strategies, and the use of algae in the mitigation of HM-associated water pollution could be a big step. Algae are found to be a very promising approach for remediation of HM water contamination. They have several advantages over chemical and other

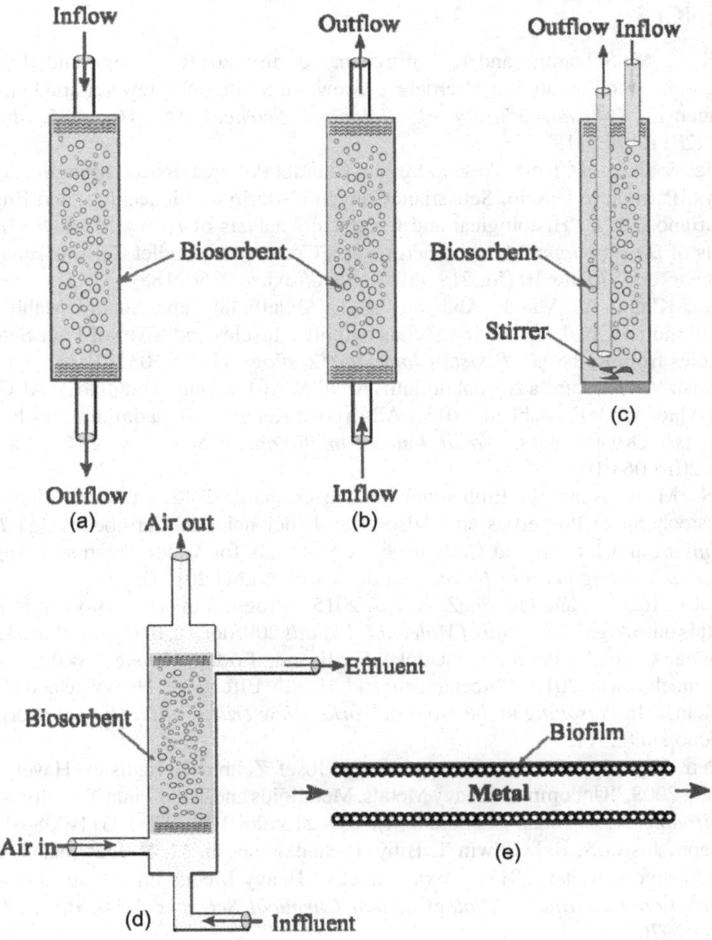

FIGURE 13.5 Different types of continuous bioreactor system for HMs removal.

types of bioremediation. The high availability, low maintenance cost, high tolerance, and extensive HM removal capacity of algae are the basic reasons for selecting them for biomonitoring and bioremediation of HMs. Several green microalgae are recognized as the most potent HM removers. Due to their high demand, they are now also cultivated in an artificial system. Along with living algae, dead algal biomass has also gained enormous popularity due to its highly efficient metal biosorption property. Moreover, different biotechnological manipulations in algal cells have demonstrated enhanced bioremediation performance against HM pollution. However, there are only a very few examples of genetically engineered algae available for use in sustainable bioremediation; thus, we expect more research and practical input from bioengineered algae for better bioremediation of HMs, which could be added in the near future.

REFERENCES

Ajayan, K. V., M. Selvaraju, and K. Thirugnanamoorthy. 2011. "Growth and Heavy Metals Accumulation Potential of Microalgae Grown in Sewage Wastewater and Petrochemical Emuents." *Pakistan Journal of Biological Sciences* 14 (16): 805. doi:10.3923/pjbs.2011.805.811.

Alesci, Alessio, Nicola Cicero, Angelo Fumia, Claudia Petrarca, Rocco Mangifesta, Vincenzo Nava, Patrizia lo Cascio, SebastianoGangemi, Mario di Gioacchino, and Eugenia Rita Lauriano. 2022. "Histological and Chemical Analysis of Heavy Metals in Kidney and Gills of *Boops boops*: Melanomacrophages Centers and Rodlet Cells as Environmental Biomarkers." *Toxics* 10 (5): 218. doi:10.3390/toxics10050218.

Al-Ghanim, Khalid A., Magda Abdelatty, Lyla Abdelfattah, and Shahid Mahboob. 2015. "Differential Uptake of Heavy Metals by Gill, Muscles and Liver of Four Selected Fish Species from Red Sea." *Pakistan Journal of Zoology* 47 (4): 1031–1036.

Al-Homaidan, Ali A., Jamila A. Alabdullatif, Amal A. Al-Hazzani, Abdullah A. Al-Ghanayem, and Aljawharah F. Alabbad. 2015. "Adsorptive Removal of Cadmium Ions by *Spirulina platensis* Dry Biomass." *Saudi Journal of Biological Sciences* 22 (6). doi:10.1016/j.sjbs.2015.06.010.

Altenor, S., M. C. Ncibi, E. Emmanuel, and S. Gaspard. 2012. "Textural Characteristics, Physiochemical Properties and Adsorption Efficiencies of Caribbean Alga *Turbinaria turbinata* and Its Derived Carbonaceous Materials for Water Treatment Application." *Biochemical Engineering Journal* 67. doi:10.1016/j.bej.2012.05.008.

Anastopoulos, Ioannis, and George Z. Kyzas. 2015. "Progress in Batch Biosorption of Heavy Metals onto Algae." *Journal ofMolecular Liquids* 209. doi:10.1016/j.molliq.2015.05.023.

Azeh Engwa, Godwill, Paschaline Udoka Ferdinand, Friday Nweke Nwalo, and Marian N. Unachukwu. 2019. "Mechanism and Health Effects of Heavy Metal Toxicity in Humans." In *Poisoning in the Modern World - New Tricks for an Old Dog?* doi:10.5772/intechopen.82511.

Babula, Petr, Vojtech Adam, Radka Opatrilova, Josef Zehnalek, Ladislav Havel, and Rene Kizek. 2008. "Uncommon Heavy Metals, Metalloids and Their Plant Toxicity: A Review." *Environmental Chemistry Letters* 6 (4): 189–213. doi:10.1007/s10311-008-0159-9.

Baby, Joseph, Justin S. Raj, Edwin T. Biby, P. Sankarganesh, M. V. Jeevitha, S. U. Ajisha, and Sheeja S. Rajan. 2011. "Toxic Effect of Heavy Metals on Aquatic Environment." *International Journal of Biological and Chemical Sciences* 4 (4). doi:10.4314/ijbcs.v4i4.62976.

Bakatula, E. N., E. M. Cukrowska, I. M. Weiersbye, L. Mihaly-Cozmuta, A. Peter, and H. Tutu. 2014. "Biosorption of Trace Elements from Aqueous Systems in Gold Mining Sites by the Filamentous Green Algae (*Oedogonium* Sp.)." *Journal of Geochemical Exploration* 144 (Pt C): 492–503. doi:10.1016/j.gexplo.2014.02.017.

Balzano, Sergio, Angela Sardo, Martina Blasio, Tamara Bou Chahine, Filippo Dell'Anno, Clementina Sansone, and Christophe Brunet. 2020. "Microalgal Metallothioneins and Phytochelatins and Their Potential Use in Bioremediation." *Frontiers in Microbiology*. doi:10.3389/fmicb.2020.00517.

Bilal, Muhammad, Tahir Rasheed, Juan Eduardo Sosa-Hernández, Ali Raza, Faran Nabeel, and Hafiz M. N. Iqbal. 2018. "Biosorption: An Interplay between Marine Algae and Potentially Toxic Elements-A Review." *Marine Drugs* 16 (2). https://doi.org/10.3390/md16020065.

Blaby-Haas, Crysten E., and Sabeeha S. Merchant. 2012. "The Ins and Outs of Algal Metal Transport." *BiochimicaetBiophysicaActa - Molecular Cell Research*. doi:10.1016/j.bbamcr.2012.04.010.

Brinza, Loredana, Matthew J. Dring, and Maria Gavrilescu. 2007. "Marine Micro and Macro Algal Species as Biosorbents for Heavy Metals." *Environmental Engineering and Management Journal* 6 (3). doi:10.30638/eemj.2007.029.

Çetinkaya Dönmez, G., Z. Aksu, A. Öztürk, and T. Kutsal. 1999. "A Comparative Study on Heavy Metal Biosorption Characteristics of Some Algae." *Process Biochemistry* 34 (9). doi:10.1016/S0032-9592(99)00005-9.

Chojnacka, Katarzyna, Andrzej Chojnacki, and Helena Górecka. 2005. "Biosorption of Cr^{3+}, Cd^{2+} and Cu^{2+} Ions by Blue-Green Algae *Spirulina* Sp.: Kinetics, Equilibrium and the Mechanism of the Process." *Chemosphere* 59 (1). doi:10.1016/j.chemosphere.2004.10.005.

Cobbett, Christopher, and Peter Goldsbrough. 2002. "Phytochelatins and Metallothioneins: Roles in Heavy Metal Detoxification and Homeostasis." *Annual Review of Plant Biology*. doi:10.1146/annurev.arplant.53.100301.135154.

Danouche, M., N. El Ghachtouli, H. El Arroussi. 2021. Phycoremediation Mechanisms of Heavy Metals Using Living Green Microalgae: Physicochemical and Molecular Approaches for Enhancing Selectivity and Removal Capacity. *Heliyon*. 7(7): e07609. doi: 10.1016/j.heliyon.2021.e07609. PMID: 34355100; PMCID: PMC8322293.

David, Elise, and Claudia Cosio. 2021. "New Insights into Impacts of Toxic Metals in Aquatic Environments." *Environments* 8(1): 1. doi:10.3390/environments8010001.

Diaconu, Mariana, Camelia Betianu, Raluca Hlihor, Maria Gavrilescu. (2013). *Biomass in Environmental Remediation – A Review*. Bulletin of the Polytechnic Institute of Iasi, Section Chemistry and Chemical Engineering, pp. 97–113. LIX.

Dwivedi, Seema. 2012. "Bioremediation of Heavy Metal by Algae: Current and Future Perspective." www.sospublication.co.in.

Figueira, Marianne M., Bohumil Volesky and Hans Jörg Mathieu. "Instrumental Analysis Study of Iron Species Biosorption by Sargassum Biomass." *Environmental Science & Technology* 33 (1999): 1840–1846.

Gupta, Asha. 2016. *Water Pollution-Sources, Effects and Control*. Pointer Publishers: Jaipur.

Huang, Chieh Chen, Meng Wei Chen, Ju Liang Hsieh, Wen Hao Lin, Pei Chung Chen, and Lee FengChien. 2006. "Expression of Mercuric Reductase from *Bacillus megaterium* MB1 in Eukaryotic Microalga *Chlorella* Sp. DT: An Approach for Mercury Phytoremediation." *Applied Microbiology and Biotechnology* 72 (1). doi:10.1007/s00253-005-0250-0.

Jais, N. M., R. M. S. R, Mohamed, A. A. Al-Gheethi, and M. K. AmirHashim. 2017. "The Dual Roles of Phycoremediation of Wet Market Wastewater for Nutrients and Heavy Metals Removal and Microalgae Biomass Production." *Clean Technologies and Environmental Policy*. doi:10.1007/s10098-016-1235-7.

Jasrotia R., Shivakshi, Arun Kansal, V.V.N. Kishore. 2014. Arsenic Phyco-remediation by Cladophora Algae and Measurement of Arsenic Speciation and Location of Active Absorption Site Using Electron Microscopy. *Microchemical Journal* 114. doi: 10.1016/j.microc.2014.01.005.

Kim, Hyun Soo, Yeo Jin Kim, and Young Rok Seo. 2015. "An Overview of Carcinogenic Heavy Metal: Molecular Toxicity Mechanism and Prevention." *Journal of Cancer Prevention* 20 (4). doi:10.15430/jcp.2015.20.4.232.

Kotrba, Pavel. 2011. "Microbial Biosorption of Metals-General Introduction." *Microbial Biosorption of Metals*. doi:10.1007/978-94-007-0443-5_1.

Leong, Yoong Kit, and Jo Shu Chang. 2020. "Bioremediation of Heavy Metals Using Microalgae: Recent Advances and Mechanisms." *Bioresource Technology*. doi:10.1016/j.biortech.2020.122886.

Li, Chonghua, Chao Zheng, Hongxuan Fu, Suhua Zhai, Fan Hu, Sadiq Naveed, Chunhua Zhang, and Ying Ge. 2021. "Contrasting Detoxification Mechanisms of *Chlamydomonas reinhardtii* under Cd and Pb Stress." *Chemosphere* 274. doi:10.1016/j.chemosphere.2021.129771.

Lim, Sing Lai, Wan Loy Chu, and Siew Moi Phang. 2010. "Use of *Chlorella vulgaris* for Bioremediation of Textile Wastewater." *Bioresource Technology* 101 (19). doi:10.1016/j.biortech.2010.04.092.

Mahdavi, Hamed, Ania C. Ulrich, and Yang Liu. 2012. "Metal Removal from Oil Sands Tailings Pond Water by Indigenous Micro-Alga." *Chemosphere* 89 (3). doi:10.1016/j.chemosphere.2012.04.041.

Mantzorou, A., E. Navakoudis, K. Paschalidis, and F. Ververidis. 2018. "Microalgae: A Potential Tool for Remediating Aquatic Environments from Toxic Metals." *International Journal of Environmental Science and Technology*. doi:10.1007/s13762-018-1783-y.

Matsunaga, Tadashi, Haruko Takeyama, Takashi Nakao, and Akira Yamazawa. 1999. "Screening of Marine Microalgae for Bioremediation of Cadmium-Polluted Seawater." *Progress in Industrial Microbiology* 35 (C). doi:10.1016/S0079-6352(99)80095-2.

Mehta, S. K., and J. P. Gaur. 2005. "Use of Algae for Removing Heavy Metal Ions from Wastewater: Progress and Prospects." *Critical Reviews in Biotechnology*. doi:10.1080/07388550500248571.

Michalak, Izabela, and Katarzyna Chojnacka. 2010. "Interactions of Metal Cations with Anionic Groups on the Cell Wall of the Macroalga *Vaucheria* Sp." *Engineering in Life Sciences* 10 (3). doi:10.1002/elsc.200900039.

Monteiro, Cristina M., Paula M. L. Castro, and F. Xavier Malcata. 2012. "Metal Uptake by Microalgae: Underlying Mechanisms and Practical Applications." *Biotechnology Progress*. doi:10.1002/btpr.1504.

Naveed, Sadiq, Chonghua Li, Xinda Lu, Shuangshuang Chen, Bin Yin, Chunhua Zhang, and Ying Ge. 2019. "Microalgal Extracellular Polymeric Substances and Their Interactions with Metal(Loid)s: A Review." *Critical Reviews in Environmental Science and Technology* 49 (19). doi:10.1080/10643389.2019.1583052.

Nazal, Mazen K. 2019. "Marine Algae Bioadsorbents for Adsorptive Removal of Heavy Metals." In *Advanced Sorption Process Applications*. Intech Open: London, UK www.intechopen.com.

Nourbakhsh, M., Y. Sağ, D. Özer, Z. Aksu, T. Kutsal, and A. Çağlar. 1994. "A Comparative Study of Various Biosorbents for Removal of Chromium(VI) Ions from Industrial Waste Waters." *Process Biochemistry* 29 (1). doi:10.1016/0032-9592(94)80052-9.

Pacheco, Diana, Ana Cristina Rocha, Leonel Pereira, and Tiago Verdelhos. 2020. "Microalgae Water Bioremediation: Trends and Hot Topics." *Applied Sciences*. doi:10.3390/app10051886.

Perales-Vela, Hugo Virgilio, Julián Mario Peña-Castro, and Rosa Olivia Cañizares-Villanueva. 2006. "Heavy Metal Detoxification in Eukaryotic Microalgae." *Chemosphere*. doi:10.1016/j.chemosphere.2005.11.024.

Podgorskii, V. S., T. P. Kasatkina, and O. G. Lozovaia. 2004. "Yeasts--Biosorbents of Heavy Metals." *MikrobiolohichnyiZhurnal (Kiev, Ukraine: 1993)* 66(1): 91–103.

Poleksic, Vesna, Mirjana Lenhardt, Ivan Jaric, Dragana Djordjevic, Zoran Gacic, Gorcin Cvijanovic, and Bozidar Raskovic. 2010. "Liver, Gills, and Skin Histopathology and Heavy Metal Content of the Danube Sterlet (*Acipenser ruthenus* Linnaeus, 1758)." *Environmental Toxicology and Chemistry* 29 (3). doi:10.1002/etc.82.

Pradhan, Subhashree, and L. C. Rai. 2001. "Copper Removal by Immobilized *Microcystis aeruginosa* in Continuous Flow Columns at Different Bed Heights: Study of the Adsorption/Desorption Cycle." *World Journal of Microbiology and Biotechnology* 17 (9). doi:10.1023/A:1013800800176.

Rajamani, Sathish, Surasak Siripornadulsil, Vanessa Falcao, Moacir Torres, Pio Colepicolo, and Richard Sayre. 2007. "Phycoremediation of Heavy Metals Using Transgenic Microalgae." *Advances in Experimental Medicine and Biology*. doi:10.1007/978-0-387-75532-8_9.

Reid, Scott D., and D. G. McDonald. 1991. "Metal Binding Activity of the Gills of Rainbow Trout (*Oncorhynchus mykiss*)." *Canadian Journal of Fisheries and Aquatic Sciences* 48 (6). doi:10.1139/f91-125.

Salama, El Sayed, Hyun Seog Roh, Subhabrata Dev, Moonis Ali Khan, Reda A. I. Abou-Shanab, Soon Woong Chang, and Byong Hun Jeon. 2019. "Algae as a Green Technology for Heavy Metals Removal from Various Wastewater." *World Journal of Microbiology and Biotechnology*. doi:10.1007/s11274-019-2648-3.

Shanab, Sanaa, Ashraf Essa, and Emad Shalaby. 2012. "Bioremoval Capacity of Three Heavy Metals by Some Microalgae Species (Egyptian Isolates)." *Plant Signaling and Behavior* 7 (3). doi:10.4161/psb.19173.

Sharma, Ashita, Mandeep Kaur, Jatinder Kaur Katnoria, and Avinash Kaur Nagpal. 2015. "Heavy Metal Pollution: A Global Pollutant of Rising Concern." In *Toxicity and Waste Management Using Bioremediation*. doi:10.4018/978-1-4666-9734-8.ch001.

Singh, Kailash, N. A. Renu, and Madhu Agarwal. 2017. "Methodologies for Removal of Heavy Metal Ions from Wastewater: An Overview." *Interdisciplinary Environmental Review* 18 (2). doi:10.1504/ier.2017.10008828.

Spain, Olivia, Martin Plöhn, and Christiane Funk. 2021. "The Cell Wall of Green Microalgae and Its Role in Heavy Metal Removal." *Physiologia Plantarum* 173 (2). doi:10.1111/ppl.13405.

Suresh Kumar, K., Hans Uwe Dahms, EunJi Won, Jae Seong Lee, and Kyung Hoon Shin. 2015. "Microalgae - A Promising Tool for Heavy Metal Remediation." *Ecotoxicology and Environmental Safety*.Academic Press. doi:10.1016/j.ecoenv.2014.12.019.

Tüzün, Ilhami, Gülay Bayramoğlu, Emine Yalçin, Gökben Başaran, Gökçe Çelik, and M. Yakup Arica. 2005. "Equilibrium and Kinetic Studies on Biosorption of Hg(II), Cd(II) and Pb(II) Ions onto Microalgae *Chlamydomonas reinhardtii*." *Journal of Environmental Management* 77 (2). doi:10.1016/j.jenvman.2005.01.028.

Ubando, Aristotle T., Aaron Don M. Africa, Marla C. Maniquiz-Redillas, Alvin B. Culaba, Wei Hsin Chen, and Jo Shu Chang. 2021. "Microalgal Biosorption of Heavy Metals: A Comprehensive Bibliometric Review." *Journal of Hazardous Materials* 402. doi:10.1016/j.jhazmat.2020.123431.

Vijayaraghavan, K., and Yeoung Sang Yun. 2008. "Bacterial Biosorbents and Biosorption." *Biotechnology Advances*. doi:10.1016/j.biotechadv.2008.02.002.

WHO/UNICEF. 2019. https://www.who.int/news/item/18-06-2019-1-in-3-people-globally-do-not-have-access-to-safe-drinking-water-unicef-who

Yang, Jin Shui, Jing Cao, Guan Lan Xing, and Hong Li Yuan. 2015. "Lipid Production Combined with Biosorption and Bioaccumulation of Cadmium, Copper, Manganese and Zinc by Oleaginous Microalgae *Chlorella minutissima* UTEX2341." *Bioresource Technology* 175. doi:10.1016/j.biortech.2014.10.124.

Yu, Zhen, Teng Zhang, RuiHao, and Yi Zhu. 2019. "Sensitivity of *Chlamydomonas reinhardtii* to Cadmium Stress is Associated with Phototaxis." *Environmental Science: Processes and Impacts* 21 (6). doi:10.1039/c9em00013e.

Zhang, Wenlin, Nicole G. J. Tan, Baohui Fu, and Sam F. Y. Li. 2015. "Metallomics and NMR-Based Metabolomics of *Chlorella* Sp. Reveal the Synergistic Role of Copper and Cadmium in Multi-Metal Toxicity and Oxidative Stress." *Metallomics* 7 (3). doi:10.1039/c4mt00253a.

Znad, Hussein, MdRabiulAwual, and Sri Martini. 2022. "The Utilization of Algae and Seaweed Biomass for Bioremediation of Heavy Metal-Contaminated Wastewater." *Molecules*. doi:10.3390/molecules27041275.

Index

Printed in the United States
by Baker & Taylor Publisher Services

Printed in the United States
by Baker & Taylor Publisher Services